Future Trends in Geomathematics

Future Trends in Geomathematics

Edited by R G Craig and M L Labovitz

Pion Limited, 207 Brondesbury Park, London NW2 5JN

ISBN 0 85086 080 6

Printed in Great Britain by Page Bros (Norwich) Limited.

John C Griffiths

Dedication

We dedicate this book to John C Griffiths, Professor of Sedimentary Petrography and Operations Research at The Pennsylvania State University. 'Grif' officially retired from teaching in 1978, capping a long career that has influenced every member of the geomathematics community. Many will know him for his pioneering text *Scientific Method in Analysis of Sediments* in which he presented the fundamentals of statistical methods and, even more importantly, the coherent philosophy required for a scientific approach to problems in the geosciences. This work is typical of the 'Griffithsian' method, riddled with examples and counterexamples designed to illustrate every facet of the points in question.

Emphasizing the empirical above all else, his insights into the implications of the Heisenberg principle, Russell's paradox, the Church–Turing thesis, and general systems theory have led him to occasional rejection of naive rationalism in preference to pure empiricism. "The Federal Government is irrational, the businesses are irrational, so why the hell should I be rational?" He has always had the disarming ability to leave fellow geoscientists dead in their tracks with comments such as "Well, I don't give a damn about your theory, if you had collected *any* data on that (as I have) you would realize ...".

In his teaching and his research, Griffiths has placed considerable emphasis on defining and helping to solve the sample problem. Taking a cue from Percy Bridgeman, he has consistently stressed the importance of operational definitions. His style is profane and profound. In the discussion of problems, his analogies cut deep. For instance, discussion of grain-size measurements based on Stokes law and the impact law are peppered with such tidbits as: "Well, if you want to measure the size of this desk you take it to the top of the building, drop it off and see how far it bounces". Few meetings pass without a caution from 'Grif' that the speaker ought to "turn it around the other way", and his arguments usually reveal as many sides as a sphere!

Always alert to new ideas, 'Grif' has moved into a new field over the last decade, that of resource estimation. Here, he has shown that the solution to the problem of resource availability will require a metalogic for exploration strategies. This led him to champion the grid-drilling idea. Put simply, this says that we can make rational decisions about resources allocation only when an inventory of our resources is available. Such an inventory is possible only if we work systematically. By a clever combination of empirical data and computer simulations, he showed that a grid with a spacing of about 20 miles applied throughout the United States could tell us about our remaining mineral endowment more efficiently than any present technique. In return, he has been granted numerous titles: "the man with 20000 holes in his head", and so on. But, as usual, it is 'Grif' who has had the last laugh. Many people, both within the Government and outside, now recognize the need for such an approach.

We hereby acknowledge our debt, a difficult task for any scientist. For a science to progress, ideas must flow freely; yet there is much about the system that discourages proper recognition of contributions. Certainly, John Griffiths has given this field many more ideas than those he published himself. Others, indeed all of us in the field of geomathematics, have been greatly influenced by this man, and must acknowledge his help in shaping our ideas and careers.

Richard G Craig
Department of Geology
Kent State University
Kent, Ohio

Mark L Labovitz
Goddard Space Flight Center
Greenbelt, Md

14 January 1981

Contributors

F P Agterberg — Geological Survey of Canada, 601 Booth Street, Ottawa, Canada K1A 0E8

C Banfield — Technical Department, Logica Ltd, 64 Newman Street, London W1A 4SE, England

F Chayes — Geophysical Laboratory, 2801 Upton Street, NW, Washington, DC 20008, USA

A D Cliff — Department of Geography, University of Cambridge, Downing Place, Cambridge CB2 3EN, England

R G Craig — Department of Geology, Kent State University, Kent, Ohio 44242, USA

M F Dacey — Department of Geography, Northwestern University, Evanston, Illinois 60201, USA

M Dagbert — Geostat Systems Inc., 10403 West Colfax Suite 450, Lakewood, Colorado 80215, USA

J C Davis — Kansas Geological Survey, 1930 Avenue "A", Campus West, The University of Kansas, Lawrence, Kansas 66044, USA

L J Drew — US Department of the Interior, US Geological Survey, Reston, Virginia 22092, USA

P J Gould — Department of Geography, 306 Walker Building, The Pennsylvania State University, University Park, Pennsylvania 16802, USA

J W Harbaugh — Department of Applied Earth Sciences, Stanford University, Stanford, California 94305, USA

K W Hipel — Department of Systems Design Engineering, University of Waterloo, Waterloo, Ontario, Canada N2L 3G1

D G Krige — Department of Mining Engineering, University of the Witwatersrand, 1 Jan Smuts Avenue, Johannesburg 2001, South Africa

W C Krumbein (deceased) — Formerly of: Department of Geography, Northwestern University, Evanston, Illinois 60201, USA

M L Labovitz — Code 923, Goddard Space Flight Center, Greenbelt, Maryland 20771, USA

A I McLeod — Department of Statistical and Actuarial Sciences, University of Western Ontario, London, Ontario, Canada N6A 5B9

D F Merriam — Department of Geology, Syracuse University, Syracuse, New York 13210, USA

K Ord — *Department of Statistics, University of Warwick, Coventry CV4 7AL, England*

R A Reyment — *Paleontologiska Institutionen, University of Uppsala, Box 558, S-751 22 Uppsala, Sweden*

J E Robinson — *Department of Geology, Syracuse University, Syracuse, New York 13210, USA*

D H Root — *US Department of the Interior, US Geological Survey, Reston, Virginia 22092, USA*

A B Vistelius — *Laboratory of Mathematical Geology, V A Steklov Institute of Mathematics, USSR Academy of Sciences, Moscow, USSR*

E H T Whitten — *Department of Geological Sciences, Northwestern University, Evanston, Illinois 60201, USA*

F E Wickman — *Geologiska Institutionen, University of Stockholm, Kungstensgatan 45—Box 6801, S-113 86 Stockholm, Sweden*

Contents

Introduction

R G Craig, M L Labovitz

Languages are flexible systems, and the English language is perhaps most flexible. Hence terms such as *geomathematics* are spawned with what the reactionary grammarian might regard as reckless abandon. Our field is an intersection of the science of the Earth and a systematic mode of description through symbols and highly structured rules.

It is as inevitable as geology itself. We are part of a larger science by which the universe contemplates itself, and it is the dynamics of the Earth (and planets) that creates the science. Space and time—these are the domain of geology. Like geographers, we deal with the disposition of items in space; unlike geographers, we must deal in four dimensions. Like physicists, we study the dynamics of these items, but we must integrate over time scales not reproducible in the laboratory. Like astronomers, we deal with time scales vast and forbidding, but our interactions are strong, our time not reversible. We integrate a process that destroys itself. It is as if we are trying to reconstruct a function whose integral is not differentiable!

Thus many of the tools we require are common to other fields, but we must employ them in different ways *and* we need more: induction and deduction; the calculus and discrete mathematics; spatial models and time-series analysis; abstract algebra and spherical trigonometry.

Quantification, per se, is nothing. The art and science of geomathematics, that is, to uncover the kernal of the Earth system and express it in symbols having generative power, this alone validates the act of quantification. As in all sciences, we began with descriptions—simple nominal assignments that yield isomorphs of isolated observations. But the field has matured fast because it has been called to perform large tasks. Exploring for energy (chapters 2 and 10), temporalizing the stratigraphic record (chapter 12), structuring our observations in space (chapters 1, 9, 13, 16, and 17), and a thousand other applications of rigorous reasoning have molded a coherent body of knowledge. We stand on the threshold of a new view of the Earth.

If the world can be comprehended—a reasonable axiom for geoscientists—then we ask for the form. Neurons alone may not be the vehicle; our minds have limits, and those 10^{11} ciphers may not suffice. We note an emerging symbiosis with our machines. The operating system of the IBM 370 computer took 1600 brains to build it. Is it conceivable that the 'operating system' of the earth is any simpler? We wonder: can the answers we seek be expressed in a language comprehensible by a single person? Consider the response.

Model our earth at any moment: 10^{27} grams, about 10^{50} elementary
particles, 10^8 events per second; now extend it to 10^{17} seconds. Even if
we ignore the remainder of the universe (an unlikely choice), our system
through time still consists of 10^{67} events. We can only represent it within
the brain by compressing by a factor of 10^{-56}. Our symbols, then, must
have that power! We have far to go; it would appear unwise to balk at
the minor complexities given by geomathematics, we should look instead
for the advances needed after this one!

But we may need to abandon this compressive strategy. IF ... THEN
stores perhaps 10^7 events; we do not 'understand' the symbol, but can
use it. A well-conditioned assemblage of such brutes holds perhaps 10^{11}
events; hundred of these behemoths in tandem yield our geomodel with
less compression, but who will ride herd? We no longer evolve our science
in isolation. We stand aghast at the fearsome output of our machines;
they move now to swallow this and other sciences in the bits and bytes of
their data bases. What new forms of mental stress arise to replace our
aged profession? Griffiths has asked "If you were a committee of dinosaurs
called at the end of the Cretaceous to discuss the future of the race would
one of the options you considered have been *the correct one*?" The
committee on geology—when we have finally finished our collective
scientific duties—this group will oversee the events. Dare we imagine such
a postscientific period?

We offer, then, some small steps toward the future. Vistelius (chapter 12)
and Rement and Banfield (chapter 13) ask how much can be stored in a
transition matrix. Hipel and McLeod (chapter 5) and Labovitz (chapter 6)
show how to extract the geological structure of time. Dacey and Krumbein
(chapter 9) shake new information from a simple network. Spaces can be
modelled and compared as functional surfaces (Merriam and Robinson,
chapter 14) or as elements (Agterberg, chapter 1). Chayes (chapter 4) asks
how these are related and Gould (chapter 17) asks what is it to relate?
All of the authors have a common theme: the geological problem contains
a mathematical entity; does that entity (which I may, incidentally, be
able to 'solve') convey the essence of the problem?

We invite you to our work!

Part 1

Spatially dependent data

Cell-value distribution models in spatial pattern analysis

F P Agterberg

Introduction

This essay reviews methods of spatial analysis for modelling the shapes of frequency distributions of discrete phenomena measured with respect to cells or blocks of variable size. The frequency distribution of presence-absence data quantified for cells is of interest in several fields of geoscience. These include chemical analysis of rock samples, petrographic modal analysis, geostatistical ore-reserve estimation, litho-stratigraphic correlation of borehole data, geochemical exploration, line-spacing problems in exploration geophysics, drilling for hidden targets, and regional resource estimation.

Many geological phenomena are at least in part characterized by their boundaries or contacts with other phenomena. For example, an igneous rock consists of crystals belonging to different mineral phases. A stratigraphic section contains lithologies such as sandstone and shale which represent distinct states with an abrupt contact between them. Geological maps show the contacts between mappable rock units. On the other hand, many types of geochemical and geophysical measurements are averages for portions of the lithosphere which contain different types of materials. For example, a chemical analysis of an igneous rock is the average value of element concentration values for separate minerals. A seismic velocity reading may be the average value for different materials in a heterogeneous geological environment. Spatial pattern analysis will be useful for correlating the geometric probability of contacts between distinct states in geological environments by means of geophysical and geochemical measurements.

Griffiths (1970) has stated that changes in the paradigms (patterns of thought) of geology are needed if we are to benefit from a more widespread application of geomathematics. He goes on to predict that such changes will be brought about only by successful demonstration that the new paradigms can solve geological problems. Spatial pattern analysis is a useful topic of study in geomathematics. As stated before, it will aid in linking the quantitative measurements in geophysics and geochemistry with geological reality. It will also contribute towards placing geological observations, which are made at discrete points, in a regional framework. The aim of the following examples is to show that geological observations should often be considered as part of a broader environment.

If a contact between two sedimentary rock types has a specific strike and dip at a given point, it is of interest to know whether this measurement reflects a small local fold or belongs to a large regional structure.

Without further information of this type, the single measurement may be meaningless. A second example comes from economic geology. Suppose that a massive sulphide deposit is enclosed by sedimentary rocks which are stratigraphically underlain by acidic volcanic rocks. The latter, in turn, cap a volcanic pile of andesitic and basaltic lavas. An economic geologist may label this type of sulphide deposit as 'volcanogenic'. From a genetic point of view the deposit is more strongly related to the volcanic rocks than to the sedimentary rocks in its immediate vicinity. It is true that geological concepts such as this one have led to numerous controversies in the past. Nevertheless, it is useful in spatial pattern analysis to assume that a feature can be significantly correlated with features which occur at some distance from it. For this reason, geological pattern analysis may require a more elaborate mathematical apparatus than is available in methods of pattern recognition, to perform the multivariate analysis of attributes observed simultaneously at the same points in space.

A third reason for developing methods of spatial analysis is that the flow of groundwater, or other types of dynamic movement of matter through heterogeneous geological environments, can be influenced by the spatial variability. An example of this effect has recently been brought out by Freeze (1977) in hydrogeology. Existing models for groundwater flow in the vertical direction through a soil consist of deterministic equations with coefficients that are averages of the relevant attributes of the soil (hydraulic conductivity, compressibility, and porosity). However, Freeze has shown by Monte Carlo simulation experiments that groundwater movement is not only determined by average values for the soil but also depends on other properties of the frequency distributions of these attributes. If these effects are significant, the deterministic models may have to be replaced by stochastic process models in order to account for the heterogeneities in the system.

Autocorrelation functions
A basic tool of spatial pattern analysis is the autocorrelation function, $r_\alpha(d)$, of a random variable, X, measured at two points which are distance d apart in direction α. The methods to be discussed in this paper can be applied to phenomena which occur in either two-dimensional or in three-dimensional space. Most examples will be for patterns on two-dimensional maps. The random variable, X, which is measured at points in space, can be a binary variable for presence or absence of a feature. A binary variable assumes the value one if the feature is present and zero if it is absent. When a binary variable is coded with respect to the cells of a grid, a continuous random variable can be produced. For example, suppose that the copper in a porphyritic deposit occurs in chalcopyrite crystals which are dispersed through the rock. For very small cells, the copper concentration value would resemble a binary variable. However, it is a continuous variable for the small blocks processed for chemical analysis. In this

situation, it is advantageous to regard the element concentration value for small blocks as a continuous random variable which is measured at points in space. On the other hand, if a pattern in black and white for a given rock unit on a geological map is scanned and digitized on an image analyzer for the purpose of spatial analysis, the thresholded grey-level values at the discrete raster points become either black or white. It is desirable that a statistical model for spatial pattern analysis is comprehensive to the extent that the binary variables for small cells are linked with the continuous variables that can be defined for larger cells.

When the autocorrelation function $r_\alpha(d)$ is known, it is possible to estimate the variance of the continuous variable for the amount of rock type contained in a cell of given size and shape. This section contains a discussion of the choice of a suitable autocorrelation function.

Suppose that the random variable X has a mean value m, then the auto-covariance function, $C_\alpha(d)$, is

$$C_\alpha(d) = E\{(X_0 - m)[X_\alpha(d) - m]\} . \tag{1}$$

In equation (1), E denotes expected value, X_0 is the value of the random variable at an arbitrary point, and $X_\alpha(d)$ is the value of X at a point at a distance d from X_0 in direction α. The variance C_0 satisfies $C_0 = C_\alpha(0)$. The autocorrelation function, $r_\alpha(d)$, is defined as

$$r_\alpha(d) = \frac{C_\alpha(d)}{C_0} . \tag{2}$$

A possible choice of function for $r_\alpha(d)$ is

$$r_\alpha(d) = c\exp(g|d|) , \tag{3}$$

where c and g are constants, although it should be noted that Matheron (1965) prefers to use the function

$$\gamma_\alpha(d) = \tfrac{1}{2}E\{[X_0 - X_\alpha(d)]^2\} , \tag{4}$$

because the use of $C_\alpha(d)$ implies the assumption that there exists a stationary mean, m. This assumption is not needed when the function displayed in equation (4) is used. In this essay, the existence of a stationary mean is assumed mainly for reasons of convenience. If a study area is delineated on a two-dimensional pattern where a rock type is shown in black, m is simply the proportion of black in this study area. It also denotes the probability that a random point in the study area belongs to the pattern.

A value of c equal to one in equation (3) gives the exponential auto-correlation function which is frequently used in time-series analysis, where it represents the solution for a sequence of data X_k ($k = ..., -2, -1, 0, 1, 2, ...$) with the Markov property $P(X_k | X_{k-1}, X_{k-2}, ...) = P(X_k | X_{k-1})$. The conditional probability that the random variable assumes a specific value depends on a single previous value only. Specific models in two dimensions, which produce exponential autocorrelation functions, have been constructed by Matérn (1960) and Switzer (1965). A Boolean scheme described by Serra (1976) for porous media also gives an exponential autocorrelation function. This scheme consists of a set of grains of different forms and shapes which are implanted at points randomly distributed in space according to a simple Poisson point process.

The situation that $0 < c < 1$ in equation (3) arises when every measured value of the random variable X can be regarded as the sum of a stationary random variable, S, with exponential autocorrelation function $\exp(g|d|)$ and a random variable, N, for which the autocorrelation function is a spike at the origin and equal to zero for any nonzero value of d. The model, $X = S + N$, is the signal-plus-noise model of the statistical theory of communication. S changes continuously from place to place and N is a superimposed source of local variability, which changes abruptly from place to place. If the variance of X is equal to C_0, the variance of S is cC_0, and that of N is $(1-c)C_0$. Thus, the constant c can be interpreted as the ratio between the variance of the continuous variability and the variance of the total variability. The purpose of introducing a noise component into the model is to allow the elimination of the effect of the local variability.

It is noted that Whittle (1963) and Bartlett (1975) have argued that the exponential autocorrelation function does not arise as naturally in spatial pattern analysis as it does in one-dimensional time-series analysis. These authors have advocated a model where the Markov property (stated above) is replaced by the nearest-neighbour property, that is:

$$P(X_k | \dots X_{k-2}, X_{k-1}, X_{k+1}, X_{k+2}, \dots) = P(X_k | X_{k-1}, X_{k+1}) .$$

This leads to another type of autocorrelation function in two dimensions:

$$r_\alpha(d) = qd\mathrm{K}_1(qd) , \tag{5}$$

where q is a constant and $\mathrm{K}_1(qd)$ is a modified Bessel function of the second kind. The choice of equation (3) instead of equation (5) is justified in the following situation.

Suppose that a straight line is drawn at random across the pattern of a specific rock type on the geological map. If presence of the rock type is coded as one, and its absence as zero, the variation of the pattern along the line will be as shown in figure 1(a). As before, m represents the mean or proportion of the study area underlain by the rock type. Suppose that the location of each contact point (change from zero to one, or from one to zero) can be regarded as random and that the probability, p_k, that a line segment of length d contains exactly k contact points, has a Poisson distribution:

$$p_k = \frac{(\lambda|d|)^k}{k!}\exp(-\lambda|d|) , \tag{6}$$

where λ is a constant.

The autocovariance function $C_\alpha(d)$ is the expected value of the product D, where $D = (X_0 - m)[X_\alpha(d) - m]$, or $C_\alpha(d) = E(D)$. D can have only one of the three possible values: (a) $D = (1-m)^2$, if $X_0 = X_\alpha(d) = 1$; (b) $D = m^2$, if $X_0 = X_\alpha(d) = 0$; and (c) $D = m(1-m)$, otherwise. The probability that situation (a) arises is equal to the probability $p(X_0 = 1)$, where $p(X_0 = 1) = m$, multiplied by the sum of the probabilities, p_{2k}, that the number of contact points per distance d is an even number. This ensures that $X_\alpha(d) = 1$, if $X_0(d) = 1$. When similar arguments are developed for the situations (b) and (c), it follows that

$$C_\alpha(d) = \sum_{k=0}^{\infty} [m(1-m)^2 p_{2k} + (1-m)m^2 p_{2k} - m(1-m)p_{2k+1}]$$

$$= m(1-m) \sum_{k=0}^{\infty} (p_{2k} - p_{2k+1}) . \qquad (7)$$

Substitution of equation (6) into equation (7) and use of the following two equalities

$$\sum_{k=0}^{\infty} \frac{(\lambda|d|)^{2k}}{(2k)!} = \tfrac{1}{2}[\exp(\lambda|d|) + \exp(-\lambda|d|)] ,$$

$$\sum_{k=0}^{\infty} \frac{(\lambda|d|)^{2k+1}}{(2k+1)!} = \tfrac{1}{2}[\exp(\lambda|d|) - \exp(-\lambda|d|)] ,$$

gives:

$$C_\alpha(d) = m(1-m)\exp(-2\lambda|d|) . \qquad (8)$$

Division of both sides of equation (8) by the variance C_0 $[= m(1-m)]$ gives the exponential autocorrelation function in equation (3), with $c = 1$, and $g = 2\lambda$. Hence the coefficient g can be interpreted as twice the number of contact points per unit distance. Empirical work on some

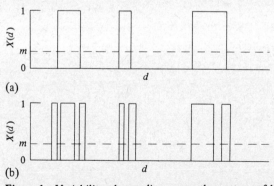

(a)

(b)

Figure 1. Variability along a line across the pattern of hypothetical rock units. Points belonging to the specified rock type have value one. Here m represents the proportion of total study area underlain by the specified rock type. (a) Number of contact points per unit of distance as a random variable with a Poisson distribution; (b) as (a), with a random variable modelling local variability superimposed on the Poisson distribution.

types of contacts between rock types on geological maps (Agterberg, 1978; Agterberg and Fabbri, 1978) has indicated that the signal-plus-noise model with $c < 1$ gives better results than an exponential function with $c = 1$. This modified model can be interpreted as the sum of the pattern of figure 1(a) and another random process whose influence is restricted to the immediate surroundings of the contacts. The resulting situation of multiple contacts is schematically represented in figure 1(b).

It may be concluded that the model for spatial autocorrelation of discrete patterns discussed in this section is characterized by three parameters $(m, g, \text{ and } c)$, all of which can be interpreted in a geometrical sense.

For a square cell, S, having a side of length a, the variance $C_0(a)$ of the proportion of the square which is underlain by the rock type is given by the formula (see Agterberg, 1978):

$$C_0(a) = \frac{C_0 c}{a^4} \int_s dx \int_s \exp(-g|x - y|)dy , \qquad (9)$$

where x and y are vectors denoting the location of a random point within S. Serra (1969) has constructed a nomogram which can be used to obtain numerical estimates of this quadruple definite integral of the exponential function. This nomogram is for the variance C_a which, after multiplication by c, gives the variance, cC_a, of the signal at a point located at random within S. If C_a is determined by this method, then $C_0(a) = c(C_0 - C_a)$, as the variance, cC_0, of the signal at a random point in the study region is the sum of cC_a and $C_0(a)$ (see Matheron, 1965). This relationship is only valid if the noise component, which has variance $(1 - c)C_0$, is strictly local.

The curve shown in figure 2 gives $C_0(a)/cC_0$ as a function of ag. A numerical solution can also be obtained by Monte Carlo simulation.

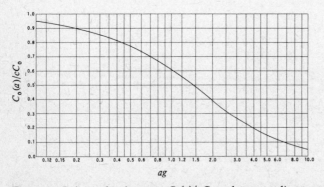

Figure 2. Relationship between $C_0(a)/cC_0$ and ag according to equation (9). $C_0(a)$ is the variance of the proportion of a square cell with side a composed of a given binary variable, C_0 is the variance of the binary variable measured at points, and a and g are coefficients of the theoretical autocorrelation function.

$C_0(a)/cC_0$ is the expected value of the random variable U with

$$U = \exp\{-g[(X_1 - X_2)^2 + (Y_1 - Y_2)^2]^{1/2}\} , \tag{10}$$

where X_1, X_2, Y_1, and Y_2 are independent random variables with uniform distributions on the interval $[0, a]$. Numerical estimates of $E(U)$ can be obtained by using the interactive graphic computer program SIMGRAI by Chung et al (1978).

Models of frequency distributions of cell values

In the previous section it was shown that by defining a mean and an auto-correlation function, it is possible to estimate the variance of the value of a random variable in a cell. The method is applicable to discrete or continuous variables in two-dimensional or three-dimensional space. Suppose that X_0 represents the random variable measured at points, and X the random variable obtained by averaging X_0 over a square cell with a side of length a. X then has mean m and variance $C_0(a)$. It is interesting to model the frequency distribution of X as a function of a and the auto-correlation function in equation (3). If X_0 is a binary variable, its density function will consist of two spikes with frequencies $1 - m$ at $X_0 = 0$, and m at $X_0 = 1$. As the cell size increases, the density function first assumes a U-shape. For $m < 0.5$, it will gradually lose its peak at $X = 1$, and start to resemble a gamma distribution for which the density tends to infinity as X tends to zero. With further increases in cell size the density function will lose the peak at $X = 0$ and become unimodal like a lognormal distribution. In the limit, for very large cells, it may assume the shape of a normal distribution. Two examples may illustrate the usefulness of modelling this type of frequency distribution.

A classical problem of ore-reserve estimation consists of predicting the shape of the frequency distribution of the average metal-grade values of mining blocks from the known frequency distribution of assay values, which are the average grade values of the much smaller blocks processed for chemical analysis. Knowledge of the shape of this frequency distribution is useful in order to determine in advance the percentage of blocks, of a given size and shape in the mineral deposit, that can be mined and milled. The average grades of these blocks should exceed a cutoff grade determined by economic considerations. If its average grade is less than this cutoff grade, a block should be left in the ground or removed as waste.

The use of methods of conditional probability for the estimation of regional mineral resources, based on data for relevant geological, geophysical, or geochemical parameters is a second example in which the frequency distributions of data observed by cell may be studied. Since it is desirable to normalize the parameters in order to correlate them with one another and with dependent variables defined for the occurrence, size, and grades of known mineral deposits, the choice of the transformations required to obtain a multivariate normal distribution is important. This transformation is dependent upon shapes of the original frequency distributions, which in turn are dependent on the size of the cells used for quantification.

Theoretical models for frequency distributions in spatial analysis have been proposed by several authors. The following four models can be used for this purpose and will be briefly reviewed: (1) the Neyman–Scott clustering model, (2) the de Wijsian model, (3) the Prokhorov model, and (4) a model based on the transfer-function theory of Matheron.

The Neyman–Scott clustering model
The earliest version of this model was proposed by Neyman (1939). An extensive review is given by Neyman and Scott (1972) as part of a comprehensive collection of papers on stochastic point processes (Lewis, 1972). This model uses a simple Poisson process for the distribution of points in space. This process can arise in the following manner. Suppose that a region A is subdivided into n cells: A_1, A_2, ..., A_n. A natural definition of randomness of points is to assume that the probability of a single point falling in A_i is given by the ratio mes(A_i)/mes(A), where mes denotes measure (the area in two-dimensional space). For an arbitrary number of points this process results in the independent Poisson variables $N(A_1), N(A_2), ..., N(A_n)$ with parameters λmes(A_1), λmes(A_2), ..., λmes(A_n). In the simple Poisson process model, the number of points in a cell of a given size and shape is, therefore, a Poisson variable controlled by the constant λ, and the size of the cell.

The basic assumptions of the Neyman–Scott model are: (a) the points occur in clusters, the centres of which are distributed according to a simple Poisson process; (b) the number of points per cluster is a random variable; and (c) if a cluster centre occurs at location u, the points belonging to this cluster are randomly distributed according to a probability distribution $f(x|u)$, where x denotes distance from the centre. An application of this model to the occurrence of massive sulphide deposits in the Bathurst area, New Brunswick, has been given by Agterberg (1976).

Neyman and Scott have derived the equation of the joint frequency distribution for numbers of points in two cells of variable size and shape. However, further work would be required to apply this type of model to the solution of the problems of spatial analysis presented at the beginning of this section.

The de Wijsian model
This model was formulated by de Wijs (1951; 1953). It has recently been advocated by several authors including Brinck (1974) and Harris (1977), who have used it to represent the average concentration value of elements such as uranium in portions of the lithosphere.

In developing his model, de Wijs (1951; 1953) argued as follows. If an ore body consists of 2^k blocks ($k = 0, 1, 2, ...$), then the logarithms of the concentration values of the blocks describe a binomial distribution. For increasing values of k, the frequency distribution of the concentration values of the blocks rapidly becomes lognormal with variance, σ^2, given by

$$\sigma^2 = \frac{k}{4}\left[\ln\left(\frac{1-Q}{1+Q}\right)\right]^2, \tag{11}$$

where Q is a constant independent of k. According to de Wijs, the constant Q is to satisfy the following condition. Suppose that the entire ore body, with average concentration value m, is divided into two sub-blocks of equal weight with concentration values W_1 and X_1, with $W_1 = (1+Q)m$ and $X_1 = (1-Q)m$; then the absolute value of the difference between these two values is $2Qm$. When each of the two blocks is again subdivided into two equal parts, there would be four blocks with concentration values given by $W_2 = (1+Q)^2m$, $X_2 = (1+Q)(1-Q)m$, $X_2 = (1-Q)(1+Q)m$, and $Y_2 = (1-Q)^2m$. The process of subdividing blocks into two equal parts can be continued k times. For each subdivision, the logarithms of the concentration values have a binomial distribution.

In practice for ore bodies the concentration values of blocks of constant size may satisfy a lognormal distribution (discussed later). This type of distribution is determined by two parameters (mean, m, and variance, σ^2), with equation (11) representing the change in shape of the lognormal distribution with changes in the block concentration values. However, the interpretation of the coefficient Q is not simple when the geometry of the ore body is considered.

Suppose that the 2^k blocks form a linear sequence as in the original example of de Wijs (1951; 1953) for the Pulacayo vein-type zinc deposit in Bolivia. At the first subdivision ($k = 1$) the two possible arrangements of the blocks with concentration values W_1 and X_1 can be written as $\{W_1 X_1\}$ and $\{X_1 W_1\}$. The four blocks at $k = 2$ with concentration values W_2, X_2, X_2, and Y_2 have four possible arrangements: $\{W_2 X_2 Y_2 X_2\}$, $\{X_2 W_2 X_2 Y_2\}$, $\{X_2 Y_2 X_2 W_2\}$, and $\{Y_2 X_2 W_2 X_2\}$. These arrangements satisfy two specific conditions: (1) the absolute value of the difference between any two successive values is equal to $2Qm$; and (2) averaging the first two values and the last two values yields one of the two possible sequences for $k = 1$. However, the total number of possible arrangements of the values W_2, X_2, X_2, and Y_2 is twelve. Eight of these arrangements do not satisfy the preceding two conditions and should be rejected as possible arrangements. From this sample, it may be concluded that the subdivision of a block into two parts of equal weight, with the absolute value of difference in concentration equal to $2Qm$, is not a simple random process but depends on the spatial configuration of the blocks. It is possible to form similar sequences for larger values of k.

Matheron (1965) has shown that the function $\gamma_\alpha(d)$ [equation (4)] becomes

$$\gamma_\alpha(d) = 3A\ln d \tag{12}$$

for the model of de Wijs. The relationship between the constants A and Q is

$$A = \left[\ln\left(\frac{1-Q}{1+Q}\right) \right]^2 \Big/ 4 \ln 2 . \tag{13}$$

The variance, var(v, V), of smaller blocks with volume v contained in a larger block with volume V satisfies the expression,

$$\text{var}(v, V) = A \ln\left(\frac{V}{v}\right) . \tag{14}$$

The model described by equations (12) and (14) has been applied to the Pulacayo vein-type zinc deposit with good results. Alternatively, Agterberg (1974a) has successfully used the autocorrelation function given in equation (3) to model the same deposit. Therefore the de Wijsian model can be used in modelling the variance of mean concentration values for cells. However, it does not provide a simple method for modelling the corresponding frequency distribution.

Thus two different statistical models have been applied to the same set of data from the Pulacayo deposit. The question of which method is best cannot be answered by a study of the discrepancies between model and observed data because both models provide a good fit. Nevertheless it is interesting to consider the implications of the different methods. Matheron (1965) has pointed out that a stationary mean and autocorrelation function do not exist in the de Wijsian model. Thus a mean value, m, computed from all data for this mineral deposit, is not a meaningful parameter. For example, within the framework of the de Wijsian model, m could not be used for extrapolation to other parts of the same mineral deposit, or to other deposits of the same type. On the other hand, a geologist might prefer to define the average concentration for a mineral deposit and estimate it as simply as possible by computing the arithmetic mean. The hope is that m has geochemical significance for a particular type of mineral deposit, and can be used to extrapolate outside the sampled area.

The Prokhorov model
Prokhorov (1965) tried to explain the widespread occurrence of approximately lognormal distributions of element concentration values in rocks. He generalized the simple Poisson process to model the spatial distribution of mineral particles.

Let $X(v)$ denote the concentration of an element in volume v. Prokhorov assumes that: (1) for the disjoint volumes $v_1, v_2 \ldots , X(v_1), X(v_2) \ldots$ are independent random variables; (2) the distribution function of $X(v)$ depends only on the size of v; and (3) all values of $X(v)$ tend to zero as v tends to zero. Under these assumptions, the distribution function of $X(v)$ satisfies Kolmogorov's definition of an infinitely divisible function of v (Gnedenko and Kolmogorov, 1954). That is, $X(v)$ will have the same form of the frequency distribution for any volume v. In an example, Prokhorov then assumes that $X(v)$ has a gamma distribution. The application of a form for canonical representation of infinitely divisible distributions (Gnedenko and Kolmogorov, 1954) then shows that the logarithm of $X(v)$ has an asymptotically normal distribution as v increases.

Although the Prokhorov model has not been developed further, it brings out several important points. The frequency distribution of $X(v)$ either has to be an infinitely divisible function in order to preserve its shape, or it has to be a good approximation to a function which is infinitely divisible. Although other infinitely divisible functions exist, the three prime candidates for consideration in cell-value distribution modelling are the Poisson, gamma, and normal distributions.

An important aspect to be considered in this type of approach is the autocorrelation of the variable $X(v)$. For example, if two adjacent volumes have values which are described by identical gamma distributions, their combined volume will have a gamma distribution which is easily parameterized only if the initial two mean values are realizations of independent gamma-distributed random variables. This condition of stochastic independence is only rarely satisfied in practical applications. In order to generalize the Prokhorov model, it would be necessary to use a multivariate gamma distribution model as described in Johnson and Kotz (1972, chapter 40).

It is noted that the lognormal distribution function is not infinitely divisible. In general, the average of several lognormally distributed variables is not another lognormal variable. In practice, however, a very close approximation to a lognormal distribution can be preserved. The preservation of a lognormal distribution in mineral deposits was first shown empirically by Krige (1951) for gold contained in blocks ranging in size from chip samples to entire mines in the Witwatersrand gold fields of South Africa. Switzer and Parker (1976) have shown by computer simulation that the lognormal model may continue to provide a good approximation for grade values of large blocks which are averages taken from autocorrelated lognormal variables for smaller blocks, and the Prokhorov model can be used to describe this phenomenon. Other models providing explanations of the rule of preservation of lognormality will be discussed later in this essay.

A model based on the transfer-function theory of Matheron
Suppose that part of an ore body with volume V is subdivided into n blocks of equal volume, v_i, where $v_i = v$ for all i. Suppose further that the random function, X, is measured at n points with values X_j ($j = 1, 2, ..., n$). According to Matheron (1976a, page 245) the 'transfer function' is defined as the conditional distribution of $X(v)$, given X_j ($j = 1, 2, ..., n$), where $X(v)$ represents the average value of a block selected at random from among the n blocks. The term 'transfer function' has a different meaning in time-series analysis and the statistical theory of communication. For this reason, Matheron's transfer functions are described as conditional

average-cell-value frequency distributions or 'cell-value distributions' elsewhere in this paper.

Matheron (1976a) and Journel and Huijbregts (1978) have considered several types of problems which can be solved by use of transfer-function theory. An important problem is the estimation of the frequency distribution of the unknown mean values of large blocks from known assay values. This problem is solved by using disjunctive Kriging (Matheron, 1973; 1976b).

Matheron also distinguishes between direct and indirect transfer functions. Indirect transfer functions are based on the observations X_j ($j = 1, 2, ..., n$) available at the time when the ore-reserve estimation is made, but they also allow for the incorporation of data that will become available in the course of future mining operations. Applications of transfer-function theory to ore-reserve estimation have been made by Maréchal (1974; 1975; 1976) and Kim et al (1977). Only direct transfer functions are considered in this paper. The assumptions that will be made are similar to those made by Matheron in his 'discrete Gaussian model'.

In this model, which resembles the statistical technique of stratified random sampling, the volume V is subdivided into n disjoint blocks, v_i, of equal size v. The random variable X is averaged for each of the blocks to obtain the cell values X_i. In general, there exists a transformation ψ_i, such that $X_i = \psi_i(Z)$, where Z is a Gaussian random variable with zero mean and unit variance. Suppose that the value, X_0, of X is measured at a single point which is randomly located within a block v_i with cell value X_i. Let X_0 be related to a Gaussian variable Z_0 by the expression $X_0 = \psi(Z_0)$. Then the conditional expectation of X_0, given X_i, is equal to X_i because the point at which X is measured has a random location within v_i, that is

$$E(X_0|X_i) = X_i . \tag{15}$$

A complete specification of the discrete Gaussian model is obtained by assuming that: (1) the joint distribution of Z_0 and Z_1 is bivariate normal with a correlation coefficient, r, which is independent of location; (2) if Z_i is given for a cell, the corresponding value of Z_{0i} is stochastically independent of the values of Z_{0j} and Z_j in other cells ($j \neq i$); and (3) Z_i ($i = 1, 2, ..., n$) has a multivariate normal distribution. Under these conditions, the random variables Z_i and Z_{0i} ($i = 1, 2, ..., n$) have a multivariate normal distribution with correlation coefficients given by $\rho_{ij} = E(Z_{0i}, Z_{0j})$; $r_{ij} = E(Z_{0i}, Z_j)$; and $R_{ij} = E(Z_i, Z_j)$. These correlation coefficients are related to one another by the expressions $\rho_{ij} = r^2 R_{ij}$ and $r_{ij} = r R_{ij}$. For example, suppose that X_0 is lognormally distributed with mean m and variance σ^2. It can then be shown that X_i also has a lognormal distribution with variance $r^2\sigma^2$ (Matheron, 1976a, page 245). The correlation coefficients ρ_{ij}, r_{ij}, and R_{ij} can then be computed from the logarithmically transformed data, and the validity of the model can be tested in practical applications (see Agterberg, 1977).

The applicability of the discrete Gaussian model depends mainly on how well assumption (2)—that the deviations $(Z_{0i} - Z_i)$ are uncorrelated—is satisfied in practice. In a similar fashion, the validity of equation (15) depends on the deviations $(X_{0j} - X_{ij})$ being uncorrelated. Because a stratified random-sampling scheme is used, the deviations of the discrete Gaussian model are uncorrelated regardless of the nature of the auto-correlation function of the stationary random variable X. Suppose, however, that X_0 is the value of X at the midpoint of v_i, or that some other method of sampling is used. Even then the deviations are approximately uncorrelated provided that X has a stationary function $\gamma_\alpha(d)$ (Matheron, 1965, page 198). The latter condition is automatically satisfied when X has both a stationary mean and a covariance function $C_\alpha(d)$. It may be concluded that the discrete Gaussian model of Matheron can provide useful methods to obtain cell-value distribution models in spatial analysis.

A useful cell-value distribution model

In an early application of the discrete Gaussian model, Matheron (1974, pages 57–61) has considered the following cell-value distribution model for binary patterns:

$$X = \Phi[(rZ_1 - b)(1 - r^2)^{-\frac{1}{2}}] . \tag{16}$$

In equation (16), X is the random variable for the cell values generated from a binary variable X_0, Z is a normally distributed random variable with zero mean and unit variance, and $\Phi(u)$ represents the fractile of the cumulative normal distribution function with

$$\Phi(u) = \frac{1}{(2\pi)^{\frac{1}{2}}} \int_{-\infty}^{u} \exp(-\tfrac{1}{2}x^2) \mathrm{d}x . \tag{17}$$

When m represents the mean value of X for the study region, the constant b in equation (16) is defined such that

$$\Phi(b) = 1 - m . \tag{18}$$

The constant r in equation (16) represents the correlation coefficient between the Gaussian random variable Z of equation (16) and a Gaussian variable Z_0 which corresponds to the binary variable X_0. The value of r can be derived from the mean, $\bar{X}$, and variance, s^2, of the random variable X by use of the expression:

$$s^2 = \phi^2(b) \sum_{j=1}^{\infty} r^{2j} [H_{j-1}(b)]^2 , \tag{19}$$

where $\phi(u) = (2\pi)^{-\frac{1}{2}} \exp(-\tfrac{1}{2}u^2)$, and $H_j(u)$ denotes the standardized Hermite polynomial:

$$H_j(u) = u^j - \frac{j^{(2)}}{2\,1!} u^{j-2} + \frac{j^{(4)}}{2^2\,2!} u^{j-4} - \frac{j^{(6)}}{2^3\,3!} u^{j-6} + \dots , \tag{20}$$

where

$$j^{(t)} = j(j-1)(j-2) \dots (j-t+1) .$$

For example, the first five standardized Hermite polynomials are:

$$H_0(u) = 1 , \qquad\qquad H_3(u) = (u^3 - 3u)(6)^{-\frac{1}{2}} ,$$
$$H_1(u) = u , \qquad\qquad H_4(u) = (u^4 - 6u^2 + 3)(24)^{-\frac{1}{2}} ,$$
$$H_2(u) = (u^2 - 1)(2)^{-\frac{1}{2}} , \qquad H_5(u) = (u^5 - 10u^3 + 15u)(120)^{-\frac{1}{2}} .$$

The series expansion for the variance in equation (19) converges as j increases. The mean $\bar{X}$ and variance s^2 represent estimates of the population mean, m, and variance, $C_0(a)$, respectively. These can be obtained by two different methods.

(a) A sampling grid with cells of side a is superimposed on the study area. The proportion of total area which is underlain by the pattern is measured in each of these cells. $\bar{X}$ and s^2 are then computed as the arithmetic mean and variance of these proportions.

(b) The pattern can be quantified and measured with a flying-spot scanner (Agterberg and Fabbri, 1978), or with the Quantimet 720 image analyzer (Imanco, Boston) with linear correlator (Agterberg, 1978). This gives a value of m and provides the geometrical covariance which can be modelled statistically to obtain values of c and g [equation (3)]. The variance $C_0(a)$ then follows from equation (9) or figure 2.

Figure 3 can be used in both methods to obtain an estimate of r from the known values of $\bar{X}$ or m and s^2 or $C_0(a)$.

Agterberg and Fabbri (in press) used method (b) for the spatial analysis of an outcrop pattern of acidic volcanic rocks in the Bathurst area, New Brunswick. For a square study area measuring 80 km on a side, they

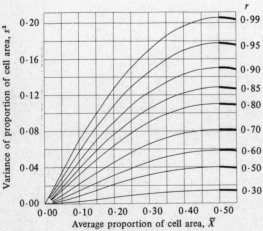

Figure 3. Relationship between estimates of the mean, $\bar{X}$, and the variance, s^2, of the proportion of cell area underlain by the rock type considered, for various values of r [equations (16) and (19)].

obtained $m = 0 \cdot 1973$, $c = 0 \cdot 87$, and $g = 0 \cdot 194$. This allowed the estimation of $C_0(a)$, and consequently of r, for cells with a equal to 4, 10, 20, and 40 km.

Unlike r in equation (16), b is independent of a. According to equation (18), b depends only on m. The corresponding frequency distributions for X are shown in figure 4. In the limiting case ($a = 0$), the density function of the binary variable X_0 consists of separate spikes at $X_0 = 1$ (with frequency m).

The frequency distribution of X in equation (16) becomes a straight line if the function is plotted on paper which has a normal probability scale both for the abcissa and for the ordinate (prob-prob line), representing the cumulative frequency percentage and the percentage cell area underlain by the pattern of interest, respectively. A straight line is also obtained by plotting the probit of the cell value (see Finney, 1971) on normal probability paper. This procedure facilitates the comparison between the model given by equation (16) and any observed frequency distribution for a sample obtained by superimposing a grid on the pattern and measuring the cell values.

Nine examples of this graphical test of the model are shown in figure 5. The first two plots are for the previously discussed acidic volcanic rocks of Middle-Upper Ordovician age in the Bathurst area. The mean $\bar{X}$ ($= 0 \cdot 1973$) has corresponding variances of (1) $s^2 = 0 \cdot 0919$ for sixty-four 10 km × 10 km [$(10 \text{ km})^2$] cells, and (2) $s^2 = 0 \cdot 0622$ for sixteen $(20 \text{ km})^2$ cells obtained by averaging values for blocks of four $(10 \text{ km})^2$ cells. These data yield the values of b and r shown in figure 5.

The third example in figure 5 is for the complement of acidic rocks in $(10 \text{ km})^2$ cells for the same study area. The other rocks which constitute

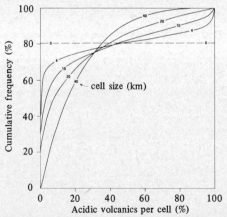

Figure 4. Sequence of cell-value distributions satisfying equation (16) for different lengths of cell sides. The example is for acidic volcanic rocks in the Bathurst area, New Brunswick (from Agterberg and Fabbri, in press).

the complement of acidic rocks have a mean value equal to $1 - \overline{X}$, and a variance equal to that of the acidic volcanics. The value of b is therefore equal to $1 - 0 \cdot 85$ or $0 \cdot 15$. From equation (19) and figure 5 it can be seen that s^2 and r are symmetric about $b = \overline{X} = 0 \cdot 5$.

The fact that percentage values for a rock type and its complement can both have a frequency distribution which satisfies equation (16) may be of interest in the study of closed number systems. If the values for a set of random variables sum to one at all observation points, these variables cannot have the same type of distribution if one or more of them has a normal, lognormal, or gamma distribution. On the other hand, each variable in a set of random variables summing to one can have the distribution function given by equation (16).

The next four examples shown in figure 5 are data from Archean acidic and mafic volcanic rocks in the Abitibi area of the Canadian Shield in east-central Ontario and western Quebec. Measurements were made for 768 $(10 \text{ km})^2$ cells. The acidic volcanics have a mean equal to $0 \cdot 0526$ and sample variances equal to $0 \cdot 0121$, $0 \cdot 0072$, and $0 \cdot 0039$ for $(10 \text{ km})^2$,

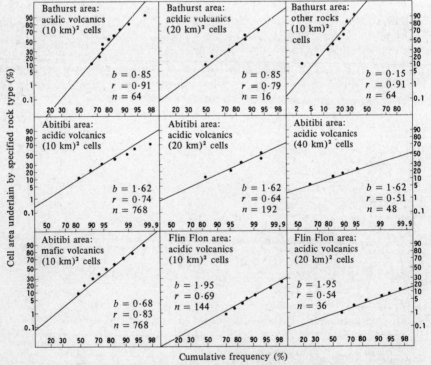

Figure 5. Experimental frequency distributions for nine samples of cell values plotted on two normal probability scales. The values of b and r were obtained from the means and variances of the cell values by using equations (18) and (19). They determine the positions and slopes of the prob-prob lines which satisfy equation (16).

(20 km)^2, and (40 km)^2 cells, respectively. This information was previously used as an example by Agterberg (1977), who fitted lognormal distributions to the data for cells with sides measuring 20 km or more. In this earlier work, it was pointed out that the lognormal distribution could not be fitted to the values for (10 km)^2 cells because there were too many cells not containing any acidic volcanics. However, a gamma distribution, with density approaching infinity as the cell value tends to zero, could be used as a model for the (10 km)^2 cells. The results shown in figure 5 indicate that the new model provides a good fit for all three cell sizes. Furthermore, the model given by equation (16) remains applicable with decreasing cell sizes until the binary variable X_0 is reached in the limit. This example also illustrates the close relationship between the new model, the gamma distribution model, and the lognormal distribution model.

The mafic volcanics in the Abitibi area have $\bar{X} = 0 \cdot 2473$, and $s^2 = 0 \cdot 0854$. They are nearly five times as abundant as the acidic volcanics in this region. Their frequency distribution can also be approximated by the distribution given in equation (16).

The final two examples of figure 5 are for the Proterozoic acidic volcanic rocks in an area of 144 (10 km)^2 cells near Flin Flon, Manitoba. These observations have an average value equal to $0 \cdot 0259$ and variances equal to $0 \cdot 0038$ and $0 \cdot 0018$ for (10 km)^2 and (20 km)^2 cells, respectively.

It may be concluded that the model given by equation (16) provides a good fit to the data used in figure 5. The observed data tend to lie about the line approximated by the theoretical line described by equation (16), with b and r estimated from $\bar{X}$ and s^2.

Some theoretical properties of the new model
Suppose, as before, that the random variables X and Z are related by $X = \psi(Z)$, where the distribution of Z is standardized normal with distribution function $\Phi(Z)$, and X has distribution function $F(X)$ and density function $f(X)$. The transformation $\psi(Z)$ satisfies

$$\int_{-\infty}^{X = \psi(z)} f(u)\,du = \int_{-\infty}^{z} \phi(u)\,du = \Phi(Z) , \qquad (21)$$

and is represented schematically in figure 6(a). This type of transformation is used by Matheron (1974; 1976a) in his transfer-function theory. It also corresponds to the Cornish–Fisher expansions of statistical theory (see Johnson and Kotz, 1970, page 33).

If X is a binary variable, it can also be related to a normally distributed random variable by the transformation $X = \psi_0(Z)$, as shown in figure 6(b).

Suppose that Z_1 and Z_2 are two random variables distributed bivariate normal with density function

$$\phi(z_1, z_2) = (2\pi)^{-1}(1-r^2)^{-\frac{1}{2}}\exp[-(z_1^2 - 2rz_1z_2 + z_2^2)/2(1-r^2)]$$

$$= (2\pi)^{-1}\exp[-\tfrac{1}{2}(z_1^2 + z_2^2)]\left[1 + \sum_{j=1}^{\infty} r^j H_j(z_1)H_j(z_2)\right] . \qquad (22)$$

Equation (22) represents the so-called 'Mehler identity' by means of which a bivariate normal distribution is expressed as the product of its marginal distributions, $\phi(z_1)$ and $\phi(z_2)$, and a series expansion using the Hermite polynomials of equation (20) (Kendall and Stuart, 1961; Lancaster, 1958).

The Mehler identity is applicable to any pair of continuous variables, say z_1 and z_2, which are not necessarily values assumed by random variables Z_1 and Z_2.

Division of equation (20) by $\phi(z_1)$ yields

$$\frac{\phi(z_1, z_2)}{\phi(z_1)} = \phi\left[\frac{z_2 - rz_1}{(1 - r^2)^{1/2}}\right] = \phi(z_2)\left[1 + \sum_{j=1}^{\infty} r^j H_j(z_1) H_j(z_2)\right] . \tag{23}$$

The Hermite polynomials have the property that

$$\int_{-\infty}^{z_2} H_j(z_2)\phi(z_2)dz_2 = H_{j-1}(z_2)\phi(z_2) . \tag{24}$$

After integration of equation (23) with respect to z_2, using the equality $\Phi(-y) = 1 - \Phi(y)$, and substitution of equation (24), it follows that

$$\Phi\left[\frac{rz_1 - z_2}{(1 - r^2)^{1/2}}\right] = 1 - \Phi(z_2) - \phi(z_2)\sum_{j=1}^{\infty} r^j H_j(z_1) H_{j-1}(z_2) . \tag{25}$$

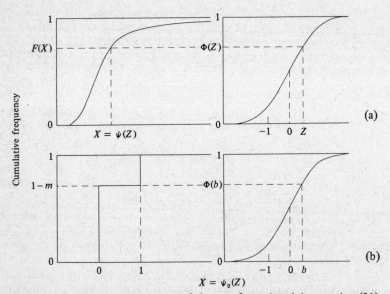

Figure 6. Graphical representation of the transformation ψ in equation (21). Distribution functions $\Phi(Z)$ on the right-hand side are for standard normal random deviates. A value $F(X)$ on the left-hand side of (a) is equal to $\Phi(Z)$ if $X = \psi(Z)$. (b) shows the distribution function of a binary random variable which can also be related to a standard normal random variable.

Setting z_2 equal to the constant b and replacing z_1 by the random variable Z, one obtains:

$$X = \Phi\left[\frac{rZ-b}{(1-r^2)^{1/2}}\right] = 1 - \Phi(b) - \phi(b) \sum_{j=1}^{\infty} r^j H_j(Z) H_{j-1}(b) . \tag{26}$$

The binary variable X, where $X = \psi_0(Z)$, shown in figure 6(b), with $\psi_0(Z) = 1$ for $Z \geqslant b$, and $\psi_0(Z) = 0$ for $Z < b$, can be obtained from equation (16) by allowing r to tend to one. Then,

$$\psi_0(Z) = \lim_{r \to 1} \Phi\left[\frac{rZ-b}{(1-r^2)^{1/2}}\right] = 1 - \Phi(b) - \phi(b) \sum_{j=1}^{\infty} H_j(Z) H_{j-1}(b) . \tag{27}$$

In general, suppose that X_1 and X_2 are random variables both arising from the single random variable X, but representing a set of smaller and a set of larger cells, respectively. Suppose further that the discrete Gaussian model applies with $E(X_1) = E(X_2) = m$, and $E(X_1|X_2) = X_2$ [see equation (15)]. Note that X_1 is the value of a cell selected at random from among the cells into which the larger cell, possessing value X_2, has been subdivided. Like other functions, the transformations ψ_1 and ψ_2 can be expanded in Hermite polynomials with

$$\left.\begin{aligned} x_1 &= \psi_1(z_1) = \sum_{j=0}^{\infty} c_j H_j(z_1) , \\ \text{and} \\ x_2 &= \psi_2(z_2) = \sum_{j=0}^{\infty} c_j^* H_j(z_2) , \end{aligned}\right\} \tag{28}$$

where x_1 and x_2 are continuous variables.

The coefficients c_j and c_j^* $(j = 1, 2, ...)$ differ from each other, except for $c_0 = c_0^* = m$, because

$$m = \int_{-\infty}^{\infty} \left[\sum_{j=0}^{\infty} c_j H_j(z_1) \phi(z_1)\right] dz_1 = c_0 . \tag{29}$$

A similar result holds for z_2.

From the assumption $E(X_1|X_2) = X_2$, it follows that

$$x_2 = \psi_2(z_2) = \int_{-\infty}^{\infty} \psi_1(z_1)[\phi(z_1, z_2)/\phi(z_2)] dz_1 . \tag{30}$$

Application of the Mehler identity [equation (22)] yields

$$x_2 = \psi_2(z_2) = \int_{-\infty}^{\infty} \psi_1(z_1)\phi(z_1)\left[\sum_{j=0}^{\infty} r^j H_j(z_1) H_j(z_2)\right] dz_1 . \tag{31}$$

The Hermite polynomials have the following property of orthogonality:

$$\int_{-\infty}^{\infty} H_i(z) H_j(z) \phi(z) dz = \begin{cases} 0 & \text{if } i \neq j , \\ 1 & \text{if } i = j . \end{cases} \tag{32}$$

Substitution of equation (32) and the first part of equation (28) into equation (31) gives

$$x_2 = \psi_2(z_2) = \sum_{j=0}^{\infty} r^j c_j H_j(z_2) \, , \tag{33}$$

or $c_j^* = r^j c_j$ for the second part of equation (28).

The variance, $\sigma^2(X_2)$, of the random variable X_2 for the larger cells is equal to $E(X_2^2) - m^2$, with

$$E(X_2^2) = \int_{-\infty}^{\infty} x_2^2 f(x_2) dx_2 = \int_{-\infty}^{\infty} \psi_2(z_2) \phi(z_2) dz_2$$

$$= \int_{-\infty}^{\infty} \left[\sum_{j=0}^{\infty} r^j c_j H_j(z_2) \right]^2 \phi(z_2) dz_2 = \sum_{j=0}^{\infty} r^{2j} c_j^2 \, .$$

Hence, because $c_0 = m$,

$$\sigma^2(X_2) = \sum_{j=1}^{\infty} r^{2j} c_j^2 \, . \tag{34}$$

From equation (33) it follows that the random variable, X, in equation (26) can be interpreted as being generated from a binary variable, X_0, by use of equation (27) and assuming $E(X_0|X) = X$. Moreover, equation (19) follows from equation (34) when the coefficients c_j, given by $c_j = \phi(b)H_{j-1}(b)$ for $j = 1, 2, ...$, are taken from equation (27).

Further use of cell-value distribution models

Three topics which can be considered in the context of cell-value distribution models are: (1) multivariate analysis of cell values, (2) compound frequency-distribution models for the regional occurrence of mineral deposits of different types, and (3) statistical models for the average amount of oil or metal per cell in a study region.

Suppose that a grid is superimposed on a region and that geological, geophysical, and geochemical parameters are observed for the cells of this grid. Correlation of the variables can be used in the estimation of (a) the conditional probability that one or more deposits of a specific type will occur in a cell, and (b) the conditional expectation of the total amount of oil or metal per cell. Several multivariate statistical techniques, such as multiple regression, may be applicable for these purposes.

For the application of multiple regression, it is desirable that the variables form a multivariate normal distribution. According to the model developed in the previous two sections, the variable representing the percentage of a rock type present can be normalized by use of the probit transformation. This is equivalent to plotting the values on paper with two probability scales (see figure 5). Approximate normality can also be obtained by replacing the data by their logits or by employing other types of transformations (Fisher and Yates, 1963).

If x is a percentage value, $\ln[x/(1-x)]$ represents the logit of x. In most practical applications there is little difference between results based on logits or probits of proportion data.

A small cell size may produce many zero values for some or all of the random variables. When a large proportion of the observations are zero, the variable cannot be normalized by a simple transformation. Suppose that, nevertheless, the probit transformation or a similar transformation is applied in preparation for multiple regression analysis. Then it would be necessary to estimate the correlation and regression coefficients by special techniques applicable for truncated and dichotomized normally-distributed random variables (Johnson and Kotz, 1972, chapter 36).

The following two methods can also be used if the cells are small.
(1) The conditional probability of occurrence of deposits in a cell can be obtained by relating the moments of the frequency distributions of the variables for cells containing deposits, with the moments of the distributions for the same variables observed in cells centred about random points in the study region. A simplified method of this type has been used by Agterberg and Fabbri (in press).
(2) The probability of occurrence of deposits in a cell can be estimated by using a version of the general qualitative response model (Amemiya, 1976).

Probit analysis (Finney, 1971) is one of several methods of modelling processes with a qualitative (yes-or-no-type) response. Agterberg (1974b) has used the logistic model to compute the probability, as a function of n variables, $p_k = f(X_{1k}, X_{2k}, ..., x_{nk})$, that a cell labelled k contains one or more deposits. In this model the function f is nonlinear, but the logit of p_k is a linear function of the variables x_i. The coefficients of this linear function can be estimated by using the method of Walker and Duncan (1967), or by the scoring method (Amemiya, 1976).

A compound frequency distribution $F_1 \, \hat{\theta} \, F_2$ arises when a parameter, θ, of a random variable with frequency distribution F_1 is itself a random variable with distribution F_2. Suppose that F_1 is a Poisson distribution with its mean distributed as a gamma distribution. Then the resulting compound Poisson distribution is a negative binomial distribution. Griffiths (1966) has shown that the frequency of mineral deposits for equal-area cells in a region can often be fitted by using the negative binomial form. Agterberg (1977) has pointed out that the negative binomial distribution of the occurrence of massive sulphide deposits for $(40 \text{ km})^2$ cells in the Abitibi area of the Canadian Shield, can be explained by the compound Poisson distribution model. The cell values [for $(40 \text{ km})^2$ cells] of acidic volcanics in this region satisfy a gamma distribution model. Further, it can be assumed that these cell values are proportional to the variable mean, μ, of a Poisson distribution for the occurrence of massive sulphide deposits.

Finally, the concept of unit regional value (Griffiths and Singer, 1971) can be considered in the context of cell-value distribution models. The unit regional value is the value, in dollars, of all commodities including

metals produced from a region averaged per unit of area. Menzie et al
(1977) concluded that in the USA the unit regional value can be considered
as a lognormal random variable when the size of the region is as small as
5000 mile^2 or 25000 km^2. Agterberg and Divi (1978) have developed a
lognormal model for the spatial distribution of copper, lead, and zinc in the
Canadian Appalachian region for which the surface area is $3\cdot6 \times 10^5$ km^2.
This approach may be useful in estimating subeconomic resources. It can
be summarized as follows: suppose that the lithosphere is sampled by
collecting blocks of rock of constant weight centred about randomly
distributed points. The spatial density of the sampling points is proportional
to the specific gravity of the material sampled. If the shape of the blocks
is kept approximately constant the average concentration by weight of a
chemical element in the aggregate of all blocks provides an unbiased
estimate of the crustal abundance, m, of the element. Although m remains
constant, the shape of the frequency distribution depends on the size of
the sampling rocks.

Suppose that the model represented by equations (28–34) applies to the
average values X_1 and X_2 for two different sizes of blocks. If X_1 for the
smaller blocks has the lognormal distribution, with mean m and variance σ^2,
it follows that (see Aitchison and Brown, 1957)

$$X_1 = m\exp[\sigma Z_1 - (\tfrac{1}{2}\sigma^2)^{\frac{1}{2}}] = m \sum_{j=0}^{\infty} \frac{\sigma^j H_j(Z_1)}{(j!)^{\frac{1}{2}}} . \tag{35}$$

Application of equation (33) gives

$$X_2 = m \sum_{j=0}^{\infty} r^j \sigma^j \frac{H_j(Z_1)}{(j!)^{\frac{1}{2}}} . \tag{36}$$

Hence, X_2 is lognormally distributed with variance $\sigma^2 r^2$ (see Matheron,
1976a, page 245). Thus the lognormality would be preserved. Likewise,
if the unit regional value according to the definition of Griffiths and
Singer (1971) has a lognormal distribution for one size of region, it would
also have a lognormal distribution for other sizes of regions if the basic
assumption formulated in equations (15) or (30) is satisfied.

References

Agterberg F P, 1974a *Geomathematics* (Elsevier, Amsterdam)
Agterberg F P, 1974b "Automatic contouring of geological maps to detect target areas
   for mineral exploration" *Journal of Mathematical Geology* 6 (4) 373–394
Agterberg F P, 1976 "New problems at the interface between geostatistics and geology"
   in *Advanced Geostatistics in the Mining Industry* Eds M Guarascio, C J Huijbregts,
   M L R David (Reidel, Dordrecht, The Netherlands) pp 403–421
Agterberg F P, 1977 "Frequency distribution and spatial variability of geological
   variables" in *Proceedings, 14th International Symposium on Computer Methods in the
   Mineral Industry* (American Institute of Mining Engineering, New York) pp 287–298
Agterberg F P, 1978 "Quantification and statistical analysis of geological variables for
   mineral resource evaluation" in *Proceedings, Goguel Colloqium on Earth Sciences
   and Measurement* (Bureau de Recherches Géologiques et Minières, Orléans, France)
   pp 399–406

Agterberg F P, Divi S R, 1978 "Statistical model for the distribution of copper, lead and zinc in the Canadian Appalachian region" *Economic Geology* **73** 230-245

Agterberg F P, Fabbri A G, 1978 "Spatial correlation of stratigraphic units quantified from geological maps" *Computers and Geosciences* **4** 285-294

Aitchison J, Brown J A C, 1957 *The Lognormal Distribution* (Cambridge University Press, Cambridge)

Amemiya T, 1976 "The maximum likelihood, the minimum chi-square and the non-linear weighted least-squares estimator in the general qualitative response model" *Journal of the American Statistical Association* **71** (7) 347-351

Bartlett M S, 1975 *The Statistical Analysis of Spatial Pattern* (Chapman and Hall, Andover, Hants)

Brinck J W, 1974 "The geochemical distribution of uranium as a primary criterion for the formation of ore deposits" in *Proceedings of a Symposium on Formation of Uranium Ore Deposits* report SM-183/19 (International Atomic Energy Agency, Vienna),pp 21-32

Chung C F, Divi S R, Fabbri A G, 1978 "An interactive graphic program for simulating the distribution of transformations of several independent random variables" in *Proceedings, 10th Annual Symposium on Interface. Computer Science and Statistics* (National Bureau of Standards, Washington, DC) pp 292-296

de Wijs H J, 1951 "Statistics of ore distribution. Part 1" *Geologie en Mijnbouw* **30** 365-375

de Wijs H J, 1953 "Statistics of ore distribution. Part 2" *Geologie en Mijnbouw* **32** 12-24

Finney D J, 1971 *Probit Analysis* 3rd edition (Cambridge University Press, Cambridge)

Fisher R A, Yates F, 1963 *Statistical Tables for Biological, Agricultural and Medical Research* 6th edition (Oliver and Boyd, Edinburgh)

Freeze R A, 1977 "Probabilistic one-dimensional consolidation" *Journal of the American Society of Civil Engineers. Geotechnical Engineering Division* **103** (GT7) 725-742

Gnedenko B V, Kolmogorov A N, 1954 *Limit Distributions for Sums of Independent Random Variables* (translated from the Russian) (Addison-Wesley, Reading, Mass)

Griffiths J C, 1966 "Exploration for natural resources" *Operations Research* **14** 189-209

Griffiths J C, 1970 "Current trends in geomathematics" *Earth-Science Reviews* **6** 121-140

Griffiths J C, Singer D A, 1971 "Unit regional value of nonrenewable natural resources as a measure of potential for development of large regions" *Geological Society of Australia. Special Publication* **3** 227-238

Harris D P, 1977 *Mineral Endowment, Resources, and Potential Supply: Theory, Methods for Appraisal and Case Studies* (Minresco, Tucson, Ariz.)

Johnson N I, Kotz S, 1970 *Distributions in Statistics* volume 2 (John Wiley, New York)

Johnson N I, Kotz S, 1972 *Distributions in Statistics* volume 4 (John Wiley, New York)

Journel A G, Huijbregts C J, 1978 *Mining Geostatistics* (Academic Press, London)

Kendall M G, Stuart A, 1961 *The Advanced Theory of Statistics* volume 2 (Hafner Press, New York)

Kim Y C, Myers D E, Knudsen H P, 1977 "Advanced geostatistics in ore reserve estimation and mine planning" research report, Department of Mining and Geological Engineering, University of Arizona, Tuscon, Ariz.

Krige D G, 1951 "A statistical approach to some basic valuation problems on the Witwatersrand" *Journal of the South African Institute of Mining and Metallurgy* **52** 119-139

Lancaster H O, 1958 "The structure of bivariate distributions" *Annals of Mathematical Statistics* **29** 719-736

Lewis P A W (Ed.), 1972 *Stochastic Point Processes: Statistical Analysis, Theory, and Applications* (John Wiley, New York)

Maréchal A, 1974 "Généralités sur les fonctions de transfert" Centre de Morphologie Mathématique, Fontainebleau, France

Maréchal A, 1975 "Analyse numérique des anamorphoses gaussiennes" Centre de Morphologie Mathématique, Fontainebleau, France

Maréchal A, 1976 "The practice of transfer functions: numerical methods and their application" in *Advanced Geostatistics in the Mining Industry* Eds M Guarascio, C J Huijbregts, M L R David (Reidel, Dordrecht, The Netherlands) pp 137-162

Matérn B, 1960 "Spatial variation" *Meddelanden fran Statens Skogsforskningsinstitut* **49** 144

Matheron G, 1965 *Les Variables Régionalisées et leur Estimation* (Masson, Paris)

Matheron G, 1973 "Le krigeage disjonctif" Centre de Morphologie Mathématique, Fontainebleau, France

Matheron G, 1974 "Les fonctions de transfert des petits panneaux" note géostatique 127, Centre de Morphologie Mathématique, Fontainebleau, France

Matheron G, 1976a "Forecasting block grade distributions: the transfer functions" in *Advanced Geostatistics in the Mining Industry* Eds M Guarascio, C J Huijbregts, M L R David (Reidel, Dordrecht, The Netherlands) pp 237-251

Matheron G, 1976b "A simple substitute for conditional expectations: the disjunctive kriging" in *Advanced Geostatistics in the Mining Industry* Eds M Guarascio, C J Huijbregts, M L R David (Reidel, Dordrecht, The Netherlands) pp 221-236

Menzie W D, Labovitz M L, Griffiths J C, 1977 "Evaluation of mineral resources and the unit regional value concept" in *Proceedings, 14th International Symposium on Computer Methods in the Mineral Industry* (American Institute of Mining, Metallurgical, and Petroleum Engineers, New York) pp 322-339

Neyman J, 1939 "On a new class of 'contagious' distributions applicable in entomology and bacteriology" *Annals of Mathematical Statistics* **10** 35-37

Neyman J, Scott E L, 1972 "Processes of clustering and applications" in *Stochastic Point Processes: Statistical Analysis, Theory, and Applications* Ed. P A W Lewis (John Wiley, New York) pp 646-681

Prokhorov Yu V, 1965 "On the lognormal distribution in geochemistry" *Teoriya Veroyatnostei i Ee Primeneniya* **10** 184-187

Serra J, 1969 "Introduction à la morphologie mathématique" in *Cahiers du Centre de Morphologie Mathématique de Fontainebleau, France* volume 3 (Centre de Morphologie Mathématique, Fontainebleau, France)

Serra J, 1976 "Stochastic models in stereology: strengths and weaknesses" in *Proceedings, 4th International Congress for Stereology* special publication 431 (National Bureau of Standards, Washington, DC) pp 83-86

Switzer P, 1965 "A random set process in the plane with a Markovian property" *Annals of Mathematical Statistics* **36** 1859-1863

Switzer P, Parker N M, 1976 "The problem of ore versus waste discrimination for individual blocks: the lognormal model" in *Advanced Geostatistics in the Mining Industry* Eds M Guarascio, C J Huijbregts, M L R David (Reidel, Dordrecht, The Netherlands) pp 203-218

Walker S H, Duncan D B, 1967 "Estimation of the probability of an event as a function of several independent variables" *Biometrika* **54** 167-179

Whittle P, 1963 "Stochastic processes in several dimensions" in *Bulletin of the 34th Session of the International Statistical Institute* (International Statistical Institute, The Hague) pp 974-985

The simulation of space-dependent data in geology

M Dagbert

Introduction

Most of the simulation studies which have been performed in geology deal with the development of deterministic or stochastic models (Harbaugh and Bonham-Carter, 1970). These models are supposed to dissect the mechanism of geological phenomena, and enable their response under various conditions to be examined. Another kind of simulation is simply the production of missing data. Of course, these data are not generated purely at random; they must be as alike as possible to available sampling data of the same variables. Since most of these data are space dependent, they show some spatial continuity and this quality has to be reproduced by the simulation. This problem of the restitution of spatial continuity has received special attention in recent years, and a practical method to achieve it has been devised (Journel, 1974). The method has been refined to cope with virtually any kind of continuity (Alfaro and Huijbregts, 1974; Journel, 1976). Its main application has been the simulation of grade or any other economic parameters in ore deposits. In this case, the efficiencies of several exploration programs (Bilodeau and MacKenzie, 1977; Bernuy and Journel, 1977) and mining methods (David et al, 1974; Dowd and David, 1976; Clark and White, 1977; Deraisme, 1977; Maréchal and Shrivastava, 1977) have been tested by using the simulated values plotted on a very fine grid. The purpose of the study presented here is to show how the method can be applied to the simulation of multielement geochemical data (Koch and Link, 1977). In this case, a correlation between the variables is added to the spatial correlation. The steps of the proposed procedure are illustrated by the simultaneous simulation of concentrations of seven major elements in the Lakeview Mountains pluton.

The data: concentrations of major elements in the Lakeview Mountains pluton

The data consist of 147 composite samples analyzed by x-ray fluorescence for seven elements: sodium, magnesium, aluminium, silicon, potassium, calcium, and iron. The samples are taken on a regular 610 m (2000 ft) grid (figure 1) that extends over the entire exposed part of the Lakeview Mountains pluton in the southern California batholith. The pluton is composed of coarse-grained hornblende–biotite quartz diorite (Morton, 1969). Composite samples have been made by taking nine individual specimens on a 15 m (50 ft) grid around each grid intersection (Baird et al, 1969; Morton et al, 1969). The original spectrographic concentrations, measured in percentages of elements, have been kept in this study.

Histograms of the distributions of the chemical concentrations (figure 2) show that most of them could be well represented by normal models. However, some asymmetry can be observed for the distributions of potassium, calcium, and iron. The computation of statistical parameters (table 1) confirms the symmetry of the distributions, with means almost equal to the median values. It also indicates a low dispersion of the concentrations around their mean: the relative standard deviation is lower than 10% except in the case of potassium, which appears to be the most variable element.

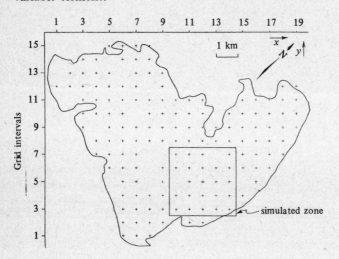

Figure 1. Sample localities on the Lakeview Mountains pluton. The position of a sample site is indicated by a cross. The limits of the exposed part of the pluton and of the squared area where simulated concentrations are produced on a 120 m (400 ft) grid are also shown.

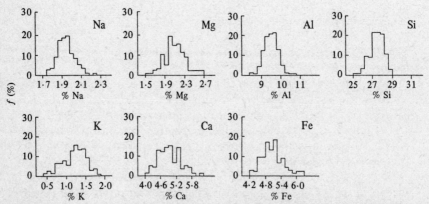

Figure 2. Histograms of the concentrations of major elements in the Lakeview Mountains pluton; f is the frequency of the values in a class.

The degree of association between the elements is assessed by computing the correlation matrix (table 2). It shows positive relationships between values for silicon and potassium on the one hand, and between values for magnesium, calcium, and iron on the other hand. The values for these two groups of elements show negative correlation, whereas the concentrations of sodium and aluminium behave rather independently.

The spatial continuity, or rather discontinuity, of the elements can be expressed by means of semivariograms. The semivariograms along the two directions of the sampling grid have been computed (figure 3). They exhibit no real anisotropy and the average difference between the values increases with the distance between the sample points, h. The elements can be classified in three groups according to the rate of increase of the difference (David and Dagbert, 1975).

The curves for potassium, calcium, and silicon present a small intercept at the origin ('nugget effect') and increase steadily towards a sill corresponding to the variance of the concentration of element. This means that, for these elements, closely-spaced samples are similar whereas samples far from each other are quite different. The concentrations are continuous on a large scale.

The nugget effect of magnesium and iron is almost one-half of their total variability as measured by the variance. The concentrations of these

Table 1. Statistics of the distributions of the concentrations of major elements in the Lakeview Mountains pluton. All the values are expressed in percentages.

Element	Mean	Median	Minimum	Maximum	Standard deviation
Na	1·94	1·93	1·75	2·23	0·086
Mg	2·11	2·10	1·56	2·69	0·205
Al	9·52	9·52	8·40	10·71	0·369
Si	27·44	27·49	25·11	28·93	0·711
K	1·20	1·24	0·42	1·80	0·293
Ca	4·94	4·90	4·09	6·15	0·428
Fe	5·08	5·03	4·24	6·25	0·414

Table 2. Correlation matrix of the concentrations of major elements in the Lakeview Mountains pluton.

Element	Na	Mg	Al	Si	K	Ca	Fe
Na	1·00						
Mg	0·30	1·00					
Al	0·63	0·14	1·00				
Si	−0·54	−0·80	−0·53	1·00			
K	−0·71	−0·66	−0·61	0·86	1·00		
Ca	0·66	0·74	0·59	−0·88	−0·93	1·00	
Fe	0·35	0·74	0·23	−0·74	−0·68	0·68	1·00

two elements have erratic variations over short distances, but they still show some continuity on a large scale.

Finally, sodium and aluminium concentrations are not continuous at all, since two samples separated by one grid interval are almost as different as two samples with more than five grid intervals between them.

The purpose of the study is to generate a set of 'possible' concentrations of the seven major elements in the Lakeview Mountains pluton. 'Possible' means that the simulated value must obey the following four conditions.
(1) The distribution of the simulated concentration values of each element must be similar to the distribution of the measured concentration values of the same element. These distributions are characterized by the histograms of figure 2 and the statistics of table 1.
(2) The relationships between the simulated concentrations must be similar to the relationships between the measured concentrations. These relationships are characterized by the correlation matrix of table 2.
(3) The simulated values are in fact simulated samples and they are assigned to specific points on the surface of the pluton. Thus the spatial continuity of the simulated values must be similar to the spatial continuity of the measured values. This continuity is characterized by the semi-variograms of figure 3.
(4) Simulated values can be generated at sample points already existing. In this case, it is desirable that the simulation procedure is able to reproduce the concentrations actually measured at these points.

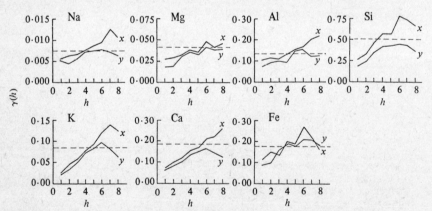

Figure 3. Semivariograms of the concentrations of major elements in the Lakeview Mountains pluton. The horizontal dotted line indicates the variance of the concentrations. For these and all subsequent semivariograms, $\gamma(h)$ is half the average quadratic difference between concentrations of samples separated by a distance h; x and y are the main directions of the grid shown on figure 1; and h is expressed in number of grid intervals (610 m or 2000 ft).

Normalization of the data

Gaussian variates are easily generated by combining random numbers. For example, if one draws series of six random numbers (each in the range 0–1) and computes the sum of them, one obtains a normal distribution of these sums with a mean of 3 and a variance of 0·5.

Since the experimental distributions at hand are not exactly normal, it is desirable to define a transformation of the original data which makes them normal. This is done by comparing the cumulative frequency of the experimental variable, Y, and the distribution function of the standard normal variable, X: Y_i is associated with X_i if there are the same frequencies of values below Y_i and X_i. This procedure defines an experimental function $Y(X)$ on which an analytical model is based (figure 4). The model is expressed as an expansion of Hermite polynomials (table 3). This form allows direct expression of absolute and conditional means or variances of functions of Y. However, this property is not really used here.

The models can be used either to derive the experimental distribution of normalized concentrations (figure 4 and table 4) or to generate possible concentration values from simulated standard normal deviates. This procedure has been called a "Gaussian anamorphosis" (Maréchal, 1976),

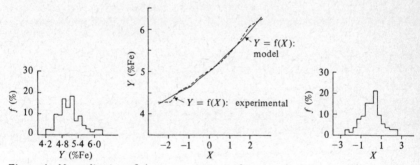

Figure 4. Normalization of the concentrations of iron in the Lakeview Mountains pluton. Y is the original concentration and X is the normalized concentration derived from the model $Y = f(X)$.

Table 3. Analytical models for the normalization functions. Y is the original variable; X is a standard normal variate; 1, X, $X^2 - 1$, $X^3 - 3X$ are Hermite polynomials.

Element	Normalization function
Na	$Y = 1·94 + 0·083X + 0·003(X^2 - 1) - 0·0009(X^3 - 3X)$
Mg	$Y = 2·10 + 0·205X + 0·004(X^2 - 1) + 0·005(X^3 - 3X)$
Al	$Y = 9·50 + 0·365X - 0·014(X^2 - 1) + 0·009(X^3 - 3X)$
Si	$Y = 27·40 + 0·719X - 0·071(X^2 - 1) + 0·010(X^3 - 3X)$
K	$Y = 1·20 + 0·286X - 0·028(X^2 - 1) - 0·004(X^3 - 3X)$
Ca	$Y = 4·94 + 0·427X + 0·030(X^2 - 1) - 0·002(X^3 - 3X)$
Fe	$Y = 5·06 + 0·404X + 0·037(X^2 - 1) + 0·001(X^3 - 3X)$

and it can be viewed as a particular form of a Monte Carlo simulation (Harbaugh and Bonham-Carter, 1970).

Table 4. Statistics of the distributions of the normalized concentrations of major elements in the Lakeview Mountains pluton. A normal model with a mean of zero and a standard deviation of one is valid for all the distributions.

Element	Mean	Median	Minimum	Maximum	Standard deviation
Na	0·009	0·0	−2·53	2·50	1·02
Mg	0·05	0·02	−2·50	2·56	1·00
Al	0·03	0·0	−2·57	3·22	1·02
Si	0·03	0·0	−2·55	2·59	0·99
K	0·04	0·02	−2·35	2·80	1·04
Ca	0·02	−0·04	−2·31	2·50	0·99
Fe	0·04	0·04	−2·45	2·48	1·00

Production of independent factors

It is also easy to reproduce the quality of independence: when one is dealing with independent variables, one can simulate them one at a time. The idea is then to transform the correlated normalized concentrations into independent factors by a principal components analysis. The eigenvalues and eigenvectors have been computed for the correlation matrix of the normalized concentrations (table 5). This correlation matrix is almost identical to the correlation matrix of the real concentrations (table 2), as it has been shown that the distributions of the original variables are already close to normality. Sample scores on the new independent factors are also normally distributed and their statistics are presented in table 6.

The geological significance of the factors has already been discussed by David and Dagbert (1975) and by Miesch and Morton (1977). Factor 1

Table 5. Eigenvalues and eigenvectors of the correlation matrix of the normalized concentrations. The eigenvalues are the variances of the scores of samples on the corresponding factors.

	Factor						
	1	2	3	4	5	6	7
Eigenvalues	4·73	1·22	0·38	0·32	0·17	0·11	0·06
Total contribution (%)	67·5	17·5	5·5	4·6	2·5	1·6	0·8
Eigenvector components on							
Na	−0·32	0·48	0·72	0·17	−0·33	0·06	0·00
Mg	−0·36	−0·48	0·08	−0·40	−0·46	−0·45	−0·25
Al	−0·29	0·59	−0·63	0·10	−0·24	−0·29	−0·10
Si	0·43	0·12	0·23	0·21	0·25	−0·79	−0·11
K	0·43	−0·08	−0·07	0·12	0·65	−0·06	0·60
Ca	−0·44	0·01	0·04	−0·20	0·37	−0·27	0·74
Fe	−0·35	−0·41	−0·08	0·84	0·00	−0·05	0·01

accounts for 67·5% of the variations of the concentrations of the elements around their mean value. It shows an antipathetic relationship of silicon and potassium to all the other elements. Thus it reflects the main characteristics of the correlation matrix. A map of the factor scores (David and Dagbert, 1975) shows a clear zonality with a mafic core (negative scores) and acid margins (positive scores). Maps of the scores on the other factors do not show the same regular pattern, and these factors seem only to account for local variations with respect to the trend expressed by the first factor.

A study of the spatial continuity of the factor scores confirms these observations. Semivariograms have been computed along the two principal directions of the sampling grid (figure 5). Factor 1 is the only variable which shows good spatial continuity. In contrast, all the other factors present very high nugget effects. In practice, they can be considered as a random noise with strictly local fluctuations superimposed on the regional variations of factor 1. Models have been fitted to the experimental curves.

Table 6. Statistics of the distributions of factor scores derived by principal components analysis of the normalized concentrations in table 4.

Factor	Mean	Median	Minimum	Maximum	Standard deviation
1	−0·03	0·16	−5·82	5·06	2·17
2	−0·02	−0·05	−3·55	3·55	1·10
3	−0·01	0·03	−1·75	1·46	0·62
4	0·03	−0·01	−1·75	1·94	0·57
5	−0·04	−0·03	−1·14	1·05	0·42
6	−0·06	−0·06	−1·00	0·84	0·33
7	0·02	0·04	−1·05	0·95	0·24

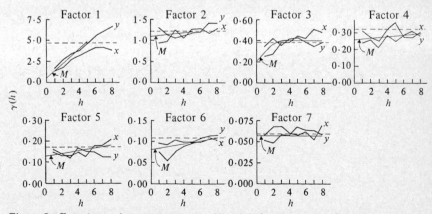

Figure 5. Experimental semivariograms and models for the scores of samples on factors derived by principal components analysis of the normalized concentrations. M is the isotropic model fitted to the curves; the horizontal dotted line indicates the variance of the factor scores.

They are valid for distances not more than five grid intervals ($3 \cdot 05$ km or
10000 ft), and consist of the sum of a nugget effect and a linear (factors 1,
2, 4, 5, 6) or spherical (factor 3) component. The model adopted for
factor 7 is a pure nugget effect.

At this point, the emphasis of the study will be directed towards the
production of random normal variates with a continuity of a specific type,
or in more precise terms, towards the simulation of a two-dimensional
stationary random function with a Gaussian distribution and a given auto-
covariance.

Simulation of factor scores

The method used to simulate the spatially dependent data is an application
of the "turning-bands" method (Journel, 1974) to two-dimensional
problems. The idea behind the method is as follows: if a two-dimensional
stationary random function $Z(x)$ with an autocovariance $C_2(h)$ is projected
onto a line, it defines a one-dimensional stationary random function $Y(t)$
with an autocovariance $C_1(h)$ related to $C_2(h)$. Then, to simulate values
of Z it is sufficient to simulate values of Y on several equally spaced lines
(figure 6). Here x is a position vector on the x–y grid, and h is a vector
denoting a distance across the grid in a particular direction. This reasoning
can be extended to three-dimensional situations. In fact, the determination
of the $C_1(h)$ corresponding to usual expressions of $C_2(h)$ has proved to be
easier in that case.

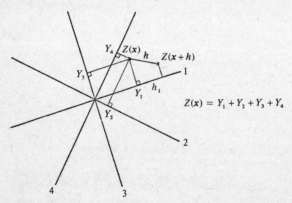

Figure 6. The simulation of a two-dimensional stationary random function, $Z(x)$,
from simulated stationary random processes, Y_1, Y_2, Y_3, and Y_4, on lines.

Relationship between $C_1(h)$ and $C_2(h)$

Stationary random functions, Y_i, are independently simulated on each
line i. The properties of the Y_i are:

$E(Y_i) = m$, for all i; expectations are the same everywhere;

$\mathrm{var}(Y_i) = \sigma^2$, for all i; variances are the same everywhere;

$$\text{cov}[Y_i(t),\ Y_i(t+h)] = C_1(h)\ , \quad \text{for all } i;\ \text{the covariance is stationary;}$$

$$\text{cov}(Y_i,\ Y_j) = 0\ , \quad \text{for all } i \neq j;\ \text{functions on two different lines are}$$
independent.

Then, if the two-dimensional random function Z is defined by

$$Z = \sum_{i=1}^{n} Y_i\ ,$$

the properties of Z are as follows:

$$E(Z) = E\left(\sum_{i=1}^{n} Y_i\right) = nm\ ,$$

$$\text{var}(Z) = \text{var}\left(\sum_{i=1}^{n} Y_i\right) = n\sigma^2\ ,$$

$$\text{cov}[Z(x),\ Z(x+h)] = \sum_{i=1}^{n} \sum_{j=1}^{n} \text{cov}(Y_i,\ Y_j) = \sum_{i=1}^{n} \text{cov}[Y_i(t_i),\ Y_i(t_i+h_i)]\ ,$$

where t_i is the projection of the point x on the line i and h_i is the projection of distance h on the same line (figure 6). If θ_i designates the angle between the vector h and the line i, then

$$h_i = h\cos\theta_i\ ,$$

and

$$\text{cov}[Z(x),\ Z(x+h)] = C_2(h) = \sum_{i=1}^{n} C_1(h\cos\theta_i)\ .$$

If the number of lines tends toward infinity, the discrete sum becomes proportional to the integral of $C_1(h\cos\theta)$ with respect to θ:

$$C_2(h) = \frac{n}{\pi} \int_{-\pi/2}^{+\pi/2} C_1(h\cos\theta)\,d\theta\ .$$

If the $C_1(h)$ function is a polynomial expression

$$C_1(h) = \sum_{k=0}^{p} a_k h^k\ ,$$

then $C_2(h)$ is a polynomial expression of the same type,

$$C_2(h) = \sum_{k=0}^{p} b_k h^k\ ,$$

with

$$b_{2k} = \frac{na_{2k}}{\pi} \int_{-\pi/2}^{+\pi/2} (\cos\theta)^{2k}\,d\theta = na_{2k}\,\frac{(2k-1)(2k-3)\ ...\ 1}{2k(2k-2)\ ...\ 2}\ ,$$

and

$$b_{2k+1} = \frac{na_{2k+1}}{\pi} \int_{-\pi/2}^{+\pi/2} (\cos\theta)^{2k+1}\,d\theta = \frac{2na_{2k+1}}{\pi}\,\frac{2k(2k-2)\ ...\ 2}{(2k+1)(2k-1)\ ...\ 3}\ .$$

Linear and spherical components are both found in the semivariograms of the factor scores of the Lakeview Mountains pluton. They correspond to covariance functions of the type

$$C_2(h) = c\left(1 - \frac{h}{a}\right)$$

for a linear component, and

$$C_2(h) = c\left[1 - \frac{3}{2}\left(\frac{h}{a}\right) + \frac{1}{2}\left(\frac{h}{a}\right)^3\right]$$

for a spherical component, where a and c are constants.

Thus the determination of the corresponding covariances on the line follows immediately from the polynomial expressions for $C_2(h)$ above, viz

$$C_1(h) = \left(\frac{c}{n}\right)\left(1 - \frac{\pi h}{2a}\right)$$

for a linear component, and

$$C_1(h) = \left(\frac{c}{n}\right)\left(1 - \frac{3\pi h}{4a} + \frac{3\pi h^3}{8a^3}\right)$$

for a spherical component.

However, it should be noted that these unidimensional covariances only reproduce the continuity of the two-dimensional function for distances less than a.

Simulation of the Y_i on the lines

At this point, the problem which remains to be solved is the simulation of a stationary random function $Y(t)$ with a Gaussian distribution of mean m, variance σ^2, and covariance $C_1(h)$. It is possible to generate Y by computing moving averages of a random noise, T (figure 7).

Let $f(u)$ designate a weighting function with

$$Y(t) = \int_{-\infty}^{+\infty} f(t-u)T(u)\,du ,$$

and

$$E(T) = \tfrac{1}{2}, \quad \mathrm{var}(T) = \tfrac{1}{12}, \qquad \mathrm{cov}[T(u), T(u+h)] = 0, \quad h \neq 0 .$$

Then the random function Y generated in this way has the following properties:

$$E(Y) = \tfrac{1}{2}\int_{-\infty}^{+\infty} f(u)\,du ,$$

$$\mathrm{var}(Y) = \tfrac{1}{12}\int_{-\infty}^{+\infty} f^2(u)\,du ,$$

$$\mathrm{cov}[Y(t), Y(t+h)] = \tfrac{1}{12}\int_{-\infty}^{+\infty} f(u)f(u+h)\,du .$$

Thus, the mean, the variance, and the autocovariance of the random function Y are stationary. Moreover, since Y is a linear combination of random numbers, it is normally distributed.

Weighting functions which produce polynomial covariances can be inferred from the results of Journel (1974) regarding the simulation of spherical covariances in three-dimensional space.

A linear function in a limited interval $[A, B]$ around the point where a value Y is generated has the form

$$f(u) = \begin{cases} Ku & \text{for } A \leqslant u \leqslant B , \\ 0 & \text{for } u < A , \quad \text{or } u > B , \end{cases}$$

with $A < 0$, $B > 0$, and K a constant.

When applied to a random noise, it produces a stationary random function with covariance

$$C_1(h) = \int_{-\infty}^{+\infty} f(u)f(u+h)\,du = K^2 \int_A^{B-h} u(u+h)\,du$$

$$= K^2 \left[\frac{h^3}{6} - \frac{h}{2}(A^2 + B^2) + \frac{(B^3 - A^3)}{3} \right] .$$

This is a polynomial expression of degree 3, which can be identified with the covariance corresponding to the spherical model with values for the parameters K, A, and B given by

$$K^2 = \frac{9\pi c}{4na^3} , \qquad A = \frac{a}{(6)^{1/2}} , \qquad B = \frac{a}{(2)^{1/2}} .$$

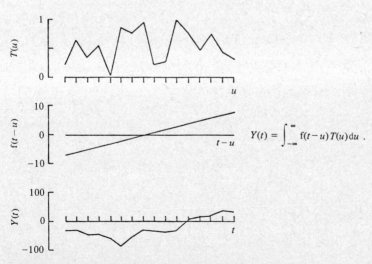

$$Y(t) = \int_{-\infty}^{\infty} f(t-u)\,T(u)\,du .$$

Figure 7. The simulation of a stationary random process on a line, $Y(t)$, by 'dilution' of a random noise, $T(u)$, with the weighting function, $f(t-u)$.

Similarly, the covariance $C_1(h)$ corresponding to the linear model can be generated by using the weighting function:

$$f(u) = \begin{cases} K & \text{for } A \leqslant u \leqslant B , \\ 0 & \text{for } u < A \quad \text{or} \quad u > B . \end{cases}$$

In this case, the values of the parameters are

$$K^2 = \frac{\pi c}{2na} , \qquad B = -A = \frac{a}{\pi} ,$$

and the procedure is reduced to that of computing the average value of the random noise within an interval centered on the point assigned to this value.

In practice, the integrals are replaced by discrete sums. Fifty random numbers are generated at equally spaced points within the interval A, B. The K coefficient must then be corrected by multiplying the previously given value by the factor $(B-A)/49$.

Simulation of a nugget effect and a covariance with several components
A two-dimensional random function with a nugget effect C_0 as auto-covariance is simply a random noise with a variance of C_0. If the random function is Gaussian, it can be generated by adding random numbers and by scaling the sum in order to obtain the appropriate mean and variance C_0.

The covariances of the factor scores at Lakeview are generally the sum of two components; a nugget effect and a linear or spherical function. In this case, each component is treated separately. First, a random function, Z_1, with a covariance scaled to that of the nugget effect is simulated; Z_1 is Gaussian with zero mean and variance C_0. Next, a second random function, Z_2, with a Gaussian distribution, zero mean, variance C, and linear or spherical covariance is generated by the method described at the beginning of this section. Then the desired random function, Z, is simply the sum of the two independent variables Z_1 and Z_2.

Simulation of the factor scores at Lakeview
Values of the factor scores have been simulated at 625 points on a 122 m (400 ft) grid. The grid covers a $3 \cdot 05$ km $\times$ $3 \cdot 05$ km [$(3 \cdot 05$ km$)^2$, or $(10\,000$ ft$)^2$] area in the eastern margin of the pluton (see figure 1). The required mean of each simulated variable is set equal to zero. The variances are derived from the semivariograms of the factor scores (table 7), since it is well known that a continuous variable shows less dispersion in a restricted area than over the entire region where it is defined.

Statistics (table 8) and semivariograms (figure 8) of the simulated values have been computed. The distributions appear to be normal and their parameters have the desired values. The experimental semivariograms generally follow the required model although some discrepancy can be observed for large distances. A common nuisance is the appearance of an anisotropy which seems to result from the restricted number of lines used in the simulation procedure. The computation of a correlation matrix (table 9) confirms the independence of the simulated factor scores.

Table 7. Total and local variances of the factor scores. The local variances are derived from the models for the semivariograms shown in figure 5.

Factor	Variance		Factor	Variance	
	experimental	theoretical [a]		experimental	theoretical [a]
1	4·60	2·68	5	0·18	0·16
2	1·20	1·15	6	0·11	0·09
3	0·40	0·35	7	0·06	0·06
4	0·29	0·27			

[a] Variance in $(3\cdot05 \text{ km})^2$ area.

Table 8. Statistics of the simulated factor scores.

Factor	Mean	Median	Minimum	Maximum	Standard deviation
1	0·007	0·262	−4·18	3·52	1·63
2	0·003	−0·020	−2·43	3·06	1·08
3	0·004	0·026	−1·96	1·63	0·59
4	0·002	−0·002	−1·68	1·35	0·52
5	0·000	−0·019	−0·99	1·14	0·40
6	0·000	0·005	0·79	0·91	0·30
7	0·001	−0·006	0·59	0·68	0·24

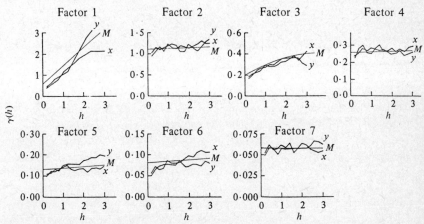

Figure 8. Semivariograms of the simulated factor scores. M is the required model derived from the curves of figure 5.

Table 9. Correlation matrix of the simulated factor scores.

Factor	1	2	3	4	5	6	7
1	1·00						
2	−0·01	1·00					
3	−0·14	0·08	1·00				
4	−0·03	−0·06	−0·08	1·00			
5	−0·16	−0·01	0·05	0·03	1·00		
6	−0·14	0·02	−0·07	0·07	0·20	1·00	
7	0·03	−0·02	0·05	−0·08	0·08	0·01	1·00

Conditional simulation of factor scores

In the area where simulated values have been produced, there are twenty-three samples on the 610 m (2000 ft) sampling grid. Simulated factor scores have also been assigned to these sample localities. They are, of course, different from the real factor scores derived from the concentrations measured at these points.

A special transformation called "conditionalization" has been proposed by Journel (1974) to eliminate this discrepancy. The transformation does not alter the distribution and continuity of the simulated values, and it uses two properties of the Kriging estimation method (David, 1977). First, when Kriging is performed, the error and the estimated value are independent. Second, Kriging reproduces exactly the values measured at sample points.

Thus, at a point x where a simulated value $Z_s(x)$ has been produced, the value of $Z(x)$ is estimated by Kriging from the *measured* values of the neighboring samples. This produces a second parameter, $Z_k(x)$. Then, the value of $Z_s(x)$ is estimated by Kriging from the *simulated* values of the neighbouring samples. This gives a third parameter, $Z_{ks}(x)$. The simulated value, $Z_c(x)$, finally retained is

$$Z_c(x) = Z_k(x) + Z_s(x) - Z_{ks}(x) \ .$$

This means that a simulated error is added to the Kriged value of the function. This simulated error is zero at sample points because the simulated and the real values are identical at these points.

The statistics of the distributions of the simulated factor scores after conditionalization are presented in table 10. From a comparison with the values in table 8, it appears that the conditionalization does not affect the dispersion of the simulated scores. The means and median values are now slightly different from zero and closer to the average factor scores in the simulated zone (table 11). Thus the conditionalization forces the local mean and dispersion of the simulated and real values to be similar.

Table 10. Statistics of the simulated factor scores after conditionalization.

Factor	Mean	Median	Minimum	Maximum	Standard deviation
1	−0·38	−0·30	−4·28	3·53	1·53
2	0·31	0·30	−2·14	3·41	1·08
3	0·37	0·34	−1·16	2·07	0·57
4	0·05	0·05	−1·63	1·38	0·52
5	−0·09	−0·09	−1·15	1·13	0·41
6	−0·10	−0·10	−0·85	0·81	0·29
7	−0·04	−0·05	−0·63	0·64	0·24

Table 11. Statistics of the real factor scores in the zone where simulation has been performed. They are computed from twenty-three sample values.

Factor	Mean	Median	Minimum	Maximum	Standard deviation
1	−0·44	0·32	−4·15	1·92	1·87
2	0·18	0·36	−0·95	1·19	0·62
3	0·11	0·25	−1·23	1·20	0·65
4	−0·03	−0·03	−0·50	0·50	0·27
5	−0·12	−0·04	−1·14	0·41	0·35
6	−0·13	−0·08	−0·79	0·34	0·28
7	−0·06	−0·09	−0·75	0·27	0·24

From simulated factor scores to simulated concentrations

The vector of the simulated normalized concentrations at a point is derived by multiplying the vector of the simulated factor scores at this point by the matrix of factor loadings (table 5). The simulated concentrations are then obtained by entering the simulated normalized concentrations into the expression of the normalization functions (table 3).

The distributions of the simulated concentrations (figure 9 and table 12) are quite similar to the distributions of the real concentrations (figure 2 and table 1). The main difference is a lower dispersion of the simulated values around their mean. Again, this phenomenon could be anticipated, because the simulated concentrations only cover a restricted part of the pluton. Differences of means and median values can also be explained by the local character of the simulated values. However, in most cases the gentle asymmetry observed in the original histograms (figure 2) is well reproduced in the distributions of the simulated values (figure 9).

The correlation matrix of the simulated concentrations (table 13) shows the same general features as the correlation matrix of the real concentrations (table 2). The associations of silicon and potassium, and of calcium, iron, and magnesium are well reproduced, although the magnitude of the correlation coefficients between these elements is slightly reduced. Small correlation coefficients of real concentrations remain small for the simulated concentrations, but their values are generally quite different.

The main spatial-continuity characteristics of the real concentrations (figure 3) are reproduced in the semivariograms of the simulated concentrations (figure 10). The curves for silicon, potassium, and calcium show a linear increase of the average difference from a small nugget effect. The semivariograms for all other elements have a similar shape; they show a more important nugget effect and a lower rate of increase of the average difference. The major differences between the two sets of real and simulated

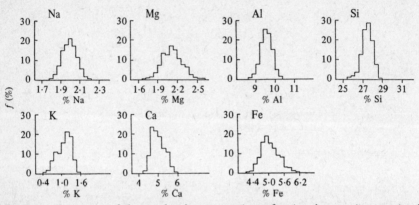

Figure 9. Histograms of the simulated concentrations of major elements in a restricted area of the Lakeview Mountains pluton. f is the frequency of the values in a class.

Table 12. Statistics of the distributions of the simulated concentrations. All the values are expressed in percentages.

Factor	Mean	Median	Minimum	Maximum	Standard deviation
Na	1·99	1·99	1·79	2·19	0·073
Mg	2·12	2·12	1·69	2·60	0·152
Al	9·55	9·56	8·65	10·36	0·301
Si	27·45	27·49	25·46	28·65	0·549
K	1·16	1·22	0·47	1·60	0·252
Ca	4·99	4·95	4·22	5·89	0·326
Fe	5·06	5·01	4·19	6·24	0·352

Table 13. Correlation matrix of the simulated concentrations.

	Na	Mg	Al	Si	K	Ca	Fe
Na	1·00						
Mg	−0·03	1·00					
Al	0·59	−0·10	1·00				
Si	−0·31	−0·68	−0·41	1·00			
K	−0·57	−0·46	−0·46	0·77	1·00		
Ca	0·52	0·56	0·43	−0·81	−0·89	1·00	
Fe	0·05	0·64	0·02	−0·63	−0·53	0·54	1·00

curves are (1) a lower nugget effect for the simulated concentrations of all the elements, and (2) the appearance of undesirable anisotropies for the simulated concentrations of potassium and calcium.

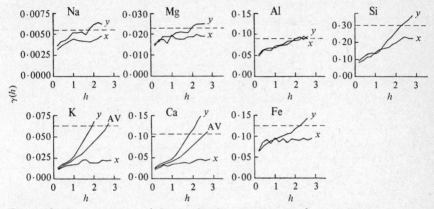

Figure 10. Semivariograms of the simulated concentrations of major elements in a restricted area of the Lakeview Mountains pluton. AV is the average semivariogram; the horizontal dotted line indicates the variance of the simulated concentrations.

Conclusions

The simulation of the concentrations of major elements in the Lakeview Mountains pluton described in this study shows that it is feasible to generate 'possible' values of geological parameters which are continuous in space and correlated with one another. Most geological variables are of this type, and the method could be generalized to other fields of application in addition to the processing of multielement geochemical data.

The main steps of the method can be summarized as follows:
(1) each variable is normalized;
(2) normalized variables are made independent by principal components analysis;
(3) factors are simulated separately by a variant of the 'turning-bands' method for two-dimensional problems;
(4) similarity between the local means of simulated and real factors scores is insured by conditionalization;
(5) the transformations of the first two steps are applied in the reverse order.

However, all conditions imposed on the simulated values are not fully respected and several research areas remain open for the improvement of results. First, the influence of operational parameters such as the number of Hermite polynomials in the expansion of the normalization function, the density of random numbers generated on the lines, and/or the number of these lines, will have to be assessed by extensive comparative studies.

Second, alternative methods could be used and their efficiency could be tested against those proposed here. For example, a moving average of a random noise within specific geometrical figures (a circle or an ellipse, for

example) could replace the 'turning-bands' method and generate the
required continuity in a better and quicker way. Also, the potential of
the expansion of normalization functions by using Hermite polynomials
has not been fully exploited.

Last, the possible applications of simulation in geochemical exploration
have to be reviewed. It is the opinion of the author that simulation
provides a more realistic view of the spatial variation of geochemical
parameters than do isoline maps derived from the interpolation of sample
values. Of course, when certain regional trends are required, these maps
provide valuable help. However, they generally do not reproduce small-
scale fluctuations. Moreover, the simulation can be repeated several times
and, in this way, an idea of the range of possible values for a concentration
at any point can be obtained.

Acknowledgements. Thanks are due to Mike Davis and Michel David for their helpful
discussion and criticism. This study was partly supported by NRCC grant A-7035.
Computer time has been kindly provided by the Université de Montréal.

References
Alfaro M A, Huijbregts C J, 1974 "Simulation of a subhorizontal sedimentary deposit"
    in *Proceedings of the 12th APCOM Symposium* (Colorado School of Mines, Golden,
    Col.) pp F65-F77
Baird A K, McIntyre D B, Welday E E, Morton D M, 1967 "A test of chemical variability
    and field sampling methods, Lakeview Mountains tonalite, Lakeview Mountains,
    Southern California batholith" special report 92, California Division of Mines and
    Geology, Sacramento, Calif.
Bernuy O, Journel A G, 1977 "Simulation d'une reconnaissance séquentielle" *Revue de
    l'Industrie Minérale* **59**(10) 472-478
Bilodeau M L, MacKenzie B W, 1977 "The drilling investment decision in mineral
    exploration" in *Application of Computer Methods in the Mineral Industry,
    Proceedings of the 14th APCOM Symposium* Ed. R V Ramani (American Institute
    of Mining Engineers, New York) pp 932-969
Clark I, White B, 1977 "Geostatistical modelling of an orebody as an aid to mine
    planning" in *Application of Computer Methods in the Mineral Industry, Proceedings
    of the 14th APCOM Symposium* Ed. R V Ramani (American Institute of Mining
    Engineers, New York) pp 1004-1012
David M, 1977 *Geostatistical Ore Reserve Estimation* (Elsevier, Amsterdam)
David M, Dagbert M, 1975 "Lakeview revisited: variograms and correspondence
    analysis—new tools for the understanding of geochemical data" in *Geochemical
    Exploration 1974* Eds I L Ellicot, W K Fletcher (American Elsevier, New York)
    pp 163-181
David M, Dowd P A, Korobov S, 1974 "Forecasting departure from planning in
    open pit design and grade control" in *Proceedings of the 12th APCOM Symposium*
    (Colorado School of Mines, Golden, Col.) pp F131-F149
Deraisme J, 1977 "Modélisations de gisement et choix d'une méthode d'exploitation"
    Revue de l'Industrie Minérale **59**(10) 483-489
Dowd P A, David M, 1976 "Planning from estimates: sensitivity of mine production
    schedules to estimation methods" in *Advanced Geostatistics in the Mining Industry*
    Eds M Guarascio, M L R David, C Huijbregts (Reidel, Dordrecht, The Netherlands)
    pp 163-184

Harbaugh J W, Bonham-Carter G, 1970 *Computer Simulation in Geology* (John Wiley, New York)

Journel A G, 1974 "Geostatistics for conditional simulation of ore bodies" *Economic Geology* **69** 673–687

Journel A G, 1976 "Ore grade distributions and conditional simulations—two geostatistical approaches" in *Advanced Geostatistics in the Mining Industry* Eds M Guarascio, M L R David, C Huijbregts (Reidel, Dordrecht, The Netherlands) pp 195–202

Koch G S, Link R F, 1977 "Anomaly recognition in exploration geochemistry through a statistical analysis of multivariate data" in *Application of Computer Methods in the Mineral Industry, Proceedings of the 14th APCOM Symposium* Ed. R V Ramani (American Institute of Mining Engineers, New York)

Maréchal A, 1976 "The practice of transfer functions: numerical methods and their applications" in *Advanced Geostatistics in the Mining Industry* Eds M Guarascio, M L R David, C Huijbregts (Reidel, Dordrecht, The Netherlands) pp 137–162

Maréchal A, Shrivastava P, 1977 "Geostatistical study of a lower proterozoic iron orebody in the Pilabara region of Western Australia" in *APCOM 77* (proceedings of the 15th APCOM symposium) (Australian Institute of Mining and Metallurgy, Parkville, Australia) pp 221–230

Miesch A T, Morton D M, 1977 "Chemical variability in the Lakeview Mountains pluton, Southern California batholith—a comparison of the methods of correspondence analysis and extended *Q*-made factor analysis" *Journal of Research of the United States Geological Survey* **5** (1) 103–116

Morton D M, 1969 "The Lakeview Mountains pluton, Southern California batholith—part 1, petrology and structure" *Geological Society of America Bulletin* **80** (8) 1539–1551

Morton D M, Baird A K, Baird K W, 1969 "The Lakeview Mountains pluton, Southern California batholith—part 2, chemical composition and variation" *Geological Society of America Bulletin* **80** (8) 1553–1563

Semivariograms and Kriging: possible useful tools in fold description[†]

E H T Whitten

Introduction

To communicate the nature of a fold to another person in an objective manner is not easy. Some fold attributes are readily quantified, for example: fold-axis orientation, fold size (half wavelength, etc), vergence, or actual orientation of the bisecting surface. Variation of the orthogonal thickness of a folded lithic unit, and of the thickness parallel to the bisecting surface, provide useful insights into fold geometry. The geometry of the fold profile, or the profiles of a fold train, are not easily described quantitatively; where a fold set is not cylindroidal so that the fold profiles change parallel to the fold axes, the three-dimensional geometry of an individual folded stratum is even more difficult to describe (see, for example, Rech, 1977).

One may question why the nature of a fold (or set of folds) requires description. First, it is a useful goal to describe and catalog the nature and variability of folds in a specified domain; such information may also have direct economic importance where, for example, there is mineralization. Second, in the consideration of a genetic process–response model for the formation of a set of folds, evaluation of appropriate models requires comparison of *significant* characteristics of actual folds with those predicted by the current model.

In an inventory, it would be possible to include an almost limitless number of different fold attributes which can be measured and recorded; by convention, a standard set of variables tends to be recorded, but many others could readily be observed and included. For example, Ramsay (1967), Whitten (1966), Hansen (1971), Hudleston (1973), and Hudleston and Stephansson (1973) all described an assortment of attributes of varying value for comparing and contrasting folds. The observed scalar values for each of these variables can be measured at successive geographical sites, and maps of each attribute can then be drawn. Although many of these variables allow comparison of different members of a fold set, actual values for variables currently available do not permit unambiguous reconstruction of the nature of the particular fold or fold set. Similarly, it is by no means certain as to which physical or geometrical fold attributes would permit quantitative discrimination between correct and incorrect genetic models; that is, positive elimination of incorrect process–response models.

[†] Adapted from a soft-copy preprint (Whitten, 1977a) of a paper read in Příbram, Czechoslovakia.

It is established practice to use measured variables for various types of samples to estimate composition, structure, and/or variability of rock units. When the whole target population of interest is not available for probability sampling, it is necessary to rely on data based on an inferior, sampled population (Whitten, 1961). Subsurface structure based on data for samples drawn from the sampled population is frequently portrayed by isolines produced by manual contouring, arbitrary interpolation techniques or surface-fitting methods such as polynomial or double Fourier trend surfaces, spatial filtering, or spline surfaces (see, for example, Whitten and Beckman, 1969; Whitten and Koelling, 1973). Such surface-fitting techniques use convenient, but arbitrary, mathematical functions of no identified genetic significance (Whitten, 1974; 1975). In each case the isolines describe the sampled, rather than the target, population (unless the sampled and target populations happen to be the same, which is rarely the case), and they represent little more than an attempt to make an inventory.

More appropriate models would be based on specified genetic processes that are thought to have produced the observed subsurface phenomena; this approach depends on the identification of the relevant process–response model (Whitten, 1964). Ideally, such a model leads to the design of a stochastic equation which could be precisely accurate, provided that the process factors were correctly identified. By contrast, the corresponding deterministic equation would only be an approximation. The formal relationship between a deterministic model (differential equation) and the stochastic model based upon it is that the former yields the expected values, or mean values, for the probability distributions (Whitten, 1974, page 190). Isoline maps constructed with such a model (hypothesis) would represent predictions based upon factors of identified genetic significance; the validity of the maps, and thus of the model, could be specifically tested with newly collected data.

Identification of all the relevant factors, and of their roles in the relevant process–response model, is commonly a very time-consuming and elusive problem. In the interim, in the face of pressing economic or scientific demands, the use of the best available prediction of the target population may be justified. To this end, the probabilistic approach known as Kriging permits (a) significantly more accurate assessment of spatial variability (than other surface-fitting techniques); and (b) statements about the confidence levels to be associated with predictions[1].

In essence, the method involves two steps: (1) production of semi-variograms; and (2) construction of a variety of moving-average isoline maps by the process of Kriging.

[1] See, for example, Brooker (1975); David (1969; 1971, 1976; 1977); Delfiner and Delhomme (1975); Huijbregts (1975); Huijbregts and Matheron (1971); Journel (1974; 1975); Maréchal and Serra (1970); Matheron (1965); Royle and Hosgit (1974); Royle et al (1972); Sinclair and Deraisme (1974); Watson (1971; 1972a; 1972b); Whitten (1977b).

"Kriging provides an optimal estimate of the surface at every point;
that is, the interpolated surface will have minimum error. This is achieved
by taking into account prior knowledge about the spatial autocorrelation
between sample data points" (Sampson, 1975a, page 258).
The technique also permits optimum location of new data points to improve
predictions of spatial composition and variability (see, for example,
Delhomme and Delfiner, 1973; Huijbregts, 1975; McCullagh, 1975).

For the more general case in which the mean value of the dependent
variable changes across the area, Matheron (1969) introduced 'universal
Kriging', which yields superior estimates; the logical adjustments to the
ordinary Kriging model have been described (for example, by Huijbregts
and Matheron, 1971; Maréchal, 1972; Watson, 1972b; Olea, 1974; 1975;
Krige and Rendu, 1975). Kriging and universal Kriging have been used
most extensively in the mining industry for the prediction of ore values.
However, the techniques appear to have potential in meteorology, forestry,
structural geology (Delfiner and Delhomme, 1975; Krige and Rendu, 1975;
McCullagh, 1975; Guarascio et al, 1976; Sampson, 1975a; 1975b; Olea,
1975), or any field in which the assessment of spatial variability is a primary
objective. This is especially true where it is important to (a) associate
confidence levels with isoline maps, and/or (b) identify the locations of
inadequate data control and the degree of improvement to be anticipated
from securing additional samples (see McCullagh, 1975; Huijbregts, 1975,
for example).

Of direct relevance to the present work are the contoured maps (and
maps of the expected errors) made by Sampson (1975a; 1975b) for the
top of the subsurface Pennsylvanian Lansing Group in Stafford and
Graham Counties, Kansas, and based on universal Kriging of orthogonalized
data. Olea (1975, page 103) illustrated similar maps for the top of the
Tobifera Series (Cullen area, Magellan Basin, Chile); these were also
estimated by universal Kriging of orthogonalized data.

These preliminary studies suggest that, in general, three-dimensional
fold-geometry analysis by Kriging methods can be useful. Again, two
phases are apparent: (1) description and inventory, and (2) use of the
quantitative knowledge gained about the fold geometry to test genetic
process–response models. A necessary preliminary to Kriging is the
preparation of semivariograms. In the course of exploring the utility of
maps based on Kriging, it has been found that the semivariograms are a
significant and powerful tool in the identification and description of fold-
shape attributes.

Subsurface folds of the Michigan Basin, USA
Because the center of the Michigan Basin has been used to compare trend-
surface analyses, spatial filtering, and spline-surface interpolation in
subsurface-structure evaluation (Whitten and Beckman, 1969; Whitten and

Koelling, 1973), the same data base for the top of the Devonian Dundee
Limestone is used here (see figure 1).

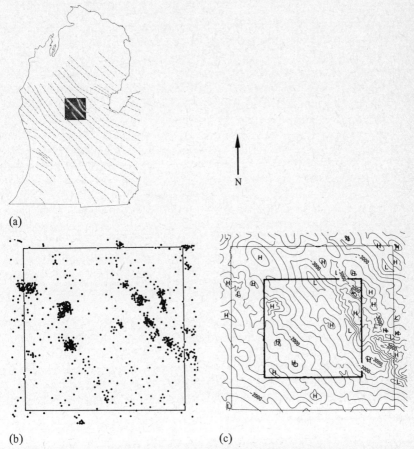

(a)

(b)   (c)

Figure 1. Top of subsurface Devonian Dundee Limestone, Michigan Basin, USA.
(a) Regional fold trends; (b) data points (drill holes) in 900 mile^2 (2333 km^2) area
shaded in (a); (c) subjective manually drawn structure contours (feet below sea level)
based on data points in (b) showing structural highs (H) and lows (L); outer frame is
that in (b), inner frame is 839 km^2 area of this study. After Whitten and Beckman
(1969, figures 1, 2, and 3).

Semivariograms

Semivariograms are graphical statements of the change in variance (of the
measured dependent variable—stratum elevation observed in well logs) at
successively increased sample spacing, h. For each different azimuth, the
semivariogram is a plot of $\gamma(h)$ versus h; h is the distance between a pair
of individual samples and $\gamma(h)$ is defined as follows. If $Y(x_i)$ is the
dependent variable, Y, at the ith sample point, x_i, then, where $N(h)$ pairs

of points occur at distance h apart, $\gamma(h)$ is approximated from the expression

$$2\gamma^*(h) = \frac{1}{N(h)} \sum_{i=1}^{N(h)} [Y(x_i+h) - Y(x_i)]^2 \ ,$$

where the superscript implies an estimated value.

Construction of semivariograms is easy when data points are regularly spaced and occur at the nodes of a grid; however, subsurface structural data points are almost always irregularly spaced. Various methods can be used to generate gridded data from an observed irregularly-spaced array. Whitten and Koelling (1973, pages 120 and following) demonstrated that each available gridding operation introduces unacceptable bias. Olea (1975, page 89) used regularly-arranged data from an area adjacent to the Cullen field to construct semivariograms for the latter area. In most cases, the extent of the difference between the structure of an adjacent area and that of the subject area would be uncertain. For example, the Dundee Limestone in the area immediately north of that studied by Whitten and Beckman (1969) in Michigan has radically dissimilar structure (Beckman and Whitten, 1969).

David (1977, pages 146 and following) provided a FORTRAN computer program for semivariogram construction that uses irregularly-distributed data. If pairs of samples are a linear distance h apart along a line of azimuth θ, semivariogram values could be estimated for each value of h and θ. In practice, it is convenient to group the pairs of samples into classes according to distance Δh and direction $\Delta\theta$. David (1977, page 146) claimed that, in most cases, an estimate of the variogram given by

$$\gamma^*(\theta, h) = \frac{1}{\Delta\theta\Delta h} \int_{-\Delta h/2}^{\Delta h/2} \int_{-\Delta\theta/2}^{\Delta\theta/2} \gamma(h+x, \theta+\psi)\mathrm{d}x\,\mathrm{d}\psi$$

introduces only limited discrepancies, provided that Δh and $\Delta\theta$ are small (that is, Δh is about the average distance between samples, and $\Delta\theta < 45°$).

The variance-variation pattern is anisotropic (that is, different for different azimuths) for most spatially-varying geological variables. In simple cases, the maximum and minimum semivariogram values of $\gamma^*(h)$ tend to be associated with orthogonal azimuths; where, in such cases, these directions are unknown, David (1977, page 145) showed that they can be idenrified by an ordinary least-squares technique. However, previous work in Michigan indicated the directions parallel to which maximum and minimum variance values are to be expected. Hence, the data for that area provide a good basis for testing the utility of semivariograms for structural and analytical purposes.

Where data comprise elevations of points on a cylindroidally (or cylindrically) folded stratum, the semivariogram for the direction parallel to the fold axis has minimal (near zero) values of $\gamma^*(h)$; the sill and range values are dependent on the amplitude and wavelength of the folds, respectively. Figure 2 shows semivariograms constructed at 15° azimuth intervals for elevations on idealized cylindroidal folds (like those on 'corrugated iron' sheeting). The harmonic nature of the original data results in harmonic semivariogram patterns for azimuths oblique to the B fold axis (for example, directions 030 to 090 in figure 2). The minima occur at h values equal to multiples of the apparent fold wavelength along each azimuth. Parallel to the B fold axis (000, or north, in figure 2), $\gamma^*(h)$ is zero. If there is a regional gradient ['drift' of Matheron (1965)] in the data set, the semivariogram tends to be parabolic (convex downwards) near the origin (see Rutledge, 1976, page 305), as shown by fold (b) in figure 3 and the semivariogram parallel to 015 in figure 2.

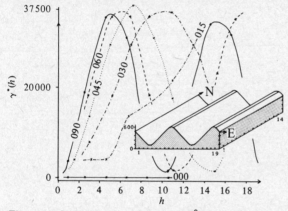

Figure 2. Semivariograms drawn at 15° azimuth intervals for 336 gridded data points on synthetic cylindroidally folded surface (inset). Bearings (030, for instance) indicate the degrees east of north of the azimuth of each semivariogram. Geographic coordinates of the inset are in units of length h.

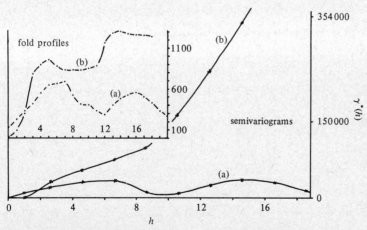

Figure 3. Semivariograms for synthetic fold profiles (inset); (a) without and (b) with strong regional gradient.

Most actual folds depart from ideal cylindroidal geometry; those in Michigan [figure 1(c)] may reflect superposed, intersecting fold trends. For the top of the Dundee Limestone in the test area [figure 1(c), inner frame], semivariograms for successive azimuths 15° apart (with $\Delta\theta = 10°$) illustrate the structural anisotropy (figure 4). In the southeast direction the variance, $\gamma^*(h)$, stabilizes rapidly with increasing h; by contrast, other $\gamma^*(h)$ traces (those parallel to 090, for instance) are strongly parabolic. The synoptic diagram (left side, figure 4) based on the set of semivariograms shows minimum $\gamma^*(h)$ values ('low') associated with directions 150 and 165 (that is, subparallel to the horizontal B fold axis); the range, although difficult to read, is some 15 miles (24·1 km). As one moves away from the direction B, the range decreases to 8 or 9 miles (12·9–14·5 km), and $\gamma^*(h)$ increases. Semivariograms parallel to directions 045–090 have comparable high values of $\gamma^*(h)$ ('high' in figure 4), and small ranges. Several of the semivariograms show harmonics (reflecting fold features crossed in that azimuth), although these are partially masked by the parabolic form resulting from the presence of significant regional gradients.

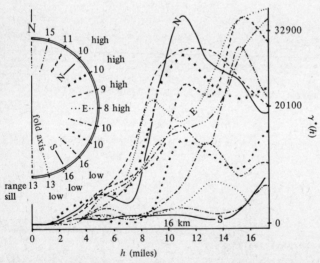

Figure 4. Semivariograms at 15° azimuth intervals for the top of Dundee Limestone [figure 1(c), inner frame]; line patterns correspond to the inset (left side) which indicates azimuth, approximate range (in miles), and sill [in $\gamma^*(h)$ units] of each semivariogram.

Isoline maps based on Kriging
Most regions analyzed for subsurface structure are likely to have a regional gradient, so, for the general case, universal Kriging is a more appropriate mapping tool than Kriging. Unfortunately, a computer program for

universal Kriging is not currently available[2]. Thus, despite the modest gradients in the Michigan data, ordinary Kriging is used here for illustrative purposes; the computations were made with a program adapted from that published by David (1977, pages 261 and following).

When the dependent variable at a point is estimated by Kriging, the 'influence' of known sample values at neighboring locations is proportional to their distance from the required point. In most cases of the use of Kriging and universal Kriging described in the literature, authors used semi-variograms to determine the anisotropy of their data, but used a single range value for all azimuths when calculating the weighting dependent on distance (see, for example, Sampson, 1975b, page 112). Hence, known data values from a circular area about the point are included in the computation.

However, the data influencing the required point should, ideally, be drawn from within the range distance, which may be different for each azimuth. That is, if maximum and minimum range values relate to orthogonal azimuths, data influencing the Krige value at each point should be drawn from an approximately elliptical area. It is complicated, although theoretically possible, to use a different range for each azimuth in modifications of available Kriging programs. However, for many data sets, it is appropriate to transform the range of the semivariogram for each different azimuth to a single value by linear transformation of h. In figure 5, if the range varies from OA to OB, transformation of the

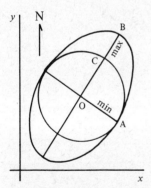

Figure 5. Orthogonal range maximum (OB) and minimum (OA) transformed to isotropic range (OA = OC) on the sampled surface.

[2] Olea (1975) and Sampson (1975b) described the logic and an available computer program for universal Kriging, but the program does not accommodate irregularly-spaced data without unacceptable preliminary orthogonalization of the observed data (see Whitten and Koelling, 1973, pages 120 and following). Delfiner (1976, page 62) drew attention to the commercially-available BLUEPACK computer package; this computer program may overcome these difficulties but it was not available for the present work.

(x, y) coordinates so that OA = OA and OB = OC means that the anisotropic ranges (OA and OB) are transformed to isotropic ranges, OA and OC. Now, when data to be used to estimate the Krige value at O are identified, use of the isotropic range OA (= OC) in transformed-coordinate space actually identifies sample points in an elliptical area of untransformed space; weighting of the identified samples in terms of isotropic range OA in transformed spatial coordinates effectively permits the estimation of a value weighted by anisotropic range in the original space.

Because of the significant anisotropy shown in figure 4, this method was used to modify the Kriging program of David (1977) to take range anisotropy into account; a simple parabola equation was used to represent the mean shape of the semivariogram of minimum range. The Krige map shown in figure 6 for the top of the Dundee Limestone was therefore based on the anisotropic variability of range. For this map, the elevations were estimated at the center of each symbol space on a continuous-symbol map produced by a line printer; isolines were drawn manually around the computer-printed symbols.

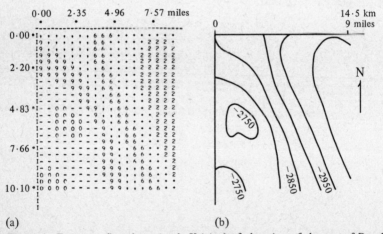

(a)   (b)

Figure 6. Estimates (based on simple Kriging) of elevation of the top of Dundee Limestone in the northwest quarter of the Michigan study area [figure 1(c)]. (a) Estimates for specific points printed by line printer; (b) isopleths (feet below sea level) drawn manually for those points.

Middle Cretaceous sedimentation rate, southeast USA

Sedimentation rates (meters/million years) for the Lower Atkinson (in northern Florida) and Tuscaloosa (in Georgia) stages at one hundred and two subsurface sites (figure 7) were derived from a study of intra-Cretaceous sedimentation rates of eastern USA (Whitten, 1976a; 1976b). For these data, the principal axes parallel to which $\gamma^*(h)$ would have maximum or minimum values were unknown *a priori*. Semivariograms were drawn along azimuths 10° apart, with $\Delta\theta = 10°$; alternate curves (at 20°

azimuth intervals) are shown in figure 8(a). The synoptic diagram (inset left) demonstrates that semivariograms parallel to directions 050–060 have minimum variance, and ranges of about 275 km (171 miles), whereas those parallel to directions 160–180 have maximum variance, and ranges of about 225 km (140 miles). Despite significant structural control of

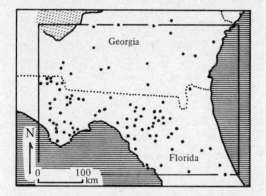

Figure 7. Data points (drill holes) used for the study of the Lower Atkinson/Tuscaloosa stages, Florida and Georgia, USA. The rocks outcrop in the shaded area to the north-west.

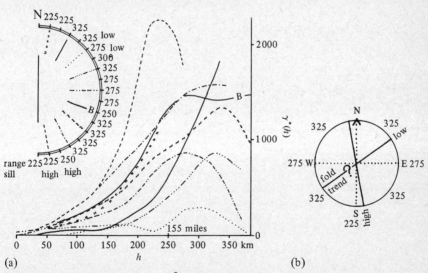

(a)

(b)

Figure 8. (a) Semivariograms at 20° azimuth intervals for sedimentation rate (meters/ million years) in Lower Atkinson/Tuscaloosa stages in parts of Florida and Georgia (see figure 7); line patterns correspond to inset at left, which indicates azimuth and approximate range (in miles) and sill [in $\gamma^*(h)$ units] of semivariograms at 10° azimuth intervals. Note that several semivariograms show 'nugget' values of $\gamma^*(h)$ of about 100. (b) Generalized synopsis of results in (a), showing where fold axis would have been if based on cylindroidal-fold data.

sedimentation rate, there is no reason to expect orthogonal principal axes (as for cylindroidal folds) for these data; the axes happen to be some 60° apart. Similarly, although the range varies systematically from about 225 km to 325 km (140–203 miles), extreme values do not coincide with extreme values of $\gamma^*(h)$. Unlike the Michigan example, a very complex transformation would be needed to accommodate the range variations summarized in figure 8(b). Hence, an isotropic mean range of 287·5 km (179·1 miles) was assumed in the computation of the isoline Krige map (figure 9). As with the Michigan data, the parabolic semivariograms imply regional gradients, and universal Kriging would yield more accurate estimates than the simple Kriging method.

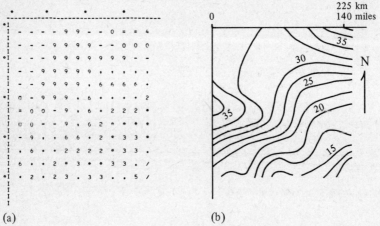

(a)   (b)

Figure 9. Estimates (based on simple Kriging) for the sedimentation rate of Lower Atkinson/Tuscaloosa stage rocks in the northwest quarter of the study area shown in figure 7. (a) Estimates for specific points printed by line printer; (b) isopleths (meters/million years) drawn manually for these points. Both maps are to the same scale.

Conclusions

The preliminary results described here suggest that semivariograms are sensitive tools for the analysis of fold geometry from the basis of irregularly-spaced subsurface data. With additional refinement, the technique should be capable of accurately identifying fold directions and perpendicular and nonperpendicular cross-fold directions, and of discriminating between folded and faulted geometry. In fact, although the Michigan structure is usually described in terms of folds, the major structure may have resulted, in part, from subparallel normal faults (Whitten and Beckman, 1969, page 1056; James, 1970). Work to determine the precision with which such structural details can be identified from semivariograms is in progress. It seems possible that further work will permit fold shape to be retrieved from the semivariograms of actual folds.

Figures 6 and 9 are based on an optimal estimate of the surface elevation at the center of each line-printer symbol, so that the interpolated surface has minimum error. In practice, more accurate results would flow from a calculation of these values by the more complex universal Kriging, rather than by simple Kriging, because of the regional gradients across the areas shown in figures 1(c) and 7. The computer cost for generating a set of semivariograms is small, but contour maps based on Kriging involve considerably more computing time than traditional trend-surface mapping techniques.

Acknowledgements. The assistance of Mrs Rose Feder with the collation and analysis of data, and grants from Northwestern University Research Committee, the Amax Foundation, and the National Science Foundation (Grant Number EAR76-22467) that have made this work possible, are gratefully acknowledged.

References
Beckman W A Jr, Whitten E H T, 1969 "Three-dimensional variability of fold geometry in the Michigan Basin" *Geological Society of America Bulletin* **80** 1629-1634
Brooker P I, 1975 "Optimal block estimation by Kriging" *Proceedings of the Australasian Institute of Mining and Metallurgy* (253) 15-19
David M L R, 1969 "The notion of 'extension variance' and its application to the grade estimation of stratiform deposits" in *A Decade of Digital Computing in the Mineral Industry—A Review of the State-of-the-Art* Ed. A Weiss (American Institute of Mining, Metallurgical, and Petroleum Engineers, New York) pp 63-81
David M L R, 1971 "Geostatistical ore estimation; a step-by-step case study" *Canadian Institute of Mining and Metallurgy, Special Volume* **12** 185-191
David M L R, 1976 "The practice of Kriging" in *Advanced Geostatistics in the Mining Industry* Eds M Guarascio, C J Huijbregts, M L R David (Reidel, Dordrecht, The Netherlands) pp 31-48
David M, 1977 *Geostatistical Ore Reserve Estimation* (Elsevier, Amsterdam)
Delfiner P, 1976 "Linear estimation of non-stationary spatial phenomena" in *Advanced Geostatistics in the Mining Industry* Eds M Guarascio, C J Huijbregts, M L R David (Reidel, Dordrecht, The Netherlands) pp 49-68
Delfiner P, Delhomme J P, 1975 "Optimum interpolation by Kriging" in *Display and Analysis of Spatial Data* Eds J C Davis, M J McCullagh (John Wiley, New York) pp 96-114
Delhomme J P, Delfiner P, 1973 "Application du krigeage à l'optimisation d'une campagne pluviométrique en zone aride" in *Symposium on the Design of Water Resources Projects with Inadequate Data* (UNESCO-WHO-IAHS, Madrid) pp 191-210
Guarascio M, Huijbregts C J, David M L R, 1976 (Eds) *Advanced Geostatistics in the Mining Industry* (Reidel, Dordrecht, The Netherlands)
Hansen E, 1971 *Strain Facies* (Springer, New York)
Hudleston P J, 1973 "Fold morphology and some geometrical implications of theories of fold development" *Tectonophysics* **16** 1-46
Hudleston P J, Stephansson O, 1973 "Layer shortening and fold shape development in buckling of single layers" *Tectonophysics* **17** 299-321
Huijbregts C J, 1975 "Regionalized variables and quantitative analysis of spatial data" in *Display and Analysis of Spatial Data* Eds J C Davis, M J McCullagh (John Wiley, New York) pp 38-53

Huijbregts C J, Matheron G, 1971 "Universal kriging (an optimal method for estimating and contouring in trend surface analysis)" *Canadian Institute of Mining and Metallurgy, Special Volume* 12 159-169

James W R, 1970 "Regression models for faulted structural surfaces" *Bulletin, American Association of Petroleum Geologists* 54 638-646

Journel A G, 1974 "Geostatistics for conditional simulation of ore bodies" *Economic Geology and the Bulletin of the Society of Economic Geologists* 69 673-687

Journel A G, 1975 "Geological reconnaissance to exploitation—a decade of applied geostatistics" *The Canadian Institute of Mining and Metallurgy Bulletin* 68 75-84

Krige D G, Rendu J M, 1975 "The fitting of contour surfaces to hanging and footwall data for an irregular ore body" in *Schriften für Operations Research und Datenverarbeitung in Bergbau* volume 4 (Glückauf, Essen) pp cv1-12

Maréchal A, 1972 "El Problema de la Curva Tonelaje—Ley de Corte y su Estimación" *Boletín Geostadística (Chile)* 1 3-20

Maréchal A, Serra J, 1970 "Random Kriging" in *Geostatistics—a Colloquium* Ed. D F Merriam (Plenum, New York) pp 91-112

Matheron G, 1965 *Les Variables Régionalisées et leur Estimation* (Masson, Paris)

Matheron G, 1969 "Le Krigeage universel" in *Cahiers du Centre de Morphologie Mathématique de Fontainebleau, France* number 1 (Centre de Morphologie Mathématique, Fontainebleau, France) pp 1-83

McCullagh M J, 1975 "Estimation by Kriging of the reliability of the proposed Trent Telemetry Network" *Computer Applications* 2 357-374

Olea R A, 1974 "Optimal contour mapping using universal kriging" *Journal of Geophysical Research* 79 695-702

Olea R A, 1975 *Optimum Mapping Techniques Using Regionalized Variable Theory* (Kansas Geological Survey, Lawrence, Kan.)

Ramsay J G, 1967 *Folding and Fracturing of Rocks* (McGraw-Hill, New York)

Rech W, 1977 "Zur Geometrie der geologischen Falten" *Geologische Rundschau* 66 352-373

Royle A G, Hosgit E, 1974 "Local estimation of sand and gravel reserves by geostatistical methods" *Institute of Mining and Metallurgy, Transactions Section A: Mining Industry* 83 A53-A62

Royle A G, Newton M J, Sarin V K, 1972 "Geostatistical factors in design of mine sampling programmes" *Institute of Mining and Metallurgy, Transactions Section A: Mining Industry* 81 A82-A88

Rutledge R W, 1976 "The potential of geostatistics in the development of mining" in *Advanced Geostatistics in the Mining Industry* Eds M Guarascio, C J Huijbregts, M L R David (Reidel, Dordrecht, The Netherlands) pp 295-312

Sampson R J, 1975a "The SURFACE II graphics system" in *Display and Analysis of Spatial Data* Eds J C Davis, M J McCullagh (John Wiley, New York) pp 244-266

Sampson R J, 1975b *SURFACE II Graphics System* (Kansas Geological Survey, Lawrence, Kan.)

Sinclair A J, Deraisme J, 1974 "A geostatistical study of the Eagle copper vein, northern British Columbia" *Transactions of the Canadian Institute of Mining and Metallurgy* 77 357-368

Watson G S, 1971 "Trend-surface analysis" *Journal of the International Association for Mathematical Geology* 3 215-226

Watson G S, 1972a "Prediction and efficiency of least squares" *Biometrika* 59 1-8

Watson G S, 1972b "Trend surface analysis and spatial correlation" *Geological Society of America, Special Paper* 146 39-46

Whitten E H T, 1961 "Quantitative areal modal analysis of granitic complexes" *Geological Society of America Bulletin* 72 1331-1360

Whitten E H T, 1964 "Process and response models in geology" *Geological Society of America Bulletin* **75** 455–464

Whitten E H T, 1966 *Structural Geology of Folded Rocks* (John Wiley, New York)

Whitten E H T, 1974 "Scalar and directional field and analytical data for spatial variability studies" *Journal of the International Association for Mathematical Geology* **6** 183–198

Whitten E H T, 1975 "The practical use of trend-surface analyses in the geological sciences" in *Display and Analysis of Spatial Data* Eds J C Davis, M J McCullagh (John Wiley, New York) pp 282–297

Whitten E H T, 1976a "Cretaceous phases of rapid sediment accumulation, continental shelf, eastern USA" *Geology* **4** 237–240

Whitten E H T, 1976b "Geodynamic significance of spasmodic, Cretaceous, rapid subsidence rates, continental shelf, USA" *Tectonophysics* **36** 133–142

Whitten E H T, 1977a "Kriging in subsurface structural analyses" in *Horniká Příbram ve vědě a Technice, Matematické Metody v Geologii* **3** 801–819

Whitten E H T, 1977b "Stochastic models in geology" *Journal of Geology* **85** 321–330

Whitten E H T, Beckman W A Jr, 1969 "Fold geometry within part of Michigan Basin, Michigan" *Bulletin, American Association of Petroleum Geologists* **53** 1043–1057

Whitten E H T, Koelling M E V, 1973 "Spline-surface interpolation, spatial filtering, and trend surfaces for geological mapped variables" *Journal of the International Association for Mathematical Geology* **5** 111–126

Estimating volumetric modal composition: the grain approach versus the area approach

F Chayes

The notion that no grain should be cut by more than one traverse in a linear analysis, or be occupied by more than one grid intersection in a point count, reflects a particular view of the nature of the sampling process. The notion that it does not matter whether either of these conditions holds reflects a very different view of the nature of the sampling process.

According to the first view, in an analysis made by counting points, the sample is to be regarded as a collection of grain sections. That the grains are themselves contained in a set of one or more thin sections is of obvious practical importance, but of no particular theoretical significance. Provided that the point interval is such that no grain contributes more than one point to the total count, the relative frequencies of points situated in different minerals are evidently considered to be an unbiased estimate of the modal composition of the rock.

In the second view, the sample is a set of one or more *areas*, the composition of each of which is to be estimated by a point count. That these areas contain varying numbers of grain sections is of obvious practical importance, but of no particular theoretical significance. The immediate analytical problem is to make the best estimate of the composition of each area. If the areas are randomly chosen, their average composition, under easily realizable conditions, is a consistent estimator of the volumetric composition of the specimen from which they were cut (Chayes, 1956, chapter 1).

To the extent that petrologists have paid any attention at all to the theoretical underpinning of modal analysis, it seems to me that their attitude toward the sampling problem as a whole is fairly represented by this second view. Indeed, something of the sort is surely implicit in the very first paper on modal analysis (Delesse, 1848), although Delesse did not use thin sections, and did not estimate areal compositions by point counting.

In their occasional attempts to examine the influence of what is loosely called 'grain size' or 'coarseness' on modal analysis, however, petrologists have almost invariably favored the first view. Here too there is a long and honorable precedent. The earliest treatment of the subject actually occurs in the very first paper on what is now called linear (or lineal) analysis (Rosiwal, 1898). After several rediscoveries, the rule established by Rosiwal—that no grain section is to be cut by more than one line—was extended to the case of point counting by van der Plas and Tobi (1965) (but see also Chayes, 1965), and has recently been reinvoked, though for a rather different purpose, by Neilsen and Brockman (1977).

The 'grain-sample' and 'areal-sample' views lead to very different interpretations of the basic statistic of all counting procedures: the count length. In the areal-sample view, the count length relates solely to the reproducibility of an individual areal estimate. In properly designed experiments, reproducibility variance is completely separable from the other components of sampling variance, which are usually, in one way or another, the principal concern of the experimenter. If the areas were all the same size, the same number of points were counted in every area, and the point sites were chosen simply at random in each area, this would be strictly true and the reproducibility error estimated from the count length would be binomially distributed. It will be approximately true if the areas and count lengths are nearly uniform. If, as is in fact the case, the point sampling is on a grid in each area, the reproducibility error will usually be less than binomial, and perhaps significantly so. However, with count lengths normally in the large hundreds or small thousands it is obvious that even the full binomial counting variance will nearly always be very small in relation to the total sample variance.

Thus, in the areal-sample view, the following precepts apply:
(1) In the interest of minimizing the reproducibility error attached to estimates of areal composition, the count length per area is to be made as large as practical, *without concern as to the number of point sites that fall in any grain.*
(2) The count length controls only the reproducibility error, which is not of geological or petrological interest and, except in cases of extreme sample homogeneity, is never more than a very small component of the total sampling variance.

In the grain-sample view, on the other hand, the count length is *not* to be extended if this will lead to multiple point-sites in individual grains; there seems to be no basic sampling unit above the level of the individual grain section; and no particular interest seems to be attached to the precision of areal estimates. Indeed, the whole question of partition and analysis of variance seems of little concern.

Which view is correct? The answer, I think, is that neither is necessarily incorrect in any absolute sense. By introducing a completely formal sampling unit of intermediate level, however, the areal sampling procedure opens up the possibility of convenient partition and analysis of variance. It seems to me that in most petrographic investigations which make extensive use of modal analysis, this approach is clearly implicit, if not actually explicit. To the extent that this is so, the rule of Rosiwal (1898) is both irrelevant and misleading.

Furthermore, with the possible application of the scanning electron microscope to modal analysis, the whole controversy seems to be about to disappear. The 'points' of the raster of a cathode-ray tube are so close together that, at magnifications commonly used in modal analysis, the distinction between a point count, the linear analysis of Rosiwal, and the direct areal analysis of Delesse, becomes academic. Modal analysis will once again become areal analysis. We shall then be obliged to think in terms of areal sampling as outlined in this essay, except that the 'count length' will be so large that the associated random component of the counting error will be negligible.

References
Chayes F, 1956 *Petrographic Modal Analysis* (John Wiley, New York)
Chayes F, 1965 "Reliability of point counting results: discussion" *American Journal of Science* **263** 719-721
Delesse A, 1848 "Procédé mécanique pour déterminer la composition des roches" *Annales des Mines* **13** 379-388
Neilsen M J, Brockman G F, 1977 "The error associated with point counting" *American Mineralogist* **62** 1238-1244
Rosiwal A, 1898 "Über geometrische Gesteinsanalysen usw." in *Verhandlungen der K. K. Geologische Reichsanstalt Wien* pp 143-175
van der Plas L, Tobi A C, 1965 "A chart for judging the reliability of point counting results" *American Journal of Science* **263** 87-90

Box-Jenkins modelling in the geophysical sciences

K W Hipel, A I McLeod

Introduction

In order to study a specified geophysical process, such as sediment deposition at the bottom of a lake, measurements of the process can be taken at equally-spaced intervals over a given time period. Observations that occur sequentially over time are called a time series; such observations are, in fact, a realization of the underlying stochastic process. In time-series analysis, it is often desirable to fit some type of stochastic model to the given data sequence.

Various types of stochastic models are available for use in geophysics. Of particular importance for the analysis of natural time series is the Box-Jenkins family of stochastic models. An original presentation of this class of models is given in a textbook that was completed in 1970 by Box and Jenkins (Box and Jenkins, 1970). This book is actually a culmination of the research of many prominent statisticians, starting with the pioneering work of Yule (1927).

Since 1970, progress has been made in extending and refining Box-Jenkins models for use in the geophysical sciences (see, for example, Hipel et al, 1977a; Hipel and McLeod, 1981; McLeod and Hipel, 1978a; 1978b). The purpose of this essay is to provide a comprehensive presentation of Box-Jenkins modelling techniques, for utilization by geophysicists. Numerous recommendations regarding important topics where further research is required are also given.

Many of the new developments in Box-Jenkins modelling that are presented here were originally published in the water-resources literature. Nevertheless, because of the fact that problems in the field of water resources possess characteristics that are common to many other areas in the natural sciences, the Box-Jenkins modelling procedures constitute useful tools that can be employed throughout the realm of geophysics. In order to emphasize further the importance of Box-Jenkins models, examples are cited where other types of stochastic models are shown not to work as well as Box-Jenkins models. For instance, McLeod and Hipel (1978c) demonstrate that Box-Jenkins models are superior in a statistical sense to fractional-Gaussian-noise (FGN) models. Furthermore, within the field of hydrology, Box-Jenkins models are shown to produce a reasonable explanation to problems related to the Hurst phenomenon (McLeod and Hipel, 1978c; Hipel and McLeod, 1978a).

Owing to the many uses of stochastic models in geophysics, a wide range of topics is covered within this essay. For example, intervention analysis (Box and Tiao, 1975) is a technique that can be employed to analyze the effects both of natural and of man-induced activities upon the environment (Hipel et al, 1975; Hipel et al, 1977b). Because the conventional seasonal Box–Jenkins model is not appropriate for the description of certain types of data, a special family of stochastic models is presented, for modelling monthly geophysical data. This class of models is extended for use in intervention analysis within this paper.

No matter what type of application a given stochastic model is to be used for, the first step is to fit the model to the given data. Consequently, in the next section a general philosophy of model-building is explained, and the way in which the construction of a stochastic model is linked to key underlying modelling principles is shown.

Modelling philosophies
This essay deals primarily with Box–Jenkins models and some extensions of them. However, whenever stochastic models are applied to a given data set, there are certain 'commonsense' modelling guidelines that should be adhered to. Of paramount importance to any stochastic analysis is a sound understanding of the physical phenomenon to be modelled, and also an appreciation of the limitations of the stochastic models that are employed to model that phenomenon. For instance, in certain types of monthly geophysical time series the individual monthly averages may have constant mean values, but the means vary from month to month. Consequently, the time series of all of the given data is by definition nonstationary. However, the fitting of certain types of nonstationary models (for example, Box–Jenkins models that contain differencing operators to remove homogeneous nonstationarity) to the data may obscure reality. If the fitted stochastic model does not allow for second-order stationarity within each month, serious problems may arise. An understanding of the mathematical design of the stochastic model, coupled with a knowledge of the physical process to be modelled, can aid in determining which kinds of stochastic models are worthy of consideration in the first place.

When entertaining appropriate stochastic models for a particular problem, certain fundamental modelling principles should be followed. One foundation stone to stochastic model-building is that the selected model should be kept as simple as possible. This is accomplished by developing a model that incorporates the minimum number of model parameters needed to describe the data adequately. The principle of model parsimony (that is, keeping the number of model parameters to a minimum) is of utmost importance in Box–Jenkins modelling.

Another important modelling tenet is that the model developed imparts a 'good statistical fit' to the data. In order to have a good statistical fit, efficient estimates must be obtained for the model parameters, and the fitted model must pass rigorous statistical tests to insure that the underlying modelling assumptions are satisfied.

In practice, the key modelling doctrines of parsimony and good statistical fit can be satisfied. This is effected by carefully pursuing the identification, estimation, and diagnostic-checking stages of model construction. At the identification stage, a plot of the given data, with other simple statistical inspections, can provide valuable insight into which type of model is suitable to fit to the time series. Efficient estimates of the model parameters can be obtained at the estimation stage by employing the method of maximum likelihood. The model should then be checked for possible inadequacies. If the diagnostic tests reveal any serious anomalies, appropriate modifications can then be made by repeating the identification and estimation stages.

A mathematical formulation of the parsimony criterion of model-building, together with maximum likelihood estimation, is given by the Akaike information criterion (AIC) (Akaike, 1974). When there are several competing models to choose from, the model that gives the minimum of the AIC, defined by

$$AIC = -2\ln(\text{maximum likelihood}) + 2k \, , \tag{1}$$

where k is the number of model parameters, should be selected.

The manner in which the AIC can be employed to enhance the three stages of model development is explained by Hipel (1981) and by Hipel and McLeod (1981), while numerous applications that demonstrate the efficacy of the AIC for selecting the most appropriate model are presented in the engineering literature (see, for example, McLeod et al, 1977; Hipel, 1981; Hipel and McLeod, 1981). In situations where fairly complicated models are required, it is often advisable first to consider fairly simple forms, and then to proceed to develop more complicated models. The simpler models often provide insight into how the more complicated models should be designed. For example, suppose that the time series for average monthly river flow and precipitation are available for modelling. Obviously, precipitation causes river flow. However, the first step in the model-building phases would be to determine simple univariate models to fit separately to each of the two data sets. Based upon this information, a more complicated transfer-function-noise model can be developed to link the two time series. If the more advanced model is preferable on statistical grounds, and also more realistically describes the physical process under consideration, then this model should be chosen in preference to the other simpler models.

Nonseasonal models

Many types of geophysical records are strictly nonseasonal, and in other situations when seasonal data are available, one could be required to consider a time series of average annual values. For example, tree-ring indices and mud-varve thicknesses are obtainable only in the form of yearly records. However, time series for mean annual river flow, temperature, and precipitation can be calculated from average weekly records. Whatever the case, in many situations it is often necessary to deal with nonseasonal geophysical time series.

Stationary Box–Jenkins models
The Box–Jenkins nonseasonal stationary models (Box and Jenkins, 1970) constitute a flexible and comprehensive family of linear stochastic processes for use in the geophysical sciences. These Box–Jenkins models are often referred to as autoregressive–moving average (ARMA) processes. In order to define an ARMA model, consider a time series of observations of the variable z: $z_1, z_2, ..., z_{t-1}, z_t, z_{t+1}, ..., z_N$, where the observations are spaced by equal time intervals (for example, one year for annual geophysical data). An ARMA model for z_t is given by

$$\phi(B)(z_t^{(\lambda)} - \mu) = \theta(B)a_t \ . \tag{2}$$

In this equation
$z_t^{(\lambda)}$ designates an appropriate transformation of z_t such as a Box–Cox transformation (McLeod, 1974; Box and Cox, 1964), no transformation is a possible option;
t is an index for the discrete time periods;
B is a backward shift operator defined by

$$Bz_t^{(\lambda)} = z_{t-1}^{(\lambda)} , \quad \text{and} \quad B^k z_t^{(\lambda)} = z_{t-k}^{(\lambda)} ;$$

μ is the mean level of the process, usually taken as the mean of the $z_t^{(\lambda)}$ series;
a_t is a normally and independently distributed white noise residual series with zero mean and variance σ_a^2 [written as $NID(0, \sigma_a^2)$];
$\phi(B)$ $(= 1 - \phi_1 B - \phi_2 B^2 - ... - \phi_p B^p)$ is a nonseasonal autoregressive (AR) operator, or polynomial, of order p, such that the roots of the characteristic equation, $\phi(B) = 0$, lie outside the unit circle for nonseasonal stationarity, and the ϕ_i $(i = 1, 2, ..., p)$ are the nonseasonal AR parameters;
$\theta(B)$ $(= 1 - \theta_1 B - \theta_2 B^2 - ... - \theta_q B^q)$ is a nonseasonal moving average (MA) operator, or polynomial, of order q, such that the roots of $\theta(B) = 0$ lie outside the unit circle for invertibility and the θ_i $(i = 1, 2, ..., q)$ are the nonseasonal MA parameters.
The notation $ARMA(p, q)$ is used to denote an ARMA model with an AR operator of order p and an MA operator of order q. If the process is a purely AR model of order p, it is often denoted by $AR(p)$, instead of $ARMA(p, 0)$. An $ARMA(0, q)$ model can be equivalently written as $MA(q)$.

When determining the type of ARMA model to fit to a given data set, Box and Jenkins (1970) recommend that the identification, estimation, and diagnostic-checking stages of model development should be followed. More recently, Hipel et al (1977a) have suggested the use of contemporary techniques to strengthen the three stages of model construction. At the identification stage, new methods are available for determining the orders of the AR and MA parameters. Improved estimates of these parameters can then be obtained by using the method of McLeod (1977a; 1977b). As a part of this method, the type of Box–Cox transformation needed may be estimated. At the diagnostic-checking stage, the residual assumptions of independence, normality, and homoscedasticity (that is, constant variance) can be verified. The independence assumption is the most crucial of all, and its violation can have drastic consequences (Box and Tiao, 1973). Sensitive tests are now available (McLeod, 1977a; 1978) to test the 'whiteness' of the residual series. However, if the normality and/or homoscedasticity assumptions are not true, they can often be reasonably well satisfied when the observations undergo a Box–Cox transformation.

McLeod et al (1977) have demonstrated the usefulness of the contemporary Box–Jenkins modelling procedures described by Hipel et al (1977a), by applying these methods to different types of time series. Hipel and McLeod (1978a) have successfully determined ARMA models for modelling twenty-three nonseasonal geophysical time series. These data sets consist of time series for river flow, mud varves, temperature, precipitation, sunspots, and tree rings. Hipel (1975) has also found that stationary ARMA models can adequately model many types of nonseasonal natural data sequences.

Nonstationarity
If the time series is nonstationary, a standard procedure is first to remove the nonstationarity and then to fit a stationary model to the resulting stationary data. In particular, Box and Jenkins (1970) suggest that nonstationary data should be differenced just enough times to remove what is termed "homogeneous nonstationarity". Then an ARMA model is fitted to the stationary series, w_t, given by

$$w_t = (1 - B)^d z_t^{(\lambda)} , \qquad (3)$$

where $(1 - B)^d$ is a differencing operator of order d.

A Box–Jenkins model that includes differencing is referred to as an autoregressive–integrated moving average (ARIMA) model. The notation ARIMA(p, d, q) is used to indicate the orders of the operators in the ARIMA process. An ARIMA$(p, 0, q)$ model is equivalent to a stationary ARMA(p, q) model.

Although the stationarity assumption is reasonably valid for mathematical models which are fitted to observed geophysical time series, it is possible that situations may arise where the data are nonstationary. The scientist should proceed with caution when deciding upon the method for removing the nonstationarity. It is, of course, possible that the differencing operation of equation (3) may be feasible. However, if there is a change in the mean level of a process because of a known external cause, then the method of intervention analysis should be used (Box and Tiao, 1975; Hipel et al, 1975; 1977b). For example, Hipel et al (1975) employ intervention analysis to model the drop in the mean level of the River Nile at Aswan, due to the completion of the Aswan Dam in 1902. The different mean levels before and after 1902 are automatically accounted for in the intervention model.

Fractional Gaussian noise

Within the field of hydrology, the FGN model, and approximations thereof, have been developed for modelling natural time series (Mandelbrot and Wallis, 1969a; 1969b; 1969c). McLeod and Hipel (1978c) have developed a method for obtaining maximum likelihood estimates (MLE) for the parameters of the FGN model, a technique for calculating the model residuals of a FGN process so that the residuals can be subjected to diagnostic checks, and also a procedure for exactly simulating FGN [refer to Hipel and McLeod (1978b) for the FORTRAN programs of these developments]. Even with these significant improvements in FGN modelling, McLeod and Hipel (1978c) have demonstrated that the ARMA family of models is superior to the FGN model. When a proper ARMA model, and also a FGN model, are fitted to each of six annual river-flow time series, the AIC indicates that the ARMA model is preferable to the corresponding FGN process for each of the data sets. As explained by McLeod and Hipel (1978c), the inability of FGN models to work as well as Box–Jenkins processes could be due to many reasons. In particular, besides the mean and variance, the FGN model possesses only one other model parameter. Estimates of the model parameters form the only actual link between the theoretical model and the real world as represented by the data. In contrast to ARMA modelling, where the proper number of AR and MA parameters are selected for modelling a given data set, FGN models are rendered highly inflexible by their dependence upon a fixed number of parameters.

Hipel and McLeod (1978a) have shown that ARMA models statistically preserve important historical statistics, such as the Hurst coefficient and the rescaled adjusted range. Because ARMA models almost always perform better than FGN models according to a statistical criterion such as the AIC, it is probably not advisable to employ FGN for modelling geophysical time series.

In summary, the ARMA model constitutes an excellent tool for modelling yearly geophysical data. The relative importance of FGN and other related models will probably decrease greatly over the next few years, and it is highly unlikely that these models will survive the evolutionary process.

Box–Jenkins seasonal models

The Box–Jenkins seasonal model is defined by the expression

$$\phi(B)\Phi(B^s)\{[(1-B)^d(1-B^s)^D z_t^{(\lambda)}] - \mu\} = \theta(B)\Theta(B^s)a_t . \tag{4}$$

Here

s is an index of the season length ($s = 12$ for monthly data);

$\Phi(B^s)$ $(= 1 - \Phi_1 B^s - \Phi_2 B^{2s} - ... - \Phi_P B^{Ps})$ is a seasonal AR operator of order P, such that the roots of $\Phi(B^s) = 0$ lie outside the unit circle for seasonal stationarity, and the Φ_i $(i = 1, 2, ..., P)$ are the seasonal AR parameters;

$(1 - B^s)^D$ is a seasonal differencing operator of order D, used to produce seasonal stationarity by the Dth difference of the data (usually $D = 0, 1,$ or 2);

$\Theta(B^s)$ $(= 1 - \Theta_1 B^s - \Theta_2 B^{2s} - ... - \Theta_Q B^{Qs})$ is a seasonal MA operator of order Q, such that the roots of $\Theta(B^s) = 0$ lie outside the unit circle for invertibility, and the Θ_i $(i = 1, 2, ..., Q)$ are the seasonal parameters;

μ is the mean level of the process, or mean of the differenced z_t series [if $(D+d) > 0$, then often $\mu \simeq 0$].

The model defined in equation (4) is also called a multiplicative seasonal ARIMA process. The notation ARIMA$(p, d, q) \times (P, D, Q)_s$ is used to represent the seasonal ARIMA model. The first set of brackets contains the orders of the nonseasonal operators, and the second pair of brackets contains those of the seasonal operators. For example, a stochastic seasonal noise model of the form ARIMA$(2, 1, 1) \times (1, 1, 2)_s$ is written as

$$(1 - \phi_1 B - \phi_2 B^2)(1 - \Phi_1 B^s)\{[(1 - B)(1 - B^s)z_t^{(\lambda)}] - \mu\}$$
$$= (1 - \theta_1 B)(1 - \Theta_1 B^s - \Theta_2 B^{2s})a_t . \qquad (5)$$

The problem with the model defined in equation (4) is that when this model is fitted to observed seasonal geophysical time series, it is usually necessary to difference the data both seasonally and nonseasonally. A model containing differencing operators implies that the process is nonstationary, and therefore if the model is used for simulation purposes the generated data are not restricted to any mean level. However, in many cases this does not reflect reality. For example, when considering average monthly river-flow data, the observations for any particular month tend to fluctuate around some fixed mean level. Even though the average varies from one month to the next, this does not justify differencing. Seasonal models are needed which preserve the monthly means in addition to other relevant monthly statistics.

Deseasonalization
To eliminate the need for differencing, water-resources engineers often incorporate a deterministic component or transformation into the model to remove seasonality. A nonseasonal ARMA model is then fitted to the resulting nonseasonal data. For monthly observations it is often appropriate to remove seasonality by standardizing the data. This is accomplished by subtracting out the appropriate monthly mean and then dividing the result by the monthly standard deviation for each of the seasonal data points. It is often advisable to take natural logarithms of the data before performing the standardization. Deseasonalization of monthly data, and the fitting of a nonseasonal model to the resulting time series, has been discussed by

numerous authors in the hydrological literature (see, for example, Hipel, 1975; Hipel and McLeod, 1979; McLeod et al, 1977; Tao and Delleur, 1976; McKerchar and Delleur, 1974; Croley and Rao, 1977).

For monthly data, deseasonalization causes the addition of twenty-four parameters to the model. For weekly and daily data, the number of parameters required for standardization prohibits the use of the procedure. Instead, the seasonality is removed by using Fourier series and other periodic polynomials (see, for example, McMichael and Hunter, 1972). Clarke (1973), and Croley and Rao (1977), present extensive descriptions of deseasonalization procedures for daily, weekly, and monthly data; Hipel and McLeod (1979) demonstrate how the AIC can be employed to select the most appropriate deseasonalized model.

A flexible family of geophysical models for monthly data

In many practical applications, it is convenient to present the data as average monthly values. For instance, when using simulation to design a reservoir for hydroelectric power generation, Hipel (1975) and Hipel et al (1979) fitted stochastic models to the historical average monthly river flows. Because of the importance of average monthly observations in certain branches of geophysics, McLeod and Hipel (1978a) have presented an improved family of stochastic models that are designed exclusively for modelling monthly geophysical time series.

Let $z_{r,m}$ denote an observation made in the rth year at month m. If there are n years of data, the observation that occurs in chronological order as observation number t is represented by z_t, where

$$t = 12(r-1) + m \tag{6}$$

for $t = 1, 2, ..., N$, and $N = 12n$. By using equation (6), it can be shown that $z_{10,0}$, $z_{9,12}$, and $z_{8,24}$ all refer to the 108th observation, given by z_{108}.

Consider a class of models where, for month m, the process is written as

$$\phi_m(B)(z_{r,m}^{(\lambda)} - \mu_{r,m}) = a_{r,m} . \tag{7}$$

Here

$\phi_m(B)$ $(= 1 - \phi_{m,1}B - \phi_{m,2}B^2 - ... - \phi_{m,p_m}B^{p_m})$ is a monthly AR operator of order p_m for month m, and $\phi_{m,i}$ $(i = 1, 2, ..., p_m)$ are monthly AR parameters for month m;

B   is a backward shift operator that operates on m such that $B^k z_{r,m}^{(\lambda)} = z_{r,m-k}^{(\lambda)}$, and $B^k \mu_{r,m} = \mu_{r,m-k}$, where k is a nonnegative integer;

$z_{r,m}^{(\lambda)}$ is the Box–Cox transformation of $z_{r,m}$, that is

$$z_{r,m}^{(\lambda)} = \begin{cases} \dfrac{z_{r,m}^{\lambda} - 1}{\lambda} & (\lambda \neq 0) , \\[2ex] \ln z_{r,m} & (\lambda = 0) ; \end{cases}$$

$\mu_{r,m}$ is the mean correction factor which, for a pure autoregression, is taken as the monthly mean, μ_m;

$a_{r,m}$ is a white-noise process, that is NID$(0, \sigma_m^2)$, where σ_m^2 is the variance of the innovations for month m.

From equation (7) it can be seen that a system of twelve equations ($m = 1, 2, ..., 12$) is required to model monthly geophysical time series. McLeod and Hipel (1978a) recommend that the identification, estimation, and diagnostic-checking stages of model construction should be followed when fitting their family of monthly AR models to a given data set. For each month, it is necessary to identify the order of the AR operator $\phi_m(B)$. This can be done by using the AIC for model discrimination, and also by adjusting the order of an AR operator for a particular month wherever diagnostic checks indicate inadequacies. Methods are given for obtaining MLEs of the model parameters. McLeod and Hipel (1978a) also present the distribution for the residual autocorrelation function and portmanteau statistics to allow for sensitive diagnostic checking of the 'whiteness' assumption for the residual series.

Previously a number of researchers have considered similar types of models to the monthly AR model given in equation (7) (see, for example, Thomas and Fiering, 1962; Jones and Brelsford, 1967; Yevjevich, 1972; Clarke, 1973; Rao and Kashyap, 1974; Tao and Delleur, 1976; Croley and Rao, 1977; Sen, 1978; Pagano, 1978; Parzen and Pagano, 1979). However, McLeod and Hipel (1978a) have developed many new results for utilization at the identification, estimation, and diagnostic-checking stages of model development. As explained later in this paper, this enhanced mathematical maturity allows the class of monthly AR models to be readily extended for use in transfer-function-noise modelling and intervention analysis. Nevertheless, research should continue to develop suitable families of stochastic models for modelling weekly, daily, and perhaps hourly geophysical data (Kibler and Hipel, 1979).

The Hurst phenomenon

For many types of geophysical processes, it is meaningful to assume that the yearly time series are stationary, and that observed seasonal data are stationary within a given season. These assumptions stem from the premise that, barring external interventions, natural processes tend to repeat themselves. For instance, river-flow sequences usually follow a similar stochastic pattern from one year to the next unless the river basin is devastated by a forest fire or altered by man-induced activities, such as cultivation and urbanization. Because of this, geophysicists wish to develop and apply stochastic models that preserve the important historical characteristics of the given time series.

Within the field of hydrology, McLeod and Hipel (1978c) have interpreted the controversies surrounding the Hurst phenomenon as being related to the preservation of historical statistics by stochastic models. In order to explain the Hurst phenomenon, some statistical definitions must first be given. For a nonseasonal time series $z_1, z_2, ..., z_N$, define the kth adjusted partial sum, S_k^*, by

$$S_k^* = S_{k-1}^* + z_k - \bar{z}_N = \sum_{i=1}^{k} z_i - k\bar{z}_N , \qquad k = 1, 2, ..., N , \tag{8}$$

where $\bar{z}_N$ is the mean of the N observations in the time series, and $S_0^* = S_N^* = 0$.

The adjusted range, R_N^*, is defined as

$$R_N^* = M_N^* - m_N^* , \tag{9}$$

where M_N^* and m_N^* are the adjusted surplus and adjusted deficit, respectively, given by

$$M_N^* = \max(0, S_1^*, S_2^*, ..., S_N^*) ,$$

$$m_N^* = \min(0, S_1^*, S_2^*, ..., S_N^*) .$$

The rescaled adjusted range (RAR), $\bar{R}_N^*$, is then calculated as

$$R_N^* = \frac{R_N^*}{\hat{\sigma}_N} , \tag{10}$$

where $\hat{\sigma}_N$ is the sample standard deviation.

The statistics defined above have direct physical interpretation in reservoir theory. If the z_t are average annual volumes of river flow, then summing z_i over k years gives the inflow into a reservoir over those years, whereas $k\bar{z}_N$ is the outflow after k years if $\bar{z}_N$ is released each year. S_k^* represents the storage after k years and R_N^* is the minimum reservoir capacity required to satisfy a constant draft $\bar{z}_N$ without experiencing shortages or spills over the period spanned by the N inflow sequences.

Hurst (1951; 1956) stimulated interest in the RAR statistic by his studies of long-term storage requirements on the River Nile. Based upon his experiences, Hurst developed the empirical relationship

$$\bar{R}_N^* = \left(\frac{N}{2}\right)^K , \tag{11}$$

where K is the Hurst coefficient.

For 690 annual time series for stream flow, river and lake level, precipitation, temperature, pressure, tree rings, mud varves, sunspots, and wheat prices, Hurst found K to have an average value of $0 \cdot 73$, with a standard deviation of $0 \cdot 09$.

Using the theory of Brownian motion, Feller (1951) proved that the following asymptotic formula is valid for the expected value of the RAR, $E(\bar{R}_N^*)$, for any sequence of identically independently distributed random variables.

$$E(\bar{R}_N^*) = 1 \cdot 2533 N^{\frac{1}{2}} . \tag{12}$$

The value of the exponent in the above asymptotic formula is 0·5. However, Hurst found K to have an average value of 0·73 for the 690 time series that range in length from 30 to 2000 years. This discrepancy is referred to as the Hurst phenomenon. The search for a reasonable explanation of the Hurst phenomenon and the need for methods whereby the statistics related to the work of Hurst can be incorporated into mathematical models, have intrigued researchers during the past quarter of a century.

A critical review of research in this area is given by McLeod and Hipel (1978c). In particular, the FGN model was developed as a means of explaining the Hurst phenomenon and preserving the Hurst coefficient. However, when compared to the ARMA model, McLeod and Hipel (1978c) have shown the FGN model to be inferior. Furthermore, by fitting the proper ARMA model to twenty-three geophysical time series and by using appropriate simulation procedures, Hipel and McLeod (1978a) have demonstrated that ARMA models do preserve the value of the Hurst coefficient or equivalently the RAR.

Mixture models

As one explanation of the Hurst phenomenon, Hurst (1957) conjectured that a nonstationary model, in which the mean of the series is subject to random changes, could account for higher values of K, and hence the Hurst phenomenon. Similar models have been suggested by Klemes (1974) and Potter (1976). More recently, Boes and Salas (1978) have developed the 'mixture model' as a process which possesses shifting means.

Although the mixture-model concept is still under development, there could be serious problems with the eventual use of this type of model in geophysics. First, the justification for assuming a model with a changing mean is questionable. Although the ARMA model in equation (2) has a constant mean, that model does allow data simulated from such a process to fluctuate about a constant mean level. Indeed, it is possible that over a moderately long time span the simulated data may be entirely above or below the mean level. If an external intervention has altered the process such that there is an actual change in mean level, then an intervention model (Box and Tiao, 1975) can be used to describe this change. In practice, the intervention model has already been shown to be successful for modelling changes in mean level both for nonseasonal and for seasonal hydrological data (Hipel et al, 1975; Hipel et al, 1977b). Therefore, there may be no need for a mixture model.

A second deficiency of the mixture model is that it may be difficult to fit it to actual geophysical time series. In particular, there could be many arduous problems in the development of suitable estimation procedures for the model parameters. Hinkley (1972) has described estimation problems in change-point analysis, which is in some ways similar to the mixture model. Consequently, the practical usefulness of the mixture model may be quite limited. Geophysicists are encouraged to use the technique of intervention analysis if there is a physical reason for a change in the mean level of a given phenomenon.

Conclusions
As mentioned previously, the RAR statistic is of interest in problems dealing with storage of water in reservoirs. However, the RAR may also be of importance to other types of geophysical problems that deal with cumulative sums. It could, for example, be used in modelling the yearly accumulations of sediment at the mouth of a river.

In other branches of geophysics there may be further statistics which are of practical importance. However, it is likely that these statistics will be preserved by an ARMA model which has been fitted according to the three stages of model construction stated previously. Simulation studies similar to those described by Hipel and McLeod (1978a) can be carried out to check if this is the case for particular statistics.

Simulation
Historical data represent one possible source for the realization of the underlying stochastic process that describes the phenomenon under consideration. After an appropriate Box–Jenkins model has been fitted to the data, it can then be used to simulate possible sequences of observations. Within the field of hydrology, the use of simulation is referred to as operational or synthetic hydrology.

Simulation has been extensively used in engineering. For example, water-resources engineers use simulation as an aid in the design and operation of reservoirs (Hipel, 1975; Hipel et al, 1979). As shown by Hipel and McLeod (1978a), simulation can also be employed for determining the mathematical properties of ARMA models. By using simulation with a specified Box–Jenkins model, the empirical cumulative distribution function (ECDF) can be developed for the statistic under consideration. Hipel and McLeod (1978a) develop the ECDF for the RAR and the Hurst coefficient for different types of ARMA models and different lengths of series. In many situations, the exact theoretical distribution is analytically intractable; simulation can then be used to obtain the distribution of a given statistic as accurately as required. These and other applications of simulation demonstrate the need for adoption of proper simulation methods.

McLeod and Hipel (1978b) have presented improved simulation procedures for generating synthetic traces from Box–Jenkins models. These techniques have been labelled WASIM 1 (Waterloo simulation procedure one), WASIM 2 (Waterloo simulation procedure two), and WASIM 3 (Waterloo simulation procedure three). A distinct advantage of these simulation methods is the fact that random realizations of the underlying stochastic process are employed as starting values. Because fixed initial values are not used, systematic bias is not introduced into the generated data.

Furthermore, WASIM 1 is an exact simulation method for pure MA processes whereas WASIM 2 is exact for ARMA(p, q) processes. WASIM 3 is a technique for incorporating parameter uncertainty into a simulation study.

In practical applications, if the Box–Jenkins model residuals (that is, the estimated a_t terms) are not normally distributed, then approximate normality can be achieved by implementing a Box–Cox transformation (Hipel et al, 1977a). It is possible that in certain instances it may be difficult to rectify nonnormality of the residuals. In other situations, it may be desirable to allow the residuals to follow some type of skewed distribution. An inherent advantage of the WASIM 1 simulation technique is that the only restriction on the residuals is that they are identically independently distributed with zero mean and variance σ_a^2 [that is, $IID(0, \sigma_a^2)$].

When dealing with seasonal geophysical data, the important statistical characteristics of the observations are often stationary within a given season. If a model is to be used for simulation, the seasonal ARIMA model should not be employed because the model usually contains a differencing operator. Consequently, the simulated data will not be restricted to any mean level within a given season. Therefore, either an ARMA model should be fitted to the deseasonalized data, or else for the case of monthly data, the process in equation (7) should be used. In order to select which seasonal model to use for simulation, one can show how to discriminate between the various types of seasonal models by using the AIC (Hipel, 1981; Hipel and McLeod, 1979; McLeod and Hipel, 1978a). In addition, exact simulation procedures are given for the family of monthly geophysical models given in equation (7) (McLeod and Hipel, 1978a).

It is the opinion of the authors that the importance of simulation will continue to expand within the geophysical sciences. Original research regarding various aspects of simulation is still required. For example, univariate simulation procedures can be extended for use with multivariate models. Finally, practitioners are encouraged to employ recent simulation developments in their simulation studies.

Forecasting
Within engineering, forecasting is often useful for operational purposes. For example, when planning a release policy for a reservoir during an approaching month, a forecast of river flow for the month would be useful. Minimum-mean-square-error (MMSE) forecasting procedures are described by Box and Jenkins (1970), for use both with nonseasonal and with seasonal ARIMA models. If certain types of nonlinear transformations have been applied to the data, Granger and Newbold (1976) suggest methods whereby MMSE forecasts can be obtained for the original data. McLeod et al (1977) discuss the use of the methods of Granger and Newbold, when either a natural logarithmic or a square root transformation is implemented.

To determine which models make the best forecasts, various models can be fitted to the first section of a data set and then one-step-ahead forecasts can be obtained for the remaining data as each new observation becomes available. By using the mean square error of the forecasts for each model, significance testing can be carried out to ascertain if one model forecasts significantly better than another. When modelling annual geophysical time series, Noakes (1979) found that ARMA models forecast significantly better than FGN models. For the case of models fitted to data for average monthly river flow, the monthly AR model in equation (7) and the seasonal ARIMA model in equation (4) performed equally well with respect to forecasting ability. However, the deseasonalized model did not forecast as well as the monthly AR and seasonal ARIMA models (Noakes, 1979). Newbold and Granger (1974) and Granger and Newbold (1977) have compared the capabilities of Box–Jenkins and other types of models to forecast economic time series.

Intervention analysis
Conventional intervention models
In their text, Box and Jenkins (1970) discuss transfer-function–noise models. These models are useful for describing dynamic relationships between geophysical time series. For instance, a model can be developed that relates precipitation and river flow.

A model closely related to the transfer-function–noise model is the intervention model. This model can be developed by a procedure labelled 'intervention analysis' (Box and Tiao, 1975), and is of particular importance in the geophysical sciences.

A major challenge in the natural sciences is to model and statistically describe the effects both of man-induced and of natural interventions upon the mean level of a geophysical time series. For example, how do changes in land use, such as urbanization and cultivation, affect the amount of soil erosion in a given area? Will a forest fire significantly alter the drainage characteristics of a river basin? Do human activities significantly change the world climate? Intervention analysis is a means by which the effects of such events can be introduced into the model of a time series.

An intervention model is made up of a dynamic component and a noise component. The estimates of transfer-function parameters for the dynamic component can statistically describe the intervention effects; they also determine the dependence of the response variable on any other inputs. An ARMA process constitutes the noise term. A detailed mathematical description of the intervention model can be found by referring to the statistical and engineering literature (Box and Tiao, 1975; Hipel et al, 1975; 1977b). However, in general terms, the intervention model may be written as

$$\text{response} = \text{dynamic component} + \text{noise} ,\qquad (13)$$

where

   dynamic component = interventions + inputs .

As well as modelling the intervention effects, the complete stochastic intervention model can also be used for forecasting and simulation. It is not necessary to disregard parts of the data or model sections of the time series separately because of the changes in mean level. Instead, all of the available information can be incorporated into a single model that inherently preserves the intervention effects.

It should be noted that the ordinary Student t-test cannot be used to check for a change in the mean level of a time series. This test is only valid if the observations before and after the intervention being considered vary about the two means normally and independently with constant, but not necessarily equal, variance. These assumptions are not satisfied because time series are usually serially dependent, often nonstationary, and frequently seasonal.

When an intervention model for describing a given time series is to be determined, it is recommended that the three stages of model development (Box and Tiao, 1975) are followed. Additional methods that assist in the model-building process are given in the engineering literature (Hipel et al, 1975; 1977b). The intervention component of the intervention model can be designed to simulate many different kinds of interventions. For example, if the change is a step, an impulse, an exponential decay, or an increase in the response variable (or any combination of these changes), the intervention model can be readily designed to handle these situations.

Intervention analysis has been successfully applied to problems in the geophysical and environmental sciences. Hipel et al (1975) used intervention analysis to demonstrate that the completion of the Aswan Dam in 1902 caused a significant decrease in the mean level of the average annual downstream flows of the River Nile. Intervention analysis has been employed to determine the statistical effects of reservoir construction and operation upon the mean level of monthly downstream river flows (Hipel et al, 1977b); to find out how a devastating forest fire can alter the seasonal flow characteristics of a river (Hipel et al, 1977c; 1978); to ascertain the effectiveness of pollution abatement procedures for improving air quality (Box and Tiao, 1975) and water quality (D'Astous and Hipel, 1979); to determine the effects of changing the type of snow gauges used for measuring the amount of snowfall in the Canadian Arctic (Baracos et al, 1981); to fill in missing data points in a time series (D'Astous and Hipel, 1979; Baracos et al, 1981); and to design data-collection procedures (Lettenmaier et al, 1978). Abraham (1980) has described a multivariate intervention model and applied it to economic time series.

A new class of intervention models
The monthly AR model in equation (7) can be extended to include
transfer-function–noise and intervention models. To accomplish this, the
$\mu_{r, m}$ term in equation (7) must be designed in a manner that allows for
the incorporation of covariate series and intervention components into the
model. Before an application of this new type of intervention model is
given, let us consider a study which uses the conventional intervention
model. Hipel et al (1977b) used intervention analysis to model the
changes in the monthly flow patterns of the South Saskatchewan River at
Saskatoon (Canada), caused by the operation of the Gardiner Dam.
Average monthly river-flow data are available from January 1912 to
December 1974. In January 1969 the Gardiner Dam came into full
operation, and caused a step change in the mean level of the monthly
flows. Downstream at Saskatoon, the dam caused average flows to
increase during the winter and to decrease during the summer. Because
the main use of the water behind the Gardiner Dam is for generation of
electricity, the reservoir is operated so as to release flows as evenly as
possible throughout the year.

By employing the traditional type of intervention model, the authors
devised a model that adequately describes the monthly data for the South
Saskatchewan River, and the intervention effects. However, the monthly
AR model given in equation (7) is better suited for modelling monthly
geophysical data. Let $\mu_{r, m}$ in equation (7) be represented by

$$\mu_{r, m} = \mu_m + \beta_m \xi_{r, m} , \tag{14}$$

where μ_m is the average flow for month m; $\xi_{r, m}$ is the intervention time
series, which has a value of zero before 1969 and a value of one from
1969 onwards; β_m is the intervention parameter for month m.

Equations (7) and (14) can be used to determine a unique type of
intervention model for the South Saskatchewan River. Parameter estimates
are obtained for the case where all of the monthly AR operators are of order
two, and the original data are transformed by taking natural logarithms.

Table 1. Estimated percentage changes in the average monthly flows of the South
Saskatchewan River at Saskatoon from 1969 to 1974.

Month	Percentage change	Month	Percentage change
January	450·09	July	−53·23
February	405·84	August	−28·26
March	180·34	September	−10·90
April	−40·34	October	35·22
May	−52·26	November	123·45
June	−63·91	December	339·85

It can be shown that the intervention parameter for each month can be converted to percentage change, c, in mean flow by using the equation

$$c = 100[\exp(\beta_i) - 1] ; \quad i = 1, 2, ..., 12 .\qquad(15)$$

Table 1 lists the estimated percentage change in mean level for each month during the period 1969 to 1974. These results are similar to findings from use of the conventional intervention model (Hipel et al, 1977b). However, when dealing with monthly geophysical data, it may be advantageous to employ the new family of intervention models.

Relationships between time series

Wiener (1956) originally formulated a definition of causality between two time series, which is suitable for empirical detection and verification of meaningful relationships. Work on the causal calculus of temporal systems has been done by Granger (1969), and Pierce and Haugh (1977). In this section, some practical aspects of this work are discussed and then illustrated by an example involving natural time series, which shows that sunspots do not cause river flows.

Let x_t and y_t $(t = ..., -2, -1, 0, 1, 2, ...)$ be two time series, and X_t and Y_t be the sets

$$X_t = \{x_b | b < t\} ,\qquad(16)$$

and

$$Y_t = \{y_b | b < t\} .\qquad(17)$$

Then, in qualitative terms, x_t causes y_t if a better prediction of y_t can be made by using the information set $\{X_t, Y_t\}$, rather than by using only $\{Y_t\}$. Also, if y_t can be better predicted by using $\{X_t, Y_t, x_t\}$ rather than just $\{X_t, Y_t\}$, then it is said that there is 'instantaneous causality' between x_t and y_t. By making appropriate assumptions about the structure of the time series x_t and y_t, it can be shown that if x_t causes y_t instantaneously, then it is also true that y_t causes x_t instantaneously. Thus, in the situation of instantaneous causality, the direction of causality must be inferred from geophysical considerations, rather than directly from the data.

Suppose that x_t and y_t are time series that can be modelled by ARMA models, such that

$$\phi_x(B)x_t = \theta_x(B)a_t ,\qquad(18)$$

and

$$\phi_y(B)y_t = \theta_y(B)e_t ,\qquad(19)$$

where

$$\phi_x(B) = 1 - \phi_{x,1}B - \phi_{x,2}B^2 - ... - \phi_{x,p_x}B^{p_x} \; ;$$

$$\theta_x(B) = 1 - \theta_{x,1}B - \theta_{x,2}B^2 - ... - \theta_{x,q_x}B^{q_x} \; ;$$

$$\phi_y(B) = 1 - \phi_{y,1}B - \phi_{y,2}B^2 - ... - \phi_{y,p_y}B^{p_y} \; ;$$

$$\theta_y(B) = 1 - \theta_{y,1}B - \theta_{y,2}B^2 - ... - \theta_{y,q_y}B^{q_y} \; ;$$

a_t is a white-noise process, that is NID$(0, \sigma_a^2)$;
e_t is a white-noise process, that is NID$(0, \sigma_e^2)$.

The cross-correlation, $\rho_{ae}(k)$, at lag k between the a_t series and the e_t series from equations (18) and (19), respectively, is

$$\rho_{ae}(k) = \frac{E(a_t e_{t+k})}{[E(a_t^2)E(e_t^2)]^{1/2}} \; . \tag{20}$$

The following results for $\rho_{ae}(k)$ can be proven:
(1) x_t causes $y_t \Leftrightarrow \rho_{ae}(k) \neq 0$ for some k, $k > 0$;
(2) y_t causes $x_t \Leftrightarrow \rho_{ae}(-k) \neq 0$ for some k, $k > 0$;
(3) there is instantaneous causality between x_t and $y_t \Leftrightarrow \rho_{ae}(0) \neq 0$.

Suppose in a practical situation that two time series x_t and y_t are given for $t = 1, 2, ..., N$. By use of equations (18) and (19), models can be fitted to the series to obtain estimates, $\tilde{e}_t$ and $\tilde{a}_t$, of the e_t and a_t sequences. This procedure is referred to as 'prewhitening'. The sample cross-correlation, $r_{ae}(k)$, at lag k between $\tilde{a}_t$ and $\tilde{e}_t$ is defined by

$$r_{\tilde{a}\tilde{e}}(k) = \begin{cases} \sum_{t=1}^{N-k} \tilde{a}_t \tilde{e}_{t+k} \bigg/ \left[\left(\sum_{t=1}^{N} \tilde{a}_t^2 \right)^{1/2} \left(\sum_{t=1}^{N} \tilde{e}_t^2 \right)^{1/2} \right], & k \geqslant 0 \; ; \\ \sum_{t=1}^{N+k} \tilde{a}_{t+k} \tilde{e}_t \bigg/ \left[\left(\sum_{t=1}^{N} \tilde{a}_t^2 \right)^{1/2} \left(\sum_{t=1}^{N} \tilde{e}_t^2 \right)^{1/2} \right], & k \leqslant 0 \; . \end{cases} \tag{21}$$

The cross-correlation $r_{\tilde{a}\tilde{e}}(k)$ can be shown to be a consistent estimator for $\rho_{ae}(k)$.

It has been shown (Haugh, 1976) that when $\rho_{ae}(k) = 0$ for all k, then $r_{\tilde{a}\tilde{e}}(k) \sim$ NID$(0, 1/N)$ in large samples. Thus, to check for independence between two time series, the residual cross-correlations, $r_{\tilde{a}\tilde{e}}(k)$, and their estimated 95% confidence intervals, equal to $\pm 1 \cdot 96(N)^{-1/2}$, can be plotted. Significant values of $r_{\tilde{a}\tilde{e}}(k)$ may indicate causal relationships.

Formal tests of significance may also be derived. Suppose that it is known *a priori* that y_t does not cause x_t, so that $\rho_{ae}(k) = 0$ for all $k < 0$. For instance, river flows do not cause precipitation. Then the statistic Q_M,

$$Q_M = N \sum_{i=0}^{M} r_{\tilde{a}\tilde{e}}^2(i) \; , \tag{22}$$

will be distributed as $\chi^2(M+1)$, if x_t and y_t are independent. In theory, any value of $M \geqslant 0$ could be chosen, but in practice, the value of M should have the smallest magnitude such that it would be expected that after M time periods, there would not be a relationship between the x_t and y_t time series.

In certain situations, there may be dependence between $\tilde{a}_t$ and $\tilde{e}_t$. McLeod (1979) has derived the distribution of $r_{\tilde{a}\tilde{e}}(k)$ ($k = 0, \pm1, \pm2, ...$), for the general case when $\rho_{ae}(i) \neq 0$ for some lag i.

To illustrate the foregoing theory, let x_t be the yearly Wolfer sunspot number (data from Waldmeier, 1961) in year t and let y_t be the average annual river flow of the Elbe at Děčín for the years 1851–1957 (data from Novotny, 1963). It has been suggested that variations in the sunspot numbers, which are related to variations in solar radiation, may be causally related to river flows. Application of the residual cross-correlation analysis did not reveal any relationship.

A limitation of the techniques explained above is that they are only useful when the relationships between two time series are to be described. If three or more time series are mutually related, then to analyze them only two at a time may lead to spurious relationships. Research on multivariate techniques for analyzing the causal relationships between three or more time series is needed. The results from this type of research should prove to be extremely useful within the geophysical sciences.

Conclusions

Because of the essential nature of many types of geophysical phenomena, they are amenable to analysis using stochastic techniques. As exemplified in this essay, Box-Jenkins models and other related stochastic models are useful for various types of applications. For instance, intervention analysis, simulation, and forecasting constitute some of the worthwhile uses of these models. Consequently, the authors maintain that the importance of these stochastic models should increase within the geophysical sciences. The material developed by the authors which has been presented here is also to be presented, along with other relevant topics, in a textbook (Hipel and McLeod, 1981). Computer programs are listed in the book so that all of the modelling methods can be immediately implemented by the reader.

Extensive research is still required within many areas of stochastic modelling in geophysics. For instance, specialized models should be developed for modelling weekly and daily data. In certain situations, it may be appropriate to study nonlinear models (Yakowitz, 1979). Multivariate techniques should be developed for correlating relationships between three or more data sets. These and other research topics pose interesting challenges for the research scientist. This essay should serve as a guideline for placing the research problems into proper perspective.

The geophysicist is cautioned against blindly following whatever is published in the research journals. For instance, within the field of stochastic hydrology, the development of the FGN model (Mandelbrot and Wallis,

1969a; 1969b; 1969c) is a prime example of the 'Pied-Piper syndrome' in research. Since about 1969, many scientists have been employing this model in hydrological studies without ever seriously questioning the usefulness and underlying statistical properties of the model. Consequently, valuable research time has been spent on the study of a model that is very limited in scope.

References

Abraham B, 1980 "Intervention analysis and multiple time series" *Biometrika* **67** (1) 73-78

Akaike H, 1974 "A new look at the statistical model identification" *IEEE Transactions on Automatic Control* **AC-19** (6) 716-723

Baracos P C, Hipel K W, McLeod A I, 1981 "Modelling hydrologic time series from the Arctic" *Water Resources Bulletin* **17** (3)

Boes D C, Salas J D, 1978 "Nonstationarity in the mean and the Hurst phenomenon" *Water Resources Research* **14** (1) 135-143

Box G E P, Cox D R, 1964 "An analysis of transformations" *Journal of the Royal Statistical Society, Series B* **26** 211-252

Box G E P, Jenkins G M, 1970 *Time Series Analysis: Forecasting and Control* (Holden-Day, San Francisco, Calif.)

Box G E P, Tiao G C, 1973 *Bayesian Inference in Statistical Analysis* (Addison-Wesley, Reading, Mass)

Box G E P, Tiao G C, 1975 "Intervention analysis with applications to economic and environmental problems" *Journal of the American Statistical Association* **70** (349) 70-79

Clarke R T, 1973 *Mathematical Models in Hydrology* Irrigation and Drainage paper, Food and Agriculture Organization of the United Nations, Rome

Croley T E II, Rao K N R, 1977 *A Manual for Hydrologic Time Series Deseasonalization and Serial Independence Reduction* report number 199, Iowa Institute of Hydraulic Research, The University of Iowa, Iowa City, Iowa

D'Astous F, Hipel K W, 1979 "Analyzing environmental time series" *Journal of the Environmental Engineering Division, American Society of Civil Engineers* **105** (EE5) 979-992

Feller W, 1951 "The asymptotic distribution of the range of sums of independent random variables" *Annals of Mathematical Statistics* **22** 427-432

Granger C W J, 1969 "Investigating causal relationships by econometric models and cross spectral methods" *Econometrica* **37** 424-438

Granger C W J, Newbold P, 1976 "Forecasting transformed series" *Journal of the Royal Statistical Society, Series B* **38** (2) 189-203

Granger C W J, Newbold P, 1977 *Forecasting Economic Time Series* (Academic Press, New York)

Haugh L D, 1976 "Checking the independence of two covariance-stationary time series: a univariate residual cross-correlation approach" *Journal of the American Statistical Association* **71** (354) 378-385

Hinkley D V, 1972 "Time-ordered classification" *Biometrika* **59** (3) 509-523

Hipel K W, 1975 *Contemporary Box-Jenkins Modelling in Hydrology* PhD thesis, Department of Civil Engineering, University of Waterloo, Waterloo, Ontario, Canada

Hipel K W, 1981 "Geophysical model discrimination using the Akaike information criterion" *IEEE Transactions on Automatic Control* **AC-26** (2) 358-378

Hipel K W, Lennox W C, Unny T E, McLeod A I, 1975 "Intervention analysis in water resources" *Water Resources Research* **11** (6) 855-861

Hipel K W, Lettenmaier D P, McLeod A I, 1978 "Assessment of environmental impacts, part one: Intervention analysis" *Environmental Management* **2**(6) 529-535

Hipel K W, McBean E A, McLeod A I, 1979 "Hydrologic generating model selection" *Journal of the Water Resources Planning and Management Division, American Society of Civil Engineers* **105** (WR2) 223-242

Hipel K W, McLeod A I, 1978a "Preservation of the rescaled adjusted range, part two, simulation studies using Box-Jenkins models" *Water Resources Research* **14**(3) 509-516

Hipel K W, McLeod A I, 1978b "Preservation of the rescaled adjusted range, part three, fractional Gaussian noise algorithms" *Water Resources Research* **14**(3) 517-518

Hipel K W, McLeod A I, 1979 *Modelling Seasonal Geophysical Time Series Using Deseasonalized Models* technical report number 52-XM-030579, Department of Systems Design Engineering, University of Waterloo, Waterloo, Ontario, Canada

Hipel K W, McLeod A I, 1981 *Time Series Modelling for Water Resources and Environmental Engineers* (Elsevier, Amsterdam)

Hipel K W, McLeod A I, Lennox W C, 1977a "Advances in Box-Jenkins modelling, part one, model construction" *Water Resources Research* **13**(3) 567-575

Hipel K W, McLeod A I, McBean E A, 1977b "Stochastic modelling of the effects of reservoir operation" *Journal of Hydrology* **32** 97-113

Hipel K W, McLeod A I, Unny T E, Lennox W C, 1977c "Intervention analysis to test for changes in the mean level of a stochastic process" in *Stochastic Processes in Water Resources Engineering* Eds L Gottschalk, G Lindh, L Maré (Water Resources Publications, Fort Collins, Col.) pp 93-113

Hurst H E, 1951 "Long-term storage capacity of reservoirs" *Proceedings of the American Society of Civil Engineers* **116** 770-808

Hurst H E, 1956 "Methods of using long-term storage in reservoirs" *Proceedings of the Institution of Civil Engineers* (1) 519-543

Hurst H E, 1957 "A suggested statistical model of some time series which occur in nature *Nature (London)* **180** 494

Jones R H, Brelsford W M, 1967 "Time series with periodic structure" *Biometrika* **54** 403-408

Kibler D F, Hipel K W, 1979 "Surface water hydrology" *Reviews of Geophysics and Space Physics* **17**(6) 1186-1209

Klemes V, 1974 "The Hurst phenomenon—a puzzle?" *Water Resources Research* **10**(4) 675-688

Lettenmaier D P, Hipel K W, McLeod A I, 1978 "Assessment of environmental impacts, part two: Data collection" *Environmental Management* **2**(6) 537-554

Mandelbrot B B, Wallis J R, 1969a "Computer experiments with fractional Gaussian noises, part 1, averages and variances" *Water Resources Research* **5**(1) 228-241

Mandelbrot B B, Wallis J R, 1969b "Computer experiments with fractional Gaussian noises, part 2, rescaled ranges and spectra" *Water Resources Research* **5**(1) 242-259

Mandelbrot B B, Wallis J R, 1969c "Computer experiments with fractional Gaussian noises, part 3, mathematical appendix" *Water Resources Research* **5**(1) 260-267

McKerchar A I, Delleur J W, 1974 "Application of seasonal parametric linear stochastic models to monthly flow data" *Water Resources Research* **10**(2) 246-255

McLeod A I, 1974 *Contributions to Applied Time Series* masters thesis, Department of Statistics, University of Waterloo, Waterloo, Ontario, Canada

McLeod A I, 1977a *Topics in Time Series Analysis and Econometrics* PhD thesis, Department of Statistics, University of Waterloo, Waterloo, Ontario, Canada

McLeod A I, 1977b "Improved Box-Jenkins estimators" *Biometrika* **64**(3) 531-534

McLeod A I, 1978 "On the distribution and applications of residual autocorrelations in Box-Jenkins models" *Journal of the Royal Statistical Society, Series B* **40**(3) 296-302

McLeod A I, 1979 "Distribution of the residual cross correlations in univariate ARMA time series models" *Journal of the American Statistical Association* **74** (368) 849-855

McLeod A I, Hipel K W, 1978a "Developments in monthly autoregressive modelling" technical report number 45-XM-011178, Department of Systems Design Engineering, University of Waterloo, Waterloo, Ontario, Canada

McLeod A I, Hipel K W, 1978b "Simulation procedures for Box-Jenkins models" *Water Resources Research* **14** (5) 969-975

McLeod A I, Hipel K W, 1978c "Preservation of the rescaled adjusted range, part one, a reassessment of the Hurst phenomenon" *Water Resources Research* **14** (3) 491-508

McLeod A I, Hipel K W, Lennox W C, 1977 "Advances in Box-Jenkins modelling, part two, applications" *Water Resources Research* **13** (3) 577-586

McMichael F C, Hunter J S, 1972 "Stochastic modelling of temperature and flow in rivers" *Water Resources Research* **8** (1) 87-98

Newbold P, Granger C W J, 1974 "Experience with forecasting univariate time series and the combination of forecasts" *Journal of the Royal Statistical Society, Series A* **137** (2) 131-146

Noakes D J, 1979 *Forecasting Geophysical Time Series* masters thesis, Department of Systems Design Engineering, University of Waterloo, Waterloo, Ontario, Canada

Novotny J, 1963 "Dve stolete hydrologicke rady prutokove na ceskych rekach" in *Sbornik Praci* volume 2 (Hydrometeorological Institute, Prague)

Pagano M, 1978 "On periodic and multiple autoregressions" *Annals of Statistics* **6** (6) 1310-1317

Parzen E, Pagano M, 1979 "An approach to modelling seasonally stationary time series" *Journal of Econometrics* **9** 137-153

Pierce D A, Haugh L D, 1977 "Causality in temporal systems" *Journal of Econometrics* **5** 265-293

Potter K W, 1976 "Evidence for nonstationarity as a physical explanation of the Hurst phenomenon" *Water Resources Research* **12** (5) 1047-1052

Rao R A, Kashyap R L, 1974 "Stochastic modelling of river flows" *IEEE Transactions on Automatic Control* **AC-19** (6) 874-881

Sen Z, 1978 "A mathematical model of monthly flow sequences" *Hydrological Sciences Bulletin* **23** (2) 223-229

Tao P C, Delleur J W, 1976 "Seasonal and nonseasonal ARMA models in hydrology" *Journal of the Hydraulics Division, American Society of Civil Engineers* **HY10** 1541-1559

Thomas H A, Fiering M B, 1962 "Mathematical synthesis of stream flow sequences for the analysis of river basins by simulation" in *Design of Water Resources Systems* Eds A Maass, M M Hufschmidt, R Dorfman, H A Thomas Jr, S A Marglin, G M Fair (Harvard University Press, Cambridge, Mass) pp 459-493

Waldmeier M, 1961 *The Sunspot Activity in the Years 1610-1960* (Schulthas Co, Zurich)

Wiener N, 1956 "Theory of prediction" in *Modern Mathematics for Engineers, Series 1* Ed. E F Beckenbach (McGraw-Hill, New York) chapter 8

Yakowitz S, 1979 "A nonparametric Markov model for daily river flow" *Water Resources Research* **15** (5) 1035-1043

Yevjevich V M, 1972 "Structural analysis of hydrologic time series" hydrology paper 56, Department of Civil Engineering, Colorado State University, Fort Collins, Col.

Yule G U, 1927 "On a method of investigating periodicities in disturbed series, with special reference to Wolfer sunspot numbers" *Philosophical Transactions of the Royal Society of London, Series A* **226** 267-298

The transfer-function models of Box and Jenkins: examples in a geosciences context

M L Labovitz

Introduction and objective

Many of the phenomena studied by geoscientists and those in related disciplines are dynamic systems that can be modelled as input–process–response models, illustrated schematically in figure 1 (Beer, 1964). In such models, the system is 'explained' or characterized as the relationship between its important properties or state variables, which are often chosen on the basis of theoretical considerations. In this model, changes in certain state variables precede changes in other state variables, and so the history of the process influences present and future states.

In these models, the state variables are introduced as sequences, $\{Z_t\}$ ($t = 1, ..., n$), with each individual element, Z_t, representing the value of the characteristic at time t. The problem is thus recast into one of formulating rules which relate the state-variable sequences to each other. There are a number of procedures for estimating these relationships, and the appropriate procedure will depend on certain properties of the sequences. If, for instance, the terms composing a sequence are independent of one another (the history of the sequence does not influence the present or future values) and the changes in the sequences are simultaneous, then standard regression procedures can be used. Violations of either condition— independence or time-coincidence—make the modelling problem more complex. However, for many geological systems the hypothesized relationship between the phenomena under study will include a dependence assumption and/or the presence of a time-ordering process. An alternative approach to modelling phenomena of this type is the set of transfer-function models developed by Box and Jenkins (1970). Matheron (1976) and his colleagues have developed a model that they call a "transfer function", but it is not similar to the set of models described in this essay.

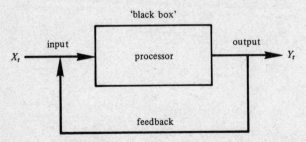

Figure 1. Schematic of general input–process–response model, with linear filter relating input to output.

The transfer function is a mathematical description of the relationship between the input and output, in other words, of the way the 'black box' of figure 1 works. Within this formalism, structures both for time delay and for time dependence can be incorporated. The models can be used either in an exploratory or in a confirmatory role. That is, the researcher can either explore for relationships or, alternatively, test the presence of hypothesized relationships. In either case, the researcher can then proceed to predict future outputs of the system.

The chief source of contention over the use of this set of models is their inability in many cases to give direct (that is, related to the subject matter) theoretical meaning to the parameters or their estimated values, in contrast to some other modelling procedures such as econometric modelling (Nelson, 1973). This disadvantage, if it is indeed a disadvantage, is compensated for by the ability of the models to simulate a great number of different stochastic processes (Kendall, 1973) with very few parameters, and at the same time to explain appreciable amounts of the variation in the system.

The purpose of this essay is to present the transfer-function approach to modelling in the context of the geosciences. A discussion of the mathematical basis of the models will be followed by a section on model-fitting and testing model adequacy. I conclude with two examples using this analysis—one in a confirmatory role, and the other in an exploratory role.

Transfer-function analysis: theoretical considerations
This section and the next one: "modelling dynamic systems", are largely a condensation and restructuring of the material presented by Box and Jenkins (1970). Other sources are noted in the acknowledgements and references.

In terms of the 'black-box' model (figure 1), a variable which measures the level of some input to the system, X, influences the level of the output, Y. Because of the inertia in the system, a change in the level of input often will not have an immediate effect upon the output "but, instead, will produce delayed response with Y eventually coming to equilibrium at a new level" (Box and Jenkins, 1970, page 335). This change in Y is referred to as a 'dynamic response' and the model which describes the response or characterizes the inertia is called a 'transfer-function model'.

If the input is being varied, and X_t and Y_t represent deviations from the equilibrium (their respective means) at time t, then it commonly happens that the inertia of the system can be represented by the linear filter

$$Y_t = v_0 X_t + v_1 X_{t-1} + v_2 X_{t-2} + ... = (v_0 + v_1 B + v_2 B^2 + ...) X_t = v(B) X_t ,$$

$$(1)$$

where v_0, v_1, v_2, ..., are a set of weights called the impulse-response function; B is the backward shift operator such that $BX_t = X_{t-1}$, and in general $B^n X_t = X_{t-n}$.

Furthermore, if one is dealing with incremental changes:

$$y_t = Y_t - Y_{t-1} = \Delta Y_t \, ,$$

and

$$x_t = X_t - X_{t-1} = \Delta X_t \, ,$$

and one wishes to relate these changes, then on differencing equation (1), it can be shown that the same transfer function which relates Y_t and X_t will relate the differenced series y_t and x_t.

However, it would be impractical to parameterize the system in terms of the impulse-response weights. Therefore Box and Jenkins developed a general linear difference-equation representation for the discrete dynamic system. This equation is given by

$$(1 - \delta_1 B - ... - \delta_r B^r)Y_t = (\omega_0 - \omega_1 B - ... - \omega_s B^s)X_{t-b} \, ,$$

or

$$\delta(B)Y_t = \omega(B)X_{t-b} \, , \qquad\qquad (2)$$

which is referred to as a transfer-function model of order (r, s, b). In this model, r is the order of the lag polynomial for the output series, and is the means by which past values of this series are allowed to influence Y_t. Similarly, s is the order of the lag polynomial for the input series, and is used to describe the manner in which values of the input series influence Y_t. Finally, b is the time delay, through which one can introduce into the model the interval of time between a change in the input and the appearance of that change in the output.

Equation (2) has the advantage of requiring fewer parameters (being more parsimonious) than if the model were parameterized by use of the impulse-response function. Substitution of equation (1) into equation (2) gives the identity

$$(1 - \delta_1 - \delta^2 B^2 - ... - \delta_r B^r)(v_0 + v_1 B + v_2 B^2 + ...) = (\omega_0 - \omega_1 B - ... - \omega_s B^s)B^b.$$

$$(3)$$

A pattern arises in the form of the impulse-response function when r, s, and b are set equal to specific values. These patterns can be derived by equating coefficients of B in equation (3). Box and Jenkins give three results which will be reflected for specific values of r, s, and b in the pattern of the impulse-response function:

(1) the first b values, v_0, v_1, ..., v_{b-1}, will be zero;
(2) the next $s - r + 1$ values, v_b, v_{b+1}, ..., v_{b+s-r}, follow no fixed pattern, with no such values occurring if $s < r$;
(3) values of v_j with $j \geqslant b + s - r + 1$ follow a pattern given by the rth-order difference equation, which has r starting values, v_{b+s}, v_{b+s-1}, ..., $v_{b+s-r+1}$. In practice the values of the v_j will decline exponentially if $r = 1$, and either exponentially or as a damped sine wave if $r = 2$. Box and Jenkins

present the form of the impulse-response function for many commonly-occurring dynamic systems (Box and Jenkins, 1970, page 349–351).

In practice a system will be subject to disturbances, or noise, which will make the observed output deviate from the predicted value of the transfer function by an amount N_t. Thus the combined transfer-function–noise model is given by

$$Y_t = \delta^{-1}(B)\omega(B)X_{t-b} + N_t .\tag{4}$$

Modelling dynamic systems

Model-building reverses the logical sequence given in the previous section. In that section, rules were set out for the generation of the impulse-response function for specific values of r, s, and b; in the model identification process, however, the objective is to select the values of r, s, and b from the examination of the plot of the impulse-response function and the plot of the related cross-correlation function (cct).

The procedure for building a transfer function is the same as that used in constructing autoregressive integrated moving-average (ARIMA) models (Box and Jenkins, 1970). The procedure is iterative, and consists of three stages. In the identification stage, tentative values for r, s, and b are chosen, as well as the coefficients for $\delta(B)$, $\omega(B)$, and an ARIMA model for the noise series. Second, in the estimation stage, a nonlinear optimization procedure is used to improve upon the initial estimates. Diagnostic checks are then performed on the residual series from the model. If the model is adequate, the researcher may proceed to the third stage, if not, the researcher returns to the model-identification stage and continues to iterate through the first and second stages until an adequate model is achieved. In the third stage, the researcher may make estimates of future outputs of the system, based upon a known sequence of inputs or estimated future inputs.

Identification

At this stage, the pairs of time series, $\{X_t\}$, $\{Y_t\}$, are regarded as a realization of a bivariate stochastic process, which, if it is not stationary, can be made so by differencing to a process $\{x_t\}$, $\{y_t\}$, where $x_t = \Delta^d X_t$; $y_t = \Delta^d Y_t$. Stationarity implies that the processes $\{X_t\}$ and $\{Y_t\}$ (or $\{x_t\}$ and $\{y_t\}$) have constant means, μ_X and μ_Y, and constant variances, σ_X^2 and σ_Y^2.

The cross-covariance and cross-correlation functions The cross-correlation function, $\rho_{XY}(k)$, which has values in the range $-1 \leqslant \rho_{XY}(k) \leqslant 1$, is used as a dimensionless measure of the cross-variance function, $\gamma_{XY}(k)$:

$$\rho_{XY}(k) = \frac{\gamma_{XY}(k)}{\sigma_X \sigma_Y}, \qquad k = 0, +1, +2, + \ldots .$$

Since only a sample is available, rather than the total population, $\gamma_{XY}(k)$ is estimated by the sample cross-covariance, $C_{XY}(k)$, and $\rho_{XY}(k)$ is estimated by the sample cross-correlation, $r_{XY}(k)$, where

$$C_{XY}(k) = \begin{cases} \dfrac{1}{n} \sum\limits_{t=1}^{n-k} (X_t - \bar{X})(Y_{t+k} - \bar{Y}), & k = 0, 1, 2, \dots, \\[2ex] \dfrac{1}{n} \sum\limits_{t=1}^{n+k} (Y_t - \bar{Y})(X_{t-k} - \bar{X}), & k = 0, -1, -2, \dots, \end{cases}$$

and

$$r_{XY}(k) = \frac{C_{XY}(k)}{s_X s_Y}, \qquad k = 0, \pm 1, \pm 2, \dots .$$

Here, $\bar{X}$ and $\bar{Y}$ are the mean values of $\{X_t\}$ and $\{Y_t\}$; s_X $[= C_{XX}^{1/2}(0)]$ and s_Y $[= C_{YY}^{1/2}(0)]$ are estimates of σ_X and σ_Y, respectively. It should be noted that unlike the autocorrelation function (acf), the ccf is not, in general, symmetrical about zero. An approximate formula for the covariance of two cross-correlation estimates $r_{XY}(k)$ and $r_{XY}(k+m)$, and the variance of a cross-correlation estimate ($m = 0$), is given by Bartlett (1955). One special case of this formula is of particular interest, and will be used later. Under the null hypothesis—that two processes are not cross-correlated, and that one process, $\{a_t\}$, is white noise—an approximation to the covariance of the cross-correlation estimates $r_{Xa}(k)$ and $r_{Xa}(k+m)$ is given by

$$\text{cov}[r_{Xa}(k), r_{Xa}(k+m)] \approx (n-k)^{-1} \rho_{XX}(m) ,$$

and the variance of the cross-correlation function is approximately equal to $(n-k)^{-1}$.

The impulse-response function, and prewhitening the input Estimation of the nonzero values of the v_k directly from the stationary processes is difficult and often inefficient. However, if the appropriate univariate model is built for the input series, and the input, output, and noise series are filtered, or 'prewhitened', by this model, the series are transformed to α_t (white noise), β_t, and ϵ_t. The impulse-response weights are now found by solution of an orthogonal set of equations of the form

$$v_k = \frac{\rho_{\alpha\beta}(k)\sigma_\beta}{\sigma_\alpha}, \qquad k = 0, 1, 2, \dots . \tag{5}$$

On substitution of estimated quantities, the approximations, $\hat{v}_k$, are obtained:

$$\hat{v}_k = \frac{r_{\alpha\beta}(k)s_\beta}{s_\alpha}, \qquad k = 0, 1, 2, \dots . \tag{6}$$

Two advantages arise from prewhitening: (1) the v_k are easily estimated; (2) the easily-computed approximate form, $(n-k)^{-1}$, for the ccf variance is applicable.

Haugh and Box (1977) have suggested that input and output series should be prewhitened by their corresponding univariate models, instead of both series being prewhitened by the univariate model of the input series. The results provided here are based on the latter procedure, as this is the only one permitted in the computer program used (Pack, 1974). Although the mathematics of this procedure remain correct, the advantage of the use of the individual prewhitening procedure is that the impulse-response function is more efficiently estimated when diagnostic checks are run on models.

Thus an estimate of the orders of the polynomials, r, s, and b, are obtained by examining the plot of the ccf for significant values of these parameters, and by using the three results given in the previous section.

The noise model The noise series, ϵ_t (that is, N_t transformed by pre-whitening), is identified and its parameters are estimated by use of the usual ARIMA techniques.

Diagnostic checking

After the model has been fitted, inadequacies are detected by examination of the acf of the residuals, $\tilde{a}_t$, from the model, and of the ccf of the prewhitened input, α_t, and the residuals, $\tilde{a}_t$. These correlation functions can be used to diagnose two types of model inadequacies. Stated in terms of the notion that the overall model is an additive combination of a transfer function and a noise model, the inadequacies are that the transfer function may be correct but the noise model is incorrect, or that the transfer function is incorrect.

These two types of model inadequacies can be detected in the acf by the following tests.

(1) If the acf, $r_{\tilde{a}\tilde{a}}(k)$, exhibits a pattern, this suggests that the transfer-function model is inadequate;

(2) if the ccf of α_t and $\tilde{a}_t$ is adequate, then it is likely that the noise model is incorrect.

As a test of model adequacy, one might be tempted to test the values of the approximate acf and ccf for significant departures from their expected values under a white-noise hypothesis. Under this hypothesis, which is equivalent to the hypothesis that an appropriate model has been fitted, a rough estimate of the variance of $r_{\tilde{a}\tilde{a}}(k)$ is $1/m$, where m is the number of residuals, which is adjusted for the number of parameters estimated. However, when estimates rather than parametric values are used in the transfer-function model, ripples are set up in the acf and ccf, as the estimates of the acf and ccf are themselves autocorrelated (Box and Jenkins, 1970).

Overall checks of the acf and ccf calculated for K lags, which take account of the distributional effects of the ripples, are given for the acf by examination of the quantity Q, where

$$Q = m \sum_{k=1}^{K} r_{\tilde{a}\tilde{a}}^2(k) ,$$

which under a white-noise null hypothesis is approximately distributed as χ^2 with $(K-p)-q$ [(order of AR linear filter) $-$ (order of MA linear filter)] degrees of freedom. For the ccf, the quantity s, where

$$s = m \sum_{k=0}^{K} r_{\alpha\hat{a}}^2(k) \; ,$$

is approximately distributed as χ^2 with $K+1-(r+s+1)$ degrees of freedom, under the hypothesis that $\hat{a}$ and α are both uncorrelated white-noise series.

Examples
The two examples given below were selected not because they illustrate transfer-function techniques with the most exotic models, but because they actually arose from hypotheses developed in other research contexts, and it became clear that fitting a transfer-function model would be one approach towards examination of the phenomena.

In the first example, the relationship between changes in population and the value of construction materials for selected Canadian provinces is examined. For the second example, the relationship between precipitation and the lake level on Lake Champlain is explored. Computations necessary for the analyses were accomplished by use of a program written by Pack (1974), implemented on an IBM 370/3033 computer at the Pennsylvania State University.

Relating the production of construction materials to population changes— an example in a confirmatory role
While the unit-regional-value approach was being used to develop areal mineral-resource estimates (Labovitz, 1978), it was suggested to the author that the level of production of construction materials is dependent on changes in population. In particular, it was suggested that one cannot establish a mineral-resource economy on construction materials because they are available everywhere; people cannot, therefore, be enticed to underdeveloped locations on the basis of these resources.

These statements can be translated into the following hypothesis. If the researcher examines the relationship between the time series for population and the time series for the value of construction materials, then (s)he will find that of the two series, changes in population will precede changes in the use of construction materials. Furthermore, because major periods of construction tend to last several years, past changes in the population will influence future usage of construction materials. In the context of the input–output model, then, population is the input, and production of construction materials is the output.

An analysis of this relationship, using the Box–Jenkins technique of transfer-function modelling, was applied to three provinces of Canada: Nova Scotia, Ontario, and Alberta.

The population time series Data on the population of Canada were
collected from the following sources:
Census of Canada (Census and Statistics Office, 1941)
Canadian Statistical Review, Historical Summary (Statistics Canada, 1970)
Canada Yearbook (Statistics Canada, 1975).

The population series were first fitted with univariate ARIMA models.
These models are used later in the analysis to prewhiten the input and
output series. For all three series, an autoregressive model of order one
on the first difference [an ARIMA(1,1,0)] was found to be appropriate.
This model has the general form

$$Z_t = (1 + \phi_1)Z_{t-1} - \phi_1 Z_{t-2} + a_t ,$$

where Z_t is the value of the series variable at time period t, ϕ_1 is a
coefficient, δ_1 is a deterministic trend, and a_t is a random noise variable
with a distribution having mean zero and variance σ_a^2, written $N(0, \sigma_a^2)$.
Specifically the models are:

$$Z_t = 1 \cdot 7302 Z_{t-1} - 0 \cdot 7302 Z_{t-2} + a_t , \qquad \text{for Alberta,} \\ 1909-1974;$$

$$Z_t = 1 \cdot 5685 Z_{t-1} - 0 \cdot 5685 Z_{t-2} + a_t , \qquad \text{for Nova Scotia,} \\ 1867-1974;$$

$$Z_t = 1 \cdot 7588 Z_{t-1} - 0 \cdot 7588 Z_{t-2} + a_t , \qquad \text{for Ontario,} \\ 1867-1974.$$

Transfer-function modelling of population and construction materials
Figure 2(a) is the ccf of the logarithm of the population of Ontario (1886
to 1974) lagged on the logarithm of the value of construction materials
[value in deflated 1967 US dollars (US67$)] produced in Ontario over the
same period. The failure of the ccf to go to zero indicates that the series
are not stationary (Box and Jenkins, 1970), and from an examination of
the series it can be seen that both are generally increasing over the period
of time under study. To continue the analysis, the series need to be
differenced. Next, the data are prewhitened with the ARIMA(1,1,0)
model for the population of Ontario (given above). The ccf of the prewhitened series is given in figure 2(b). The cross-correlation function has
two significant spikes at lags one and two. There are no other significant
spikes or discernable systematic patterns of variation in the first twenty
lags of the ccf. Thus, the values $r = 0$, $s = 1$, and $b = 1$, are chosen as
an initial estimate.

The plot of the acf, $r_{\epsilon\epsilon}(k)$, of the noise series, ϵ_t, is given in figure 3.
Analysis of the noise series indicates that it is not significantly different
from white noise.

For the initial run of the estimation and forecasting program, the values
of r and s are set at one and two, respectively. This is a procedure
known as 'overfitting', which is recommended by Box and Jenkins (1970).

The results of the run indicate that the values of the coefficients δ_1, ω_2, and θ_0 (a measure of the deterministic trend of nonstationarity in the mean of the transfer-function model) are not significantly different from zero.

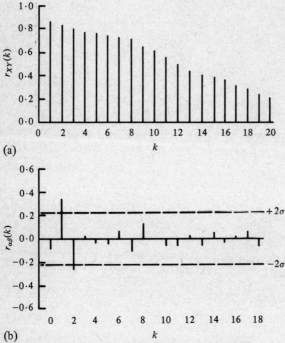

(a)

(b)

Figure 2. Plots of cross-correlation functions of series for lg(population), X, lagged by k years on lg(deflated value of construction materials), Y, for Ontario, 1886–1974. (a) Observed series; (b) prewhitened series ($X \to \alpha$, $Y \to \beta$).

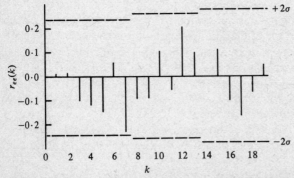

Figure 3. Plot of the acf, $r_{\epsilon\epsilon}(k)$, of the prewhitened noise series, ϵ_t, from the transfer function between population and production of construction materials in Ontario, 1886–1974.

The next run of the estimation routine used the values $(0, 1, 0)$ and $(0, 1, 1)$ for (r, s, b). The sum of squared errors was at a minimum for $b = 1$. The plots of the acf, $r_{\bar{a}\bar{a}}(k)$ [figure 4(a)], and the ccf, $r_{\alpha\bar{a}}(k)$ [figure 4(b)] do not exhibit any discernable pattern or nonrandomness. The values of Q and S are $10 \cdot 014$ and $11 \cdot 810$, respectively. When compared with their critical values of χ^2 at the $0 \cdot 95$ level [$\chi^2(20; 0 \cdot 95) = 31 \cdot 40$ for Q, and $\chi^2(19; 0 \cdot 95) = 30 \cdot 144$ for S], there is no evidence on which to reject the null hypothesis that the residuals are white noise. Thus the transfer-function model is given by

$$(1 - B)Y_t = (26 \cdot 42 - 21 \cdot 32B)(1 - B)X_{t-1}$$
$$+ [(1 - 0 \cdot 7588B)(1 - B)]^{-1}(a_t + 0 \cdot 00165) .$$

The observed series and the model are given in figure 5. Since the delay parameter, b, is small ($b = 1$), the production of construction materials responds fairly rapidly to changes in population. Furthermore, because $\omega_0 > \omega_1$, a monotonically increasing sequence of population values will force an increase in the production of construction materials.

Forecasts of the value of construction materials produced for the period 1975–1984 were obtained by first forecasting population values over the period. Table 1 gives forecasts for the population of Ontario, which is

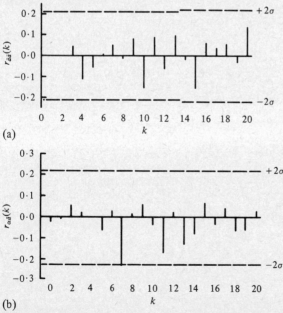

(a)

(b)

Figure 4. Correlation functions of the residuals from the transfer-function model relating population to production of construction materials in Ontario, 1886–1974. (a) acf; (b) ccf.

projected to increase from $8 \cdot 09394 \times 10^6$ people in 1974 to $9 \cdot 54089 \times 10^6$ people in 1984. It is forecasted that the increase in population will force an increase in the value of construction materials, as shown in table 2. Over the period 1975 to 1984, the construction-materials sector in Ontario will be valued at US67\$ $2 \cdot 76582 \times 10^9$. The actual value of construction materials for the years 1975 and 1976 are also given in table 2. It can be seen that these values fall well within the 75% confidence limits about the forecasted values.

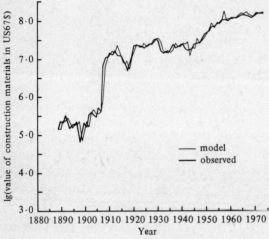

Figure 5. Plot of observed and model series for value of construction materials produced in Ontario, 1886-1974. The model series is based on the transfer function between population and production of construction materials.

Table 1. Forecasts of the population of Ontario, 1975-1984.

Year	lg(population)		
	forecast	75% confidence limits	
		lower	upper
1975	6·91613	6·91311	6·91915
1976	6·92377	6·91766	6·92988
1977	6·93117	6·92183	6·94050
1978	6·93838	6·92584	6·95092
1979	6·94544	6·92978	6·96110
1980	6·95240	6·93375	6·97105
1981	6·95928	6·93777	6·98078
1982	6·96609	6·94186	6·99032
1983	6·97286	6·94605	6·99967
1984	6·97959	6·95032	7·00885

A similar transfer-function model is found to be adequate for representing the dynamic system relating the series for the logarithm of population of Nova Scotia as the input, and for the logarithm of the US67$ value of stone production as the output. For this model, the values $(0, 1, 8)$ are chosen for (r, s, b) (see the plot of the ccf, figure 6), and the prewhitened noise series, ϵ_t, is not significantly different from white noise. Thus the model is given by

$$(1 - B)Y_t = (-28 \cdot 094 + 33 \cdot 230B)(1 - B)X_{t-8}$$
$$+ [(1 - 0 \cdot 56850B)(1 - B)]^{-1}(a_t + 0 \cdot 0014) .$$

Forecasts of the population of Nova Scotia for the period 1975 to 1984 were calculated. As in Ontario, a steady increase in the population of Nova Scotia is projected, with the population increasing by approximately

Table 2. Forecasts of the deflated value of construction materials produced in Ontario, 1975-1984.

Year	lg(value in US67$)			
	forecast	75% confidence limits		actual
		lower	upper	
1975	8·29520	8·10716	8·48323	8·20040
1976	8·32460	8·04720	8·60200	8·18590
1977	8·35469	8·01138	8·69800	
1978	8·38522	7·98739	8·78305	
1979	8·41608	7·97075	8·86140	
1980	8·44731	7·95936	8·93527	
1981	8·47870	7·95174	9·00566	
1982	8·51021	7·94706	9·07335	
1983	8·54179	7·94475	9·13884	
1984	8·57342	7·94437	9·20248	

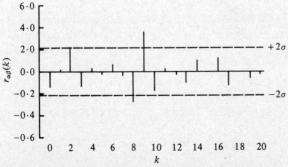

Figure 6. Plot of ccf of prewhitened series for lg(population) lagged by k years on lg(value of stone in US67$), in Nova Scotia, 1867-1974.

66000 people from 813000 to 879000. However, because the magnitude of the delay parameter, b, ($b = 8$) forces the use of the observed population values for the first eight output forecasts, and because $\omega_0 < \omega_1$, the forecasted values do not increase smoothly (see table 3). In fact there are several decreases in the forecasted values of stone production. Large fluctuations in the forecasts continue until 1982.

The actual values of stone production in Nova Scotia for 1975 and 1976 fall well within the 75% confidence limits about the forecasted values. The cumulative value of stone production in Nova Scotia over the period 1975 to 1984 is projected to be US67\$ $2 \cdot 80866 \times 10^7$.

Table 3. Forecast of the deflated value of stone produced in Nova Scotia, 1975–1984.

Year	lg(value in US67$)			
	forecast	75% confidence limits		actual
		lower	upper	
1975	6·40208	6·08204	6·72213	6·46670
1976	6·36623	5·91361	6·81884	6·41600
1977	6·37167	5·81734	6·92601	
1978	6·41168	5·77159	7·05177	
1979	6·43263	5·71699	7·14828	
1980	6·48398	5·70003	7·26793	
1981	6·40726	5·56050	7·25402	
1982	6·48508	5·57985	7·39030	
1983	6·51984	5·55533	7·48434	
1984	6·54680	5·52997	7·56363	

Additional transfer-function modelling Table 4 gives the important results from other transfer-function models constructed in this study of mineral production and population. Two important facts can be discerned from this table. First, most of the pairs of series provided no evidence by which to reject the null hypothesis that the prewhitened series are uncorrelated. Second, for several of the series, particularly those for sand and gravel production, the form of the ccf calculated with population as the forcing function indicates that the reverse analysis should be examined; that is, the commodity series should be taken as the forcing function, and the population series then modelled as the output series. This involves prewhitening both series with the model which reduces the commodity series to white-noise, first-difference models. This method of analysis leads to statistically appropriate models in two out of the five cases examined. The two cases (production in Alberta and in Ontario) both have sand and gravel production as the mineral-commodity series. Furthermore, both transfer-function models contain only a delay parameter, b ($b = 1$ for Alberta and $b = 5$ for Ontario).

A word of caution is necessary at this juncture. In this analysis, eighty-nine is the maximum number of observations used by the author for building any of the models. Although Box and Jenkins (1970) indicate that transfer-function analysis is appropriate for pairs of series with at least fifty sets of observations, other researchers have indicated, in personal communications, their reluctance to use such a small number of observations.

Table 4. Statistics from the transfer-function analyses used in this example. The last year for all series is 1974; the commodity-series values are all expressed as lg(production value in US67$).

X^a	Y^a	Series begins	(r, s, b)	Noise series	Model coefficients	Q^b	S^b
ALB	CM	1909	–				
ALB	CEM	1909	–				
ALB	S-G	1922	–				
S-G	ALB	1922	$(0, 0, 1)$	ARMA $(1,0,0) \times (0,0,1_2)$	$\tilde{\omega}_0 = 0\cdot0052$ $\tilde{\phi}_1 = 0\cdot9678^c$ $\tilde{\Theta}_2 = 0\cdot3938^c$	$12\cdot948*$ (13)	$15\cdot380*$ (15)
ALB	ST	1909	–				
NSC	CM	1886	–				
NSC	S	1922	–				
NSC	ST	1886	$(0, 1, 8)$	$a_t = \epsilon_t$	$\tilde{\omega}_0 = -28\cdot094$ $\tilde{\omega}_1 = -33\cdot230$	$17\cdot585*$ (20)	$17\cdot612*$ (19)
ONT	CM	1886	$(0, 1, 1)$	$a_t = \epsilon_t$	$\tilde{\omega}_0 = 26\cdot142$ $\tilde{\omega}_1 = 21\cdot324$	$10\cdot014*$ (20)	$11\cdot810*$ (19)
ONT	CEM	1908	–				
ONT	S-G	1920	–				
S-G	ONT	1920	$(0, 0, 5)$	AR(1)	$\tilde{\omega}_0 = 0\cdot0056$ $\tilde{\phi} = 0\cdot9680^c$	$11\cdot478*$ (14)	$16\cdot620*$ (15)
ONT	ST	1908	–				

[a] Key: ALB ≡ Alberta population; NSC ≡ Nova Scotia population; ONT ≡ Ontario population; S-G ≡ sand and gravel; CM ≡ construction materials; CEM ≡ cement; ST ≡ stone; S ≡ sand.

[b] The degrees of freedom are shown in brackets below values of Q and S; the asterisks indicate that there is no evidence by which the white-noise hypothesis can be rejected.

[c] Estimates of coefficients from a seasonal univariate model used to represent residuals or 'noise' from fitting the transfer-function model.

Conclusions It is evident from the results of the transfer-function analyses that the models do not clearly support the hypothesis set forth at the beginning of this example. The relationship between population and construction materials is more complex than that originally suggested. The observed result may be restated in terms of the feedback component illustrated in figure 1: positive feedback becomes the dominant element in the system during stages of rapid growth. This hypothesis will not be examined here. However, the results indicating that changes in production of construction materials precede changes in population should be given considerably more attention (and more analysis, including verification), with a view to the development of schemes for manipulating industrial growth and development.

Relating changes in lake level to precipitation in the drainage basin of Lake Champlain—an example in an exploratory role
A study of the relationship between the lake levels in Lake Champlain and the precipitation occurring in the drainage basin, was conducted by Barnett (1978). Because of the obvious theoretical connection between the two variables, and because of the likely time-dependence of the variables, an attempt to model the relationship with a transfer-function seems appropriate.

Univariate time series Four series of yearly observations are analyzed. These series and their univariate (ARIMA) models are given below (each sequence is composed of deviations from its mean value).

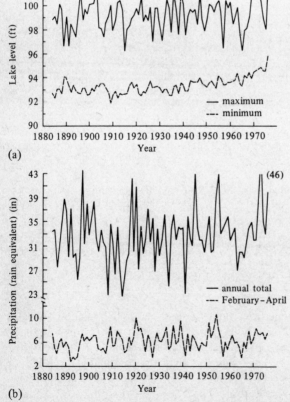

Figure 7. Plots of observed time series for (a) annual lake levels of Lake Champlain, (b) precipitation at Burlington, Vt.

Maximum lake level, 1871–1976:

$$Z_t = Z_{t-1} + a_t + 0 \cdot 2337 a_{t-1} \qquad [\text{MA}(1); \ \bar{\mu} = 99 \cdot 065 \ \text{ft}] \ .$$

Minimum lake level, 1871–1976:

$$Z_t = 0 \cdot 76045 Z_{t=1} + a_t \qquad [\text{AR}(1); \ \bar{\mu} = 93 \cdot 380 \ \text{ft}] \ .$$

Total precipitation (rain equivalent), 1884–1976:

$$(\text{white noise}; \ \bar{\mu} = 32 \cdot 767 \ \text{in}) \ .$$

February–April precipitation (rain equivalent), 1884–1976:

$$Z_t = Z_{t-1} + a_t + 0 \cdot 3414 a_{t-1} \qquad [\text{MA}(1), \ \bar{\mu} = 6 \cdot 241 \ \text{in}] \ .$$

Figure 7 displays plots of the observed data.

Transfer-function modelling of precipitation and lake level
Figure 8 shows the cross-correlation functions for four pairs of prewhitened series, with the prewhitened precipitation series as the forcing series in each case. From these graphs it is clear, first, that there is no delay factor, b, in the system; that is, the effects of changes in the precipitation are immediately realized in the lake level, if they are realized at all. Further examination of the ccfs yields the information below.
(1) The February–April precipitation and annual minimum lake level are not related.
(2) The February–April precipitation is strongly correlated with the annual maximum lake level only at zero lag; at other lags, the correlation is non-significant. A model with r, s, and b all equal to zero will therefore be tried.
(3) The ccf of total precipitation and annual minimum lake level has significant correlation at lags of zero and one; there is also a hint of an exponential decline in the ccf with increasing k. The noise series exhibits an AR(1) or AR(2) behavior. Thus several models, involving r and s values of zero, one, and two, will be tried.
(4) The series for total precipitation and annual maximum lake level are significantly correlated at zero lag, and possibly correlated at lag one. This suggests that models (0, 0, 0) or (0, 1, 0) should be tried for (r, s, b).

Results Table 5 lists the results of the estimation step (all analyses are for the years 1884 to 1976). The complete forms of these models are as follows

February–April precipitation and maximum lake level:

$$Y_t = 0 \cdot 4621 X_t + (1 - \text{B})^{-1} (a_t + 0 \cdot 3414 a_{t-1}) \ .$$

Total precipitation and maximum lake level:

$$Y_t = 0 \cdot 0688 X_t + a_t \ , \qquad \text{or} \qquad Y_t = (0 \cdot 0654 + 0 \cdot 0477 \text{B}) X_t + a_t \ .$$

Total precipitation and minimum lake level:

$$Y_t = (0 \cdot 0241 - 0 \cdot 0393 \text{B}) X_t + (1 - 0 \cdot 6247 \text{B} - 0 \cdot 2383 \text{B}^2)^{-1} a_t \ .$$

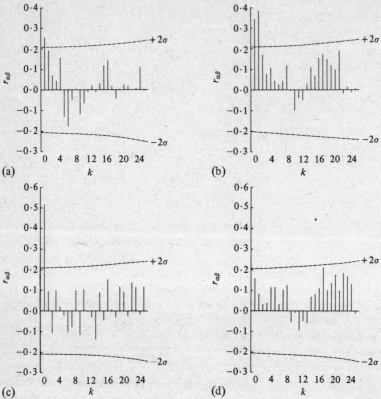

Figure 8. Plots of cross-correlation functions of prewhitened series for precipitation, lagged by k years on prewhitened series for lake levels. (a) Annual total precipitation on maximum lake level; (b) annual total precipitation on minimum lake level; (c) February-April precipitation on maximum lake level; (d) February-April precipitation on minimum lake level.

Table 5. Statistics from the transfer-function analyses used in this second example.

X	Y	(r, s, b)	Noise series	Model coefficients	Q [a]	S [a]
Febr.-Apr. precipitation	max. lake level	$(0, 0, 0)$	$a_t = \epsilon_t$	$\bar{\omega}_0 = 0 \cdot 4621$	$12 \cdot 230*$ (24)	$14 \cdot 218*$ (25)
Total precipitation	max. lake level	$(0, 0, 0)$	$a_t = \epsilon_t$	$\bar{\omega}_0 = 0 \cdot 0688$	$13 \cdot 573*$ (25)	$27 \cdot 405*$ (24)
Total precipitation	max. lake level	$(0, 1, 0)$	$a_t = \epsilon_t$	$\bar{\omega}_0 = 0 \cdot 0654$ $\bar{\omega}_1 = 0 \cdot 0477$	$14 \cdot 817*$ (24)	$25 \cdot 794*$ (24)
Total precipitation	min. lake level	$(0, 1, 0)$	AR(2)	$\bar{\phi}_1 = 0 \cdot 6247$ $\bar{\phi}_2 = 0 \cdot 2383$ $\bar{\omega}_0 = 0 \cdot 0241$ $\bar{\omega}_1 = 0 \cdot 0393$	$17 \cdot 984*$ (22)	$21 \cdot 610*$ (22)

[a] Notation as for table 4.

The coefficient $\hat{\omega}_1$ (= $0\cdot0477$) for the model relating total precipitation and maximum lake level, is not significantly different from zero at the $P < 0\cdot05$ level. However, the tail probability for this confidence region is equal to $0\cdot046$; and so this model is presented for consideration, as it seems to fit with the physical interpretation given below.

Physical interpretations Clearly, the amount of precipitation in the drainage basin has an impact on lake levels. The February–April precipitation, largely in the form of snow, constitutes the spring runoff, and has a fairly large impact on the annual maximum lake levels. However, the influence of this precipitation upon the maximum lake level is a short-term event; that is, the effects of changes in the February–April precipitation do not carry over from year to year. Thus, the water entering the lake from the February–April precipitation is discharged in less than one year. The effect of total annual precipitation on true maximum lake level is of longer duration, but weaker. This seems reasonable in light of the fact that the February–April precipitation is included in the total precipitation, and on average represents only one fifth of the total yearly precipitation, so that the total precipitation smooths out the yearly changes in spring precipitation. The influence of the total precipitation on the maximum lake level, however, is not completely dissipated in one year.

On the other hand, total precipitation has a greater impact upon the minimum lake level. This is interpreted to mean that the base level of the lake (represented by the annual minimum lake levels) is strongly influenced by the total precipitation falling in the basin, and that this influence lasts over a few years. Lastly, there does not appear to be a relationship between February–April precipitation and minimum lake level. Evidently the perturbation produced by the spring thaw is out of the system by the time of the occurrence of the minimum lake level. Thus, the period of time between the spring snowmelt and the minimum lake level (August) can be transformed into a measure of the storage capacity of the lake.

The models were not sufficient, however, to predict the maximum lake levels with the degree of precision (six inches) needed by Barnett (1978). The mean square of the residuals for the model relating February–April precipitation to maximum annual lake level is $1\cdot01$ ft^2. Thus the frequency of observation (one per year) is not sufficient, and transfer-function models using data collected monthly, or even weekly, should be constructed.

Summary and conclusions

Two examples of transfer-function modelling which use data from the geosciences have been given. The first example, in mineral economics, was used to test a specific hypothesis about the relationship between population and production of construction materials. It was concluded that the relationship is far more complex than was originally postulated.

Furthermore, it opens the realms of manipulation of population growth by use of mineral-resource-development policies. The second example examines the relationship between precipitation and lake levels in Lake Champlain. From this analysis, a reasonably easy physical interpretation of the relationship can be developed. The lake empties its spring runoff within one year, and the total precipitation has a two-year influence on the minimum level of the lake.

Workers in other process-oriented research areas, such as sedimentology, earthquake prediction, general geophysical exploration, and experimental geochemistry, could also make use of these modelling techniques.

Acknowledgements. The author would like to acknowledge Drs J C Griffiths and J Wiorkowski for instruction in time-series techniques. Drs M Hallberg, D Pack, and C Leisenring discussed specific topics in transfer-function analysis with the author. Dr S G Barnett graciously allowed the author to use his precipitation and lake-level data. The applications staff of The Pennsylvania State University Computation Center are thanked for rapidly adapting the program used in this analysis. Ms A W Siegrist and Drs R G Craig and J C Griffiths are acknowledged for their careful reviews of the manuscript.

References
Barnett S G, 1978 "Man induced changes in Lake Champlain hydrology" Regional Studies Report 7, Institute for Man and Environment, State University of New York, Plattsburgh, NY
Bartlett M S, 1955 *Stochastic Processes* (Cambridge University Press, Cambridge)
Beer S, 1964 *Cybernetics and Management* (John Wiley, New York)
Box G E P, Jenkins G M, 1970 *Time Series Analysis—Forecasting and Control* (Holden-Day, San Francisco, Calif.)
Census and Statistics Office, 1941 *Census of Canada* (Census and Statistics Office, Ottawa, Canada)
Haugh L D, Box G E P, 1977 "Identification of dynamic regression (distributed lag) models connecting two time series" *Journal of the American Statistical Association* **72**(357) 121-130
Kendall M G, 1973 *Time Series* (Hafner Press, New York)
Labovitz M L, 1978 "Unit regional value of the Dominion of Canada" PhD thesis, The Pennsylvania State University, University Park, Pa
Matheron G, 1976 "Forecasting block grade distributions: the transfer functions" in *Advanced Geostatistics in the Mining Industry* Eds M Guarascio, M L R David, C J Huijbregts (Reidel, Dordrecht, The Netherlands) pp 237-251
Nelson C R, 1973 *Applied Time Series Analysis for Managerial Forecasting* (Holden-Day, San Francisco, Calif.)
Pack D J, 1974 "Computer programs for the analysis of univariate time series models and single input transfer function models using the methods of Box and Jenkins" Department of Computer Sciences, The Ohio State University, Columbus, Ohio
Statistics Canada, 1970 *Canadian Statistical Review, Historical Symmary* (Statistics Canada, Ottawa, Canada)
Statistics Canada, 1975 *Canada Yearbook* (Statistics Canada, Ottawa, Canada)

The effects of spatial autocorrelation on geographical modelling

A D Cliff, J K Ord

Introduction

It is now fairly widely accepted that the analysis of geographically located data is made difficult by the following two facts.

(1) Most regionally located data exhibit systematic spatial variation, or spatial autocorrelation. This breaks a basic assumption upon which many statistical methods are founded: the requirement that the variates being analyzed be *independently* distributed.

(2) Many methods are *scale-dependent*: different results are obtained depending upon the spatial scale at which the analysis is conducted, with the result that it is difficult (a) to determine 'the most appropriate scale' for the analysis, and (b) to disentangle scale effects from results of genuine scientific interest.

It is the purpose of this essay briefly to review definitions and measures of spatial autocorrelation, and then to explore in some detail the ways in which an understanding of problems of spatial autocorrelation can help to alleviate the two difficulties described above.

Autocorrelation

Concepts of spatial autocorrelation

Consider a study area which has been exhaustively partitioned into n nonoverlapping subareas, such as the counties of England and Wales, or the states of the United States. Suppose that a random variable, X, has been measured in each of the subareas, and that the value of X in the typical subarea, i, is x_i. X could describe either

(1) a single population from which repeated drawings are made to give the x_i;

(2) n separate populations, one for each subarea; or

(3) a partition of a finite population among the n subareas. The underlying population model used will depend upon the problem in hand. For example, the incidence of lung cancer in the English counties might be considered under population model (1), gross national product of the countries of Western Europe under population model (2), and the distribution of the Democrat vote by states in a US presidential election under population model (3). Although the examples we have given are for area-based data, variate values collected at points could be analyzed equally well by the methods to be described. Thus we might be interested in the spatial variability of rainfall or temperature at meteorological stations.

A basic property of such spatially located data is that the values x_i are likely to be related over space. This idea underlies the concept of the *region*, and has been reiterated by Tobler (1970) as "the first law of geography: everything is related to everything else, but near things are more related than distant things", by Gould (1970), and by Cliff and Ord (1973). If the $\{x_i\}$ display interdependence over space, we say that the data are *spatially autocorrelated*. From Cliff and Ord (1973, Appendix 1), a possible model of the spatial interdependence among the $\{x_i\}$ is the scheme

$$X_i = \rho \sum_j w_{ij} X_j + \epsilon_i , \qquad i = 1, 2, ..., n .$$ (1)

Here the $\{\epsilon_i\}$ are independent and identically distributed variates with common variance, σ^2. An alternative model is the conditional expectation scheme of Besag (1974), where

$$E(X_i \mid X_j = x_j, j \neq i) = \rho \sum_j w_{ij} x_j ,$$ (2)

$$\text{var}(X_i \mid X_j = x_j, j \neq i) = \sigma^2 .$$ (3)

These two models differ for estimation purposes (see Besag, 1974), although they carry the same implications for tests of spatial dependence.

The weights $\{w_{ij}\}$ in equation (1) are any set of nonnegative constants that specify which j subareas in the study area have variate values directly spatially related with X_i. On occasion, the $\{w_{ij}\}$ are scaled so that

$$\sum_j w_{ij} = 1 , \qquad i = 1, 2, ..., n .$$

The constant ρ is a measure of the overall level of spatial autocorrelation among the $\{X_i X_j\}$ pairs for which $w_{ij} > 0$; it is the spatial autocorrelation parameter. For example, we might put $w_{ij} = 1$ (unscaled) if j is physically contiguous to i, and $w_{ij} = 0$ otherwise. More general sets of weights may, however, be constructed, and the reader is referred to Cliff and Ord (1973, section 1.4.2; 1975a) for a further discussion of the choice of weights. Thus, the "first law of geography" given by Tobler (1970) above might be captured by the relation

$$w_{ij} = d_{ij}^{-\alpha} ,$$ (4)

where d_{ij} is the distance between points or areas, i and j, and α is a 'friction of distance' parameter as used in many gravity and interaction models. Finally, when $\rho > 0$ in model (1), we say that there is *positive* spatial autocorrelation among the $\{X_i\}$, whereas $\rho < 0$ implies *negative* spatial autocorrelation. The former case is characterized by similar $\{x_i\}$ values in subareas with nonzero $\{w_{ij}\}$ values, and the latter by very different (for example, +/−) relationships (see figure 1). If $\rho = 0$ in model (1), there is said to be no spatial autocorrelation in the study area on X.

Kooijman (1976) has demonstrated how 'optimal' weights may be estimated from the data, given a certain amount of basic structure such as distance decay. His methods require the data to relate to a regular grid, but extensions to irregularly spaced locations may be possible.

0·02	0·37	2·78	-3·24	0·11
-2·34	0·00	2·27	2·89	2·33
1·79	2·04	2·13	-1·87	1·90
-0·18	1·56	-1·18	4·08	-0·64
4·93	-1·86	2·08	4·40	-4·00

$\rho = 0·00$

(a)

4·70	6·67	9·35	4·69	6·79
3·73	6·95	10·56	10·28	10·16
8·59	9·97	10·26	7·40	9·04
8·97	9·43	8·20	11·62	6·24
12·50	7·83	10·40	11·69	2·29

$\rho = 0·9$

(b)

7·17	-6·68	10·9	-11·83	8·02
-9·22	6·22	-5·87	10·52	-5·75
9·52	-5·90	9·33	-10·45	8·36
-10·65	10·21	-9·78	9·93	-5·35
15·17	-12·11	8·78	-0·43	2·63

$\rho = -0·9$

(c)

Figure 1. Effects of spatial autocorrelation on a geographical distribution. (a) Map of original observations which are spatially independent. Maps of same observations (b) positively autocorrelated on the rows and columns by the amount $\rho = 0·9$, and (c) negatively autocorrelated by the amount $\rho = -0·9$. In (b), note the similarity among the adjacent observation levels compared with those of (a); in (c), note the alternating pattern characteristic of negative spatial autocorrelation. (Source: Haggett et al, 1977, page 332.)

Space-time autocorrelation
Model (1) is the spatial equivalent of the time-series model

$$X_t = \rho X_{t-k} + \epsilon_t , \qquad t = 1, 2, ..., T , \tag{5}$$

where t denotes the typical time period, T is the total number of time periods, and k is a positive integer, $1 \leqslant k < T$. Economists have long recognised that economic time series display temporal autocorrelation. But the fundamental difference between equations (1) and (5) is that dependence in time can only extend backwards, whereas in the spatial case the dependence is multidirectional (Whittle, 1954). Often, however, the human geographer is concerned with understanding the processes which have produced regional patterns in economic data. That is, as Bennett (1974) has noted, (s)he is not concerned solely with the analysis of cross-sectional (spatial) data. His or her task is to unravel the complex patterns of autocorrelation in both time and space, to gain some insight into the functional dependencies between areas implied by the presence of autocorrelation. Thus we might postulate that

$$X_{it} = f(x_{i,\,t-k}, x_{j,\,t-k}), \qquad k = 1, 2, \dots ; \quad j \neq i . \tag{6}$$

If our time–space matrix is as shown in figure 2, the model given in equation (6) would imply a complex 'cone' of dependencies between regions, going back through time. In addition, the time interval between our data recording points is often sufficiently long, compared with the rate of operation of the geographical process, that what appear as simultaneous effects between regions may occur. The dependency of X_{it} on $x_{i,\,t-1}, \dots, x_{i,\,t-k}$ in equation (6) is, as shown in figure 2, a purely *temporal* autoregressive element; the simultaneous effects represent a purely *spatial* autoregressive component; the remaining terms in the historical part of the 'cone' of figure 2 represent general space–time covariances. The cone of dependencies which stretches from the present into the past will also project into the future. That part of the cone shows which regions, j, will be affected by area i in the future, as multiplier effects work themselves through the time–space system. Although the measures to be developed in the next subsection are defined in the context of spatial autocorrelation, we shall see that they are analogous to time-series measures of autocorrelation, and that they can be used for that purpose.

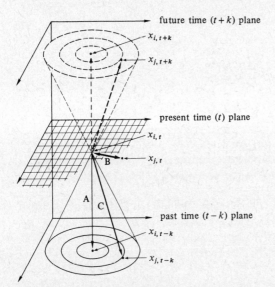

A purely temporal autoregressive component

B purely spatial autoregressive component

C mixed space–time covariances

Figure 2. Pattern of dependencies between regions in time and space. (Source: Haggett et al, 1977, page 355.)

Measures of autocorrelation

The simplest way in which the set of observations, $\{x_i\}$, can be measured is on a binary nominal scale. Conventionally, we put $x_i = 1$ if an event has occurred in the ith subarea, and $x_i = 0$ if it has not. Moran (1948) has called the former black, B, subareas, and the latter white, W, subareas; thus a two-color choropleth map is created. Two measures of spatial auto-correlation among the B and W subareas are the test statistics

$$BB = \tfrac{1}{2} \sum_{(2)} w_{ij} x_i x_j , \tag{7}$$

and

$$BW = \tfrac{1}{2} \sum_{(2)} w_{ij} (x_i - x_j)^2 . \tag{8}$$

Here

$$\sum_{(2)} = \sum_{\substack{i=1 \\ i \neq j}}^{n} \sum_{j=1}^{n} .$$

BB and BW are the (weighted) numbers of BB and BW joins in the study area. BW also reduces to the 'number of runs' statistic in one dimension (Siegel, 1956, pages 52–58), and provides a test of randomness in a time series (see Kendall, 1976). Cliff and Ord (1973, chapter 1) discuss the extensions in the literature of equations (7) and (8) to k-color choropleth maps.

If X is ranked or interval scaled, BB and BW will not be efficient measures of spatial autocorrelation, as information will be lost in reducing the data to a binary classification. Instead of doing this, we might use one or other of the following coefficients

$$I = \frac{n}{W} \sum_{(2)} w_{ij} z_i z_j \bigg/ \sum_{i=1}^{n} z_i^2 , \tag{9}$$

or

$$c = \frac{n-1}{2W} \sum_{(2)} w_{ij} (x_i - x_j)^2 \bigg/ \sum_{i=1}^{n} z_i^2 , \tag{10}$$

where $z_i = x_i - \bar{x}$, and $W = \sum_{(2)} w_{ij}$. Cliff and Ord (1973, chapter 1) discuss the origins and development of these coefficients. Both coefficients are in the classic form for a measure of autocorrelation; that is, I and c are ratios of measures of covariance to measures of variance, among the variate values.

In terms of temporal autocorrelation, we note that I reduces in one dimension to the usual serial correlation coefficient; c corresponds in form to (a) the Durbin and Watson d statistic (Durbin and Watson, 1950; 1951; 1971) used to search for temporal autocorrelation in regression residuals, and (b) the von Neumann ratio (von Neumann, 1941).

So far, we have descriptive measures of the degree of spatial auto-correlation in a map, and table 1 summarizes (in qualitative terms), the kinds of values that each statistic (7)–(10) will take on for various values of ρ in model (1).

Suppose, however, that we wish to test formally the hypothesis H_0: $\rho = 0$, against H_1: $\rho \neq 0$, in equation (1). We now indicate how this can be done.

Table 1. Qualitative variation in values of autocorrelation coefficients for different values of ρ [equation (1)].

ρ	Autocorrelation coefficient			
	BB	BW	I	c
<0	low	high	low	high
>0	high	low	high	low

Distribution theory
Cliff and Ord (1973, section 2.4) have shown that BB, BW, I, and c are all asymptotically normally distributed as $n \to \infty$, when $\rho = 0$ in equation (1), provided that in the weights matrix, $\mathbf{W} = \{w_{ij}\}$, no definite set of subareas dominates the study region as, for example, does the articulation point in a star lattice. (See also Sen, 1976; 1977; Cliff and Ord, 1977.) Thus, to test H_0: $\rho = 0$ against H_1: $\rho \neq 0$, we may treat

$$z = \frac{s - \mu_1'(s)}{\sigma(s)} \tag{11}$$

as (approximately) a standard normal deviate, where $\mu_1'(s)$ and $\sigma(s)$ denote the mean and standard deviation under H_0 of the statistic, s, which might be any one of those given in equations (7)–(10). The null hypothesis, H_0, is rejected whenever the observed value of z falls in the critical region [that is, for a two-tailed test, if $|z| \geq z_\alpha$, where $P(|z| \geq z_\alpha | H_0) = \alpha$, and α is the size of the test]. Serious inferential error is unlikely unless $n \leq 20$, when certain correction factors should be used (Cliff and Ord, 1973, chapter 2). Expressions for $\mu_1'(s)$ and $\sigma(s)$ are given for all the statistics defined in equations (7)–(10) by Cliff and Ord (1973, chapter 1).

Now that we have established what spatial autocorrelation is, and how its presence among regionally located variate values may be detected, we can consider the implications of autocorrelation for statistical models in the next two sections and the scale problem in the subsequent section.

Spatial autocorrelation and the independence assumption
Student's t test
To illustrate the effect of failure to meet the independence assumption frequently made in statistical analysis, we consider Student's t test for differences between means. Suppose that we have drawn samples of sizes N_1 and N_2 respectively from populations X_1 and X_2 which are distributed $N(\mu_1, \sigma_1^2)$ and $N(\mu_2, \sigma_2^2)$, respectively. If we take the simplest model and assume that $\sigma_1^2 = \sigma_2^2 = \sigma^2$, then under the null hypothesis, H_0, of no

difference between the population means, μ_1 and μ_2, the quantity

$$t = \frac{\bar{x}_1 - \bar{x}_2}{\tilde{\sigma}_{\bar{x}_1 - \bar{x}_2}} \tag{12}$$

will follow the Student t distribution with $N_1 + N_2 - 2$ degrees of freedom, *provided that the observations are independent.* Here $\bar{x}_1$ and $\bar{x}_2$ are the means of the samples drawn from populations X_1 and X_2, and $\tilde{\sigma}_{\bar{x}_1 - \bar{x}_2}$, the estimated standard error of the differences between sample means, is given by

$$\tilde{\sigma}_{\bar{x}_1 - \bar{x}_2} = \left(\frac{N_1 s_1^2 + N_2 s_2^2}{N_1 + N_2 - 2}\right)^{\!1/2} \left(\frac{N_1 + N_2}{N_1 N_2}\right)^{\!1/2}, \tag{13}$$

where s_1^2 and s_2^2 are the variances of samples 1 and 2. Cliff and Ord (1975b) constructed a Monte Carlo experiment to examine the sampling distribution of the t statistic (12) when X_1 and X_2 are autocorrelated by the amounts ρ_1 and ρ_2, respectively, according to model (1) among first nearest neighbours on regular lattices of various sizes. Figure 3(a) (in the next section) shows the results obtained for a 7×7 lattice and confirms the following points. First, when $\rho_1 = \rho_2 = 0$, X_1 and X_2 are uncorrelated, and the assumptions of the t test outlined for equations (12) and (13) are met. We would therefore expect a high degree of correspondence between the tabulated (light lines) and empirical (heavy lines) percentage points. As figure 3(a) shows, this is the case. Second, positive autocorrelation both in X_1 and in X_2 will result in *overestimation* of the significance of any analysis, if the conventional formulae and tables for t are used; that is, the probability of a type I error (the probability of rejecting H_0 when it is in fact true) exceeds the nominal level. The degree of overestimation increases as the level of autocorrelation increases. When the autocorrelation is negative, the situation is reversed and *underestimation* of the significance of results occurs; that is, the true probability of a type I error is smaller than the nominal level. Third, when one of the variables is positively autocorrelated, and the other is negatively autocorrelated, the effects are not self-cancelling; the true probability of a type I error again exceeds the nominal level.

These difficulties with the t test arise because positively autocorrelated observations produce a downwards bias (underestimation) in the estimated standard error which appears in the denominator of t [equation (13)]. Negatively autocorrelated observations bias the estimate of the standard error upwards (overestimation).

Spatial distortion in other statistical models
The sorts of results outlined in the previous section are well known to econometricians (see Johnston, 1972, section 8.2) from their work on regression models. A basic assumption of the familiar linear regression model

$$\underset{n \times 1}{Y} = \underset{n \times k}{X} \underset{k \times 1}{\beta} + \underset{n \times 1}{\epsilon} \tag{14}$$

is that the error terms, ϵ, are uncorrelated. For data with a geographical ordering, this requirement implies that the errors must not be spatially autocorrelated. If ordinary least squares procedures are used to estimate the model when the errors are positively autocorrelated, the consequences are as follows (Johnston, 1972, pages 246–249).

(1) The sampling variances of the regression coefficients, var($\hat{\beta}$), will be seriously underestimated;

(2) the variance of the errors, σ_ϵ^2, will be underestimated.

As this second estimate, $\bar{\sigma}_\epsilon^2$, appears in the denominator of the t and F tests used to examine the significance of the regression [Johnston, 1972, equations (2-30) and (2-36)], overstatement of the significance of the regression is likely. In addition, an inflated value for F means that the value of the coefficient of multiple correlation, R^2, will also be inflated [Johnston, 1972, equation (5-60)]. If the error terms are negatively correlated, the reverse occurs (understatement of the significance of any regression). Johnston (1972, section 8.2) has shown additionally, that the level of spatial autocorrelation in the $\{X\}$ variables in the regression is important as well. Positive (or negative) autocorrelation in those variables works in the same direction as positive (or negative) autocorrelation in the errors, thus compounding the problems already noted. Johnston (page 248) also suggests that if the $\{X\}$ variables are spatially independent, "then even if [the $\{\epsilon\}$ are] autocorrelated the bias is not likely to be serious". This is correct for those regression coefficients which correspond to X variables measured about their means, but the coefficient corresponding to the constant term will have its variance underestimated. This can be seen by analogy with the argument appearing later, where from equation (15) we see that the variance of the sample mean is of the form

$$\frac{\sigma^2}{n(1-p)},$$

so that positive autocorrelation increases the sampling variance.

What happens when other models are used? Cliff et al (1975) have demonstrated for a particular application of the X^2 goodness-of-fit test, that failure to meet the assumption of independent observed frequency counts will generally lead to substantially increased risks of a type I error if conventional statistical tables are used. For principal components and factor analysis, where independence is again assumed, the picture is more complicated. Cliff and Ord (1975c) have shown that if the variables in the analysis are equicorrelated (that is, have a single value for ρ), then the same results will be obtained, scale factors apart, as if the independence assumption were met. Whether this conclusion is of practical value remains unexplored; we have little idea of the relative levels of spatial autocorrelation in geographical variables.

However, the general lesson to be learned from the above discussion is clear. If geographers continue to apply the usual forms of many of the basic statistical models to spatially autocorrelated data, they run a very severe risk of reaching misleading conclusions. It would also indicate that the substantive results reported to date in studies which have not taken this problem into account should be interpreted with great caution. Regional concepts mean that positive autocorrelation is likely to be much more common than negative autocorrelation. We will therefore tend to think that the results of any analysis of data are substantially more significant than they really are.

Spatial autocorrelation: solutions

What approaches are available which might enable the difficulties discussed above to be overcome? An appropriate strategy might be as follows.

(1) Determine, by use of one of the measures given in equations (7)–(10), whether there is any spatial autocorrelation in the data, before carrying out any analysis. If there is not, that is, if the data are spatially independent, proceed with the analysis by using the conventional models.

(2) If we conclude that the independence assumption cannot be sustained, two basic alternatives are open to us. We can (a) modify the appropriate technique to allow for the correlated observations, or (b) try to remove the spatial autocorrelation from the data—make them spatially independent— so that the conventional models may be applied.

The formally correct procedure is always to use modified methods. However, the strategy outlined is much simpler if the independence assumption is met, and the loss of efficiency resulting from application of the strategy will generally be slight. We now consider alternatives (a) and (b) in turn.

Allowing for correlated observations

Gould (1970) has argued that understanding the pattern of autocorrelation over space in geographical variables is so basic an aim of geographical inquiry that any analysis should allow for this fact. He believes that 'cleaning up' the data to make them independent "represents a throwing out of the baby with bathwater". However, we should be quite clear that adapting the models to allow for the autocorrelation is an extremely lengthy procedure. Thus Cliff and Ord (1975b) have modified the t test for difference between means by using the autocorrelation measure, I, defined in equation (9) as the basis for a correction procedure. With reference to the earlier discussion, we note that, when the variates X_1 and X_2 are first-order spatially autocorrelated by the amounts ρ_1 and ρ_2, respectively, the estimated standard error, $\tilde{\sigma}_{\bar{x}_1 - \bar{x}_2}$, is given by

$$\tilde{\sigma}_{\bar{x}_1 - \bar{x}_2} = s \left[\frac{1}{N_1(1 - \tilde{\rho}_1)^2} + \frac{1}{N_2(1 - \tilde{\rho}_2)^2} \right]^{1/2}, \tag{15}$$

where

$$s^2 = \frac{(x_1 - \bar{x}_1 1)^T V_1^{-1} (x_1 - \bar{x}_1 1) + (x_2 - \bar{x}_2 1)^T V_2^{-1} (x_2 - \bar{x}_2 1)}{N_1 + N_2 - 2}. \tag{16}$$

In equations (15) and (16), the notation of the previous section has been followed. In addition, $\tilde{\rho}_1$ and $\tilde{\rho}_2$ are the estimates of ρ_1 and ρ_2, respectively, and $\mathbf{V}_i^{-1}$ is given by

$$\mathbf{V}_i^{-1} = (\mathbf{I} - \tilde{\rho}_i\mathbf{W}_i^{\mathrm{T}})(\mathbf{I} - \tilde{\rho}_i\mathbf{W}_i) , \qquad i = 1, 2 . \tag{17}$$

The elements of $\mathbf{W}_i$ are taken to be scaled so that $\sum_j w_{ij} = 1$. Thus, if we assume, without loss of generality, that the variates X_1 and X_2 are both $N(0, 1)$, then $\mathbf{V}_1$ and $\mathbf{V}_2$ are simply the (spatial) variance–covariance matrices for X_1 and X_2, respectively; that is, they specify the spatial covariance structure among the variate values, and so equation (15) is feeding this information explicitly into the t-test formula. It is evident that when $\tilde{\rho}_1$ and $\tilde{\rho}_2$ in equation (15) are both zero, this equation reduces to equation (13), as it should, since both sets of variate values are then spatially uncorrelated.

Strictly speaking, the estimates, $\tilde{\rho}_1$ and $\tilde{\rho}_2$, in the above equations should be obtained by maximum likelihood (ML estimation). However, the statistic I, given in equation (9), can be used to provide an acceptable approximation. As an estimator of ρ, I is inconsistent. Its great virtue, compared with maximum likelihood, is that it can be evaluated very simply. A further problem in using I is that the coefficient does not range over $[-1, +1]$ as does the ML estimator. However, $\max|I|$ can be calculated for any given lattice (Cliff and Ord, 1969, page 53) as

$$\max|I| = \left[\operatorname{var}\left(\sum_j w_{ij}z_j \right) \Big/ \operatorname{var}(z_i) \right]^{\frac{1}{2}} , \tag{18}$$

and so we can estimate ρ_i by

$$\tilde{\rho}_i = \frac{I}{\max|I|} , \tag{19}$$

which provides a more reasonable assessment on the usual $[-1, +1]$ interval.

The bias present in equation (19) used as an estimator is evident from table 2, which suggests $|\tilde{\rho}| \leqslant |\rho|$. To obtain table 2, an X variate auto-

Table 2. Comparison of $\tilde{\rho}$ and ρ for various lattices. (Source: Cliff and Ord, 1975b, page 730.)

Degree of auto-correlation in the pattern analysed, ρ	Average [a] estimate of the degree of autocorrelation, $\tilde{\rho}$			
	3×3 lattice	5×5 lattice	7×7 lattice	10×10 lattice
−0·9	−0·72	−0·81	−0·83	−0·86
−0·5	−0·52	−0·49	−0·47	−0·48
0·0	−0·26	−0·06	−0·03	0·00
0·5	0·00	0·34	0·42	0·47
0·9	0·05	0·61	0·76	0·83

[a] Based on 600 experiments.

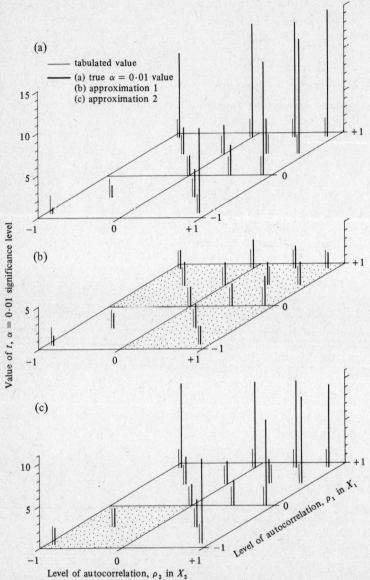

Figure 3. Impact of spatial autocorrelation upon Student's t test for differences between means. (a) Comparison of tabulated percentage point with true percentage point ($\alpha = 0\cdot01$) determined from Monte Carlo experiment for various levels of autocorrelation in X_1 and X_2. Note the highly variable cirque-shaped pattern of true values in contrast to the (by definition) horizontal plane of the tabulated values. (b) Values of t yielded by approximation 1 [equations (15), (19), and (20)] and (c) approximation 2 [equations (15), (16), and (19)]. (Source: Haggett et al, 1977, page 376.)

correlated by a known amount, ρ, was generated for each of the lattices shown. $\bar{\rho}$ was then calculated. The experiment was repeated 600 times for each value of ρ and each lattice to yield a mean value for $\bar{\rho}$ (say $\bar{\bar{\rho}}$) as $\Sigma \bar{\rho}/600$. Table 2 also shows that despite the bias, equation (19) is a reasonable approximation provided that $n \gtrsim 25$.

We may simplify approximation (15) even further by taking s in that equation as given in the usual form of the t test, namely

$$s = \left(\frac{N_1 s_1^2 + N_2 s_2^2}{N_1 + N_2 - 2}\right)^{\frac{1}{2}} ; \tag{20}$$

see equation (13). Using a Monte Carlo experiment, approximation (15) was evaluated with s given both by expression (16) and expression (20). $\bar{\rho}_i$ was calculated from equation (19). The results that Cliff and Ord (1975b) obtained for the 7×7 lattices and significance level $\alpha = 0 \cdot 01$ (one-tailed test) are reproduced in figures 3(b) and 3(c).

This figure suggests two general conclusions.
(1) If either or both of $\bar{\rho}_1$ and $\bar{\rho}_2$ are positive, take the statistic t given in equation (12) with $\tilde{\sigma}_{\bar{x}_1 - \bar{x}_2}$ given by equations (15), (19), and (20) to follow Student's t distribution with $(N_1 + N_2 - 2)$ degrees of freedom [approximation 1, say—shaded area in figure 3(b)].
(2) If both $\bar{\rho}_1$ and $\bar{\rho}_2$ are negative, take t with $\tilde{\sigma}_{\bar{x}_1 - \bar{x}_2}$ given by equations (15), (16), and (19) to follow Student's t distribution with $(N_1 + N_2 - 2)$ degrees of freedom [approximation 2, say—shaded area in figure 3(c)]. The use of these rules will provide a test of significance which is unlikely to result in serious inferential error. Cliff and Ord (1975b) have shown that the test thus modified can, in fact, be safely applied to any lattice for which N_1 and N_2 both exceed twenty-five.

Thus we have seen that, although it is possible to modify the t test to allow for correlated observations, the procedure is not straightforward. It is also possible to modify other models in a similar manner, but the task is no simpler and the theory required is often poorly developed. Given these difficulties, it is of value to consider whether the data can be transformed in some way to remove the correlation. Conventional models may then be employed. This idea is discussed in Johnston (1972, chapter 8) for the case of time series, and it has been considered by Martin (1974) for spatially located data (see also Student, 1914; Curry, 1971).

Spatial variate differencing
As before, we denote the random variable in subarea i by X_i, and suppose that the X_i are spatially autocorrelated according to the scheme given in equation (1), that is

$$X_i = \rho \sum_j w_{ij} X_j + \epsilon_i , \qquad i = 1, 2, ..., n . \tag{21}$$

The multilateral dependence among observations implied by equation (21) means that spatial difference operators can be constructed in several different ways. One of the most commonly used (Lebart, 1969; Martin, 1974) is to define

$$\Delta X_i = X_i - \sum_j w_{ij} X_j . \tag{22}$$

This difference filter is most effective in eliminating spatial autocorrelation when the autocorrelation parameter, ρ, in equation (21) is approximately one. Figure 4 shows a simple example of this difference operator applied to a set of contiguous quadrats.

If the variate differencing is used, instead of modified models, as a means of satisfying the independence assumption in analyses where hypothesis testing is the aim, the limited results available to date suggest that the following guidelines may be employed:

(a) as noted in connection with figure 3(a), if $\rho \leqslant 0$, use of the original observations will result in tests of hypotheses which are conservative, in the sense that the true risk of a type I error will be less than the nominal (tabulated) value;

(b) if ρ is slightly greater than zero, use of the original observations will result in a slightly liberal test, whereas variate differences may produce a very conservative test;

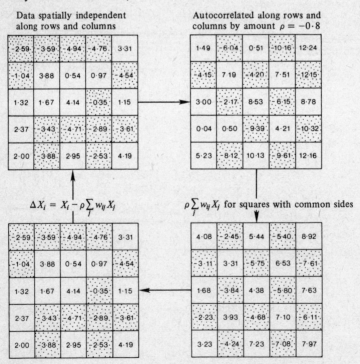

Figure 4. The use of difference operators to reduce spatial autocorrelation in a set of observations. See equation (22).

(c) as ρ approaches one, the use of data which have been differenced may result in a slightly conservative test, but the conclusions will be much more reliable than those based upon the original observations. Tests based upon the original observations will carry a risk of a type I error greatly in excess of the nominal level.

These guidelines imply that in hypothesis testing, a strategy which will always tend to err on the safe side, in terms of the risk of a type I error, is to use the original observations when $\rho \leqslant 0$, and to use variate differences when $\rho > 0$.

It should be noted that spatial differencing cannot be used for a test of mean because, when $\sum_j w_{ij} = 1$, the mean of the ΔX_i is *zero*, no matter what the common mean of the X_i may be.

The only paper in which spatial variate differencing is used, and in which substantial empirical results are given, is that of Martin (1974). His findings illustrate many of the points we have made. Martin postulated the simple population regression model

$$Y_i = 1 \cdot 0 + 0 \cdot 5 X_i + \epsilon_i = \alpha + \beta X_i + \epsilon_i , \qquad i = 1, 2, ..., n . \tag{23}$$

Table 3. Effect of spatial autocorrelation upon the variance of the estimated regression coefficient, $\hat{\beta}$, in a simple linear regression, for a 5 × 5 regular lattice. (Source: Martin, 1974, page 190.)

Level of spatial autocorrelation		Estimated regression coefficient, $\hat{\beta}$	Variance	
in X, ρ	in errors, λ		estimated, $\widetilde{\text{var}}(\hat{\beta})$	true, $\text{var}(\hat{\beta})$
0·0	0·0	0·515	0·127	0·136
	0·2	0·577	0·126	0·139
	0·4	0·534	0·131	0·145
	0·6	0·436	0·190	0·217
	0·8	0·514	0·246	0·381
	1·0	0·631	0·504	1·197
0·4	0·0	0·534	0·113	0·124
	0·2	0·481	0·113	0·153
	0·4	0·479	0·119	0·159
	0·6	0·488	0·154	0·207
	0·8	0·433	0·251	0·552
	1·0	0·481	0·635	1·264
0·8	0·0	0·519	0·071	0·080
	0·2	0·538	0·077	0·099
	0·4	0·568	0·079	0·124
	0·6	0·502	0·088	0·140
	0·8	0·495	0·153	0·402
	1·0	0·476	0·454	0·965

He assumed that the X_i and the errors, ϵ_i, were autocorrelated spatially according to the first-order schemes

$$X_i = \rho \sum_j w_{ij} X_j + v_i \, , \qquad (24)$$

and

$$\epsilon_i = \lambda \sum_j w_{ij} \epsilon_j + \mu_i \, , \qquad (25)$$

where ρ and λ are parameters expressing the level of autocorrelation in the same way as ρ_1 and ρ_2 earlier. The standard assumptions were made about v and μ. Using a Monte Carlo experiment, Martin empirically estimated the regression given in equation (23) by ordinary least squares for 5×5, 6×6, and 7×7 regular lattices, and for varying values of λ and ρ. Some of the results that he obtained are shown in table 3. It is immediately evident that the formulae led to underestimation of the true variance of $\bar{\beta}$. We use $\mathrm{var}(\bar{\beta})$ to denote this true variance, and $\widetilde{\mathrm{var}}(\bar{\beta})$ to denote the estimated variance. This confirms the discussion in the previous section, and the theoretical work in Johnston (1972, chapter 8). Martin repeated his analysis after applying the operator ΔX_i [equation (22)] to equations (24) and (25). The revised results are given in table 4. They show that $\widetilde{\mathrm{var}}(\bar{\beta})$ is much closer to $\mathrm{var}(\bar{\beta})$ when $(\rho, \lambda) \gtrsim 0.4$. It is for such values of ρ and λ that the errors in $\widetilde{\mathrm{var}}(\bar{\beta})$ are worst in table 3.

Table 4. Effect of spatial autocorrelation upon the variance of $\bar{\beta}$; regression estimated by first using spatial differences for $\{X_i\}$ and $\{\epsilon_i\}$. (Source: Martin, 1974, page 191.)

Level of spatial autocorrelation		Estimated regression coefficient, $\bar{\beta}$	Variance	
in X, ρ	in errors, λ		estimated, $\widetilde{\mathrm{var}}(\bar{\beta})$	true, $\mathrm{var}(\bar{\beta})$
0·0	0·0	0·547	0·132	0·147
	0·2	0·559	0·107	0·145
	0·4	0·510	0·106	0·144
	0·6	0·450	0·100	0·130
	0·8	0·470	0·085	0·086
	1·0	0·489	0·082	0·082
0·4	0·0	0·555	0·172	0·236
	0·2	0·507	0·138	0·179
	0·4	0·479	0·117	0·145
	0·6	0·516	0·116	0·140
	0·8	0·496	0·104	0·138
	1·0	0·506	0·104	0·103
0·8	0·0	0·495	0·185	0·191
	0·2	0·536	0·157	0·163
	0·4	0·532	0·142	0·143
	0·6	0·482	0·130	0·131
	0·8	0·481	0·127	0·127
	1·0	0·500	0·124	0·123

The account in this subsection would therefore suggest that, provided simple and reliable spatial differencing schemes can be defined, this may prove to be a more practical solution to the independence problem in hypothesis testing and estimation than the methods discussed earlier. Finally, we note that for hypothesis testing, a procedure which may be employed, instead of the correctly modified methods or variate differencing, is the randomisation (or Monte Carlo) test procedure discussed in Cliff and Ord (1973, page 81), namely:
(1) generate m sample values of the test statistic under H_0 for the appropriate level of spatial dependence in the data;
(2) reject H_0 at the $100[(j+1)/(m+1)]$ per cent level (one-tailed test) if the observed value of the test statistic exceeds at least all but the j largest generated values.

Spatial autocorrelation and the scale problem
In earlier sections, we have considered observations on a single random variable for which the 'individuals' observed are specified *a priori*, be they individual consumers, counties, boreholes, or whatever. We now consider situations where the 'individual' is not well-defined and where interrelations between random variables are of interest. The simplest measure of dependence in common use is the linear correlation coefficient, and we focus upon this measure in the rest of this section.

Usual definitions of a sample correlation coefficient assume the availability of n independent (in the sense of the two preceding sections) pairs of observations which are not *modifiable*. Yule and Kendall (1965, page 310) define nonmodifiability as meaning that each observation is based upon a natural unit, so that aggregation or subdivision of units does not produce units of the same kind. Thus, observations recorded for individual human

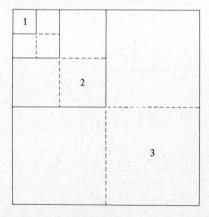

Figure 5. Aggregation scheme for cells to define higher-order levels as a function of lower-order levels ($k = 3$).

beings are not modifiable, whereas data for rectangular areas laid out in a field are modifiable.

As discussed in the introduction, geographically located data are generally both nonindependent and modifiable, and the question which therefore arises is whether a natural measure of correlation between variates can be defined for spatially dependent observations based on modifiable units. We must also ask whether such a measure can be estimated from sample data. We concentrate upon a formulation of the problem which involves contributions to the variance through common factors at different spatial scales, although other approaches are possible. To simplify the analysis, we shall consider a set of square cells (each with side of length one) laid out as a larger square of size 2^k, where k is an integer (see figure 5).

The aggregation process could involve the combination of adjacent cells to form larger units of size 2×1, 2×2, 4×2 and so on. If the square shape is preserved, the units will be 2×2, 4×4, ..., which produces k sizes of unit [excluding the single cell of size n $(= 2^{2k})$]. This scheme has the advantage of defining cells which are regular, which cover the whole study area, and which allow aggregation into larger areal units of similar shape.

Scale analysis for a single variable

The use of contiguous quadrats to search for spatial pattern at different scales was first developed in the ecological literature by Greig-Smith (1952; 1964). His analysis dealt with rectangular grids, although many ecologists now prefer to use line-transect data (Kershaw, 1957). The analysis of transects remains a possibility in some areas of geography (for example, in street surveys), but two-dimensional analysis seems to arise more naturally.

If we combine cells two at a time, then we arrive at a table of the usual analysis-of-variance form, such as table 5 (Mead, 1974). In table 5, $\Sigma(\text{cell})^2$ denotes the initial sum of squares; $\Sigma(\text{pair total})^2$ denotes the sum of squares obtained after combining the observations in adjacent cells; and so on. If the analysis if done as suggested in figure 5, then the levels will be fours, sixteens, etc.

Table 5. Analysis of variance for square cells combined in powers of two.

Level of variation	Sum of squares	Degrees of freedom
Cells within pairs	$\Sigma(\text{cell})^2 - \frac{1}{2}\Sigma(\text{pair total})^2$	2^{2k-1}
Pairs within fours	$\frac{1}{2}\Sigma(\text{pair total})^2 - \frac{1}{4}\Sigma(\text{fours total})^2$	2^{2k-2}
Fours within eights	$\frac{1}{4}\Sigma(\text{fours total})^2 - \frac{1}{8}\Sigma(\text{eights total})^2$	2^{2k-3}
$\vdots$	$\vdots$	$\vdots$
Quarter totals within half totals	$2^{2-2k}[\Sigma(\text{quarter total})^2 - \frac{1}{2}\Sigma(\text{half total})^2]$	2
Half totals	$2^{1-2k}[\Sigma(\text{half total})^2 - \frac{1}{2}(\text{total})^2]$	1
Total	$\Sigma(\text{cell})^2 - 2^{-2k}(\text{total})^2$	$2^{2k}-1$

However, a conventional analysis of variance may not be very helpful, since, as discussed earlier, positive spatial dependence at smaller scales will inflate the later sums of squares. To overcome this difficulty, Mead (1974) has provided a random-permutations test which allows separate tests to be made at each hierarchical level. Alternatively, we may postulate a model for the random variable, X_i, corresponding to the ith cell, as

$$X_i = \mu + \epsilon_1(i) + ... + \epsilon_m(i) ,\qquad (26)$$

where
(a) the means of the $\{X_i\}$ are taken to be equal to μ;
(b) there are m levels in the hierarchy, excluding the 'grand total' level;
(c) the random variables $\epsilon_j(i)$ have zero means and variances $\sigma_\epsilon^2(j)$, denoting the variance attributable to the jth level of the hierarchy.

The estimated variance components, $\bar{\sigma}_\epsilon^2(j)$, may be examined to find the scales to which most variation is attributable. If s cells are combined at each stage of the hierarchy, we may write

$$X_i(j) = \sum_{r \in A_{ij}} X_r(j-1) ,\qquad (27)$$

where A_{ij} denotes the set of s cells at level $(j-1)$ which are combined into one cell (the ith) at level j. The jth sum of squares, SS_j, corresponding to the jth row of the analysis-of-variance table, may be written as

$$SS_j = \sum_i [X_i(j) - s^{-1}X_i(j+1)]^2 ,$$

whence the mean square at level j, MS_j, is

$$MS_j = \frac{SS_j}{(n_j - n_{j+1})} ,$$

where $n_j = s n_{j+1}$. Finally, the jth component variance may be estimated as

$$\bar{\sigma}_\epsilon^2(j) = \frac{(sMS_j - MS_{j+1})}{s-1} , \qquad j = 1, ..., m .\qquad (28)$$

Although the parameters $\sigma_\epsilon^2(j)$ are nonnegative, it is possible for the estimates to be negative unless restricted. Several rules are possible, and we present one such set in the next section.

The analysis may be extended to cover unequal groups at each stage of the hierarchy, either for the conventional analysis of variance (Moellering and Tobler, 1972) or for the variance components models (Kendall and Stuart, 1975, chapter 35). When $n = s_1 s_2 ... s_m$, all that is required is to replace s by s_j in equation (28). It is evident that the variance components at each scale level reflect the form of the spatial autocorrelation function for the variate analysed, since the *average* distance between points increases each time cells are combined. Thus, a spatial autocorrelation function reflecting exponential decay will yield diminishing component variances as j increases (see the example given in Haggett et al, 1977, pages 387–391 in confirmation of this).

Correlation and scale

Now we suppose that two random variables, X_i and Y_i, are to be observed for each cell. The same hierarchical pattern is employed as before (figure 5), but we must extend the model to allow for (1) variation which is common to X and Y, and (2) for variation which affects only one of the two variables. To achieve this, we partition the jth-level random component into three parts:

Z_j the jth level *factor* which is common to both X and Y,

ϵ_j the '*noise*' component which affects X only, and

δ_j the '*noise*' component which affects Y only.

The use of the term *factor* is consistent with normal usage in factor analysis. Thus, we suppose that X_i and Y_i may be written in terms of the linear expressions

$$\left. \begin{aligned} X_i &= \mu_x + \phi_1 Z_1(i) + ... + \phi_m Z_m(i) + \epsilon_1(i) + ... + c_m(i) , \\ Y_i &= \mu_y + \theta_1 Z_1(i) + ... + \theta_m Z_m(i) + \delta_1(i) + ... + \delta_m(i) , \end{aligned} \right\} \tag{29}$$

where

μ_x and μ_y are constant (over all i);

ϕ_j, θ_j, $j = 1, ..., m$ are unknown parameters;

the factors $Z_j(i)$, $i = 1, ..., n$, $j = 1, ..., m$ have zero means, unit variances, and are uncorrelated with one another;

the random noise components $\epsilon_j(i)$, $\delta_j(i)$, $i = 1, ..., m$ have zero means, variances $\sigma_\epsilon^2(j, i)$ and $\sigma_\delta^2(j, i)$, respectively; and

(a) $\epsilon_j(i)$ and $\delta_j(i)$ are uncorrelated with $Z_j(i)$ for all j and i;

(b) $\epsilon_j(i)$ is uncorrelated with $\epsilon_k(l)$ whenever $k \neq j$ or $l \neq i$, with a similar condition for $\delta_j(i)$;

(c) $\epsilon_j(i)$ is uncorrelated with $\delta_k(l)$ for all j, k, i, and l.

This framework allows a wide variety of models to be developed. In particular, unless restrictions of the sort imposed in models 1 and 2 below are specified, the framework implies that dependence is built into the model by 'sharing' errors as shown in figure 6. For example, from

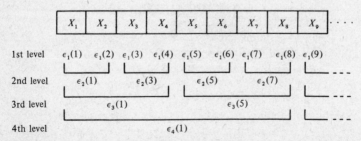

Figure 6. Illustration of way dependence is built into a scale-components model by sharing errors as cells are aggregated. Note that from level 2 upwards, the number in brackets is purely notional and indicates the first member of each group combined.

equations (29) and dealing only with the $\{\epsilon\}$ terms,

$$X_1 = \epsilon_1(1) + \epsilon_2(1) + \dots ,$$

$$X_2 = \epsilon_1(2) + \epsilon_2(2) + \dots ,$$

but $\epsilon_2(1) \equiv \epsilon_2(2)$ by specification of the model; see also equation (26).

We now examine some special cases of the model and indicate their relationship to the existing literature.

Model 1 Suppose we impose the following additional restrictions:

$$\theta_j = \phi_j = 0 , \qquad j = 2, \dots, k ,$$

$$\sigma_\epsilon^2(1, i) = \sigma_\epsilon^2 , \qquad\qquad \sigma_\delta^2(1, i) = \sigma_\delta^2 ,$$

and

$$\sigma_\epsilon^2(j, i) = \sigma_\delta^2(j, i) = 0 , \qquad j \geqslant 2 ,$$

for all i. That is, the model reduces to

$$\left.\begin{array}{l} X_i = \mu_x + \phi\xi_i + \epsilon_1(i) , \\ Y_i = \mu_y + \theta\xi_i + \delta_1(i) , \end{array}\right\} \qquad (30)$$

where $\xi_i \equiv Z_1(i)$, and we have dropped the subscripts on θ and ϕ for convenience. The model is in the form of a structural relationship (Kendall and Stuart, 1979, chapter 29) if we regard θ and ϕ, but not the $\{\xi_i\}$, as unknowns to be estimated. It is a functional relationship if we also wish to estimate the $\{\xi_i\}$. Alternatively, the model may be interpreted as a single (common) factor underlying two variates with θ and ϕ as factor loadings. We shall examine the model in the factor context with the constraint $\Sigma\xi_i = 0$. It follows directly that

$$\left.\begin{array}{ll} E(X_i) = \mu_x , & E(Y_i) = \mu_y , \\[2mm] \mathrm{var}(X_i) = \phi^2 + \sigma_\epsilon^2 , & \mathrm{var}(Y_i) = \theta^2 + \sigma_\delta^2 , \end{array}\right\}$$

and

$$\mathrm{cov}(X_i, Y_i) = \theta\phi . \qquad\qquad (31)$$

When X_i and Y_i are bivariate normal, the parameters $\{\theta, \phi, \sigma_\delta^2, \sigma_\epsilon^2\}$ cannot be estimated separately without additional information (Kendall and Stuart, 1979, pages 403–404). However, the correlation between X and Y can be estimated using the conventional sample correlation coefficient in the usual way.

Model 2 The same restrictions are imposed as in model 1, except that $\sigma_\epsilon^2(1, i) = B_i\sigma_\epsilon^2$, and $\sigma_\delta^2(1, i) = B_i\sigma_\delta^2$, where B_i is known. We also take $\mathrm{var}(\xi_i) = B_i$, which extends beyond the assumptions of equations (29). As this is the only model for which we make this extension, it seems easier to treat it on an ad hoc basis, rather than to increase further the generality of equations (29). The resulting estimators are:

$$\tilde{\mu}_x = \frac{\sum A_i X_i}{\sum A_i} , \qquad\qquad \tilde{\mu}_y = \frac{\sum A_i Y_i}{\sum A_i} ,$$

$$\widetilde{\text{cov}}(X, Y) = \frac{\sum A_i (X_i - \tilde{\mu}_x)(Y_i - \tilde{\mu}_y)}{\sum A_i} , \qquad \widetilde{\text{var}}(X) = \frac{\sum A_i (X_i - \tilde{\mu}_x)^2}{\sum A_i} ,$$

with a similar expression for $\widetilde{\text{var}}(Y)$. Here, $A_i = B_i^{-1}$, and the expectations for these estimators are given by expressions (31). These results hold for any weighting function, B_i, but a natural choice for data recorded as an average over a spatial unit of area a_i is $B_i = a_i^{-1}$. This form of weighting function has been used by Robinson (1956) and by Thomas and Anderson (1965) to weight the conventional correlation coefficient in their discussion of the ecological correlation problem. Clearly the extension to regression analysis should employ the same weighting function.

Models 1 and 2 are only of limited interest because they require that there should be no spatial autocorrelation between geographical units larger than the first level of the hierarchy. The descriptions given here are intended firstly to provide a link with earlier investigations, and secondly to pave the way for the more realistic model discussed below, which does not carry this restriction.

Model 3 We now consider all m components, so that the only restrictions are

$$\sigma_\epsilon^2(j, i) = \sigma_\epsilon^2(j) , \qquad \sigma_\delta^2(j, i) = \sigma_\delta^2(j) , \qquad (\text{all } i \text{ and } j) .$$

The variance component for X at level j is given by $\sigma_\epsilon^2(j) + \phi_j^2$, which may be estimated by

$$\tilde{\sigma}_\epsilon^2(j) + \tilde{\phi}_j^2 = \frac{\{s\text{MS}_j(x) - \text{MS}_{j+1}(x)\}}{s - 1} , \qquad j = 1, ..., m . \qquad (32)$$

This is the same expression as (28), save that an x has been inserted to indicate the variable of interest. The reason that expressions (28) and (32) are identical is that the factor and noise components collapse into one when only a single random variable is considered. In similar vein, the jth-level sum of cross products may be written as

$$\text{SCP}_j = \sum_i \{X_i(j) - s^{-1} X_i(j+1)\}\{Y_i(j) - s^{-1} Y_i(j+1)\} , \qquad j = 1, ..., m .$$

whence the mean cross product is given as

$$\text{MCP}_j = \frac{\text{SCP}_j}{n_j - n_{j+1}} ,$$

with n_j defined as before. Finally, the covariance component, $\theta_j \phi_j$, is estimated by

$$\theta_j \tilde{\phi}_j = \frac{\{s\text{MCP}_j - \text{MCP}_{j+1}\}}{s - 1} , \qquad j = 1, ..., m . \qquad (33)$$

At any given level, we have three statistics (two variances and one covariance) but four parameters $[\theta_j, \phi_j, \sigma_\epsilon^2(j), \text{ and } \sigma_\delta^2(j)]$. However, this does not matter provided we confine attention to the estimation of the three *components*.

Example Suppose there are k (= 4) levels, s (= 2) areas are combined, and $\theta_j = \phi_j = 1$ for $j = 1, ..., 4$. Also let $\sigma_\epsilon^2(1) = \sigma_\delta^2(1) = 6$, and $\sigma_\epsilon^2(j) = \sigma_\delta^2(j) = 0$, $j \geqslant 2$. Then we obtain

	$j = 1$	$j = 2$	$j = 3$	$j = 4$
$\text{var}\{X(j)\} = \text{var}\{Y(j)\}$	10	6·5	4·25	2·625
$\text{corr}\{X(j), Y(j)\}$		0·4	0·54	0·65	0·71

The reduction in variance and increase in correlation (with k) was noted in the spatial case by Yule and Kendall (1965, pages 310–312 and earlier editions). It is evident that there is no single measure of correlation that can be regarded as *the* correlation between X and Y. Rather, the correlation between the variables is a function of the size of area considered, as discussed, in the context of geography, by Robinson (1956), and Thomas and Anderson (1965). What is supplied by equations (28) and (32) is a spectral decomposition of the total (single unit) variances and covariance into components for the particular cell sizes considered. If a suitable functional form describing the change in correlation with size can be found, it might be calibrated from the estimated values.

Negative estimates To avoid estimates which imply negative values for any $\sigma_\delta^2(j)$ or $\sigma_\epsilon^2(j)$, we may modify the estimators given in equations (28) and (32) as follows.

(a) If both variance estimators are greater than zero but the modulus of the correlation exceeds unity, scale the covariance down by a factor c_1 and each variance up by a factor c_2. These factors c_1 and c_2 are chosen such that the modulus of the correlation is reduced to unity.

(b) If one variance is negative, replace that variance (corresponding to X, say) by

$$\widetilde{\text{var}}[X(j)] = \frac{[\text{cov}\{X(j), Y(j)\}]^2}{\widetilde{\text{var}}\{Y(j)\}}.$$

(c) If both variances are negative, set both variances and covariances to zero.

In all cases, the investigator may prefer to set the appropriate estimators to zero or to aggregate these values with the next level of the hierarchy. These negative estimates arise from the use of equations (28) and (32) and can be avoided by deriving the maximum likelihood estimators subject to a series of boundary constraints. The numerical details tend to be very messy, which is why ad hoc rules such as the above are sometimes used. In the following section we have quoted negative estimates as they arise, without any attempts to 'doctor' the analysis.

An alternative approach
In place of equations (29) we might employ the usual regression model for X and Y; that is

$$X_i = Z_i^T \beta + \epsilon_i \; ; \qquad Y_i = Z_i^T \gamma + \delta_i \; . \tag{34}$$

The trend-surface models (with polynomial or sinusoidal terms) are special cases of model (34). Clearly the covariance between the residuals will depend upon the value specified for $\text{cov}(\epsilon_i, \delta_i)$, leading to Zellner's (1962) 'seemingly unrelated' regression equations. When ϵ_i and δ_i are uncorrelated, the spatial variation in X and Y can be 'explained' entirely by other variables. If a regression model is analysed using model 3 given above, and the Z are spatially autocorrelated, the results will, quite reasonably, reflect correlation at different distances (or for different sized units). The all-important question of interpretation is whether the correlation is intrinsic or whether it is the result of a common dependence upon other variables. The answer to this question must depend upon geographical theory rather than upon statistical methods.

It is possible to envisage other schemes which interrelate spatial auto-correlation and factor analytic methods. However, this development is beyond the scope of the present discussion.

Wheat yields and land use
The original example of the effect of different sized areal units upon values of the correlation coefficient, r, is given in early editions of *An Introduction to the Theory of Statistics* by Yule and Kendall. They computed the correlation between yields of wheat (X) and potatoes (Y) in the 48 English and Welsh counties in 1936, and obtained a value of $0 \cdot 22$ for the sample product–moment correlation coefficient. When counties were combined on an index of productivity to give 24, 12, 6, and 3 subareas, the revised values of r obtained were $0 \cdot 30$, $0 \cdot 58$, $0 \cdot 76$, and $0 \cdot 99$, respectively, as shown in the last column of table 6.

The jth level of aggregation corresponds to groupings of 2^{j-1} counties. Table 6 also gives the variances and covariance for each level, and the estimated components of variance and covariance evaluated by use of

Table 6. Scale analysis of data for wheat yield (X) and yield of potatoes (Y). Aggregation steps are $2 \times 2 \times 2 \times 2 \times 3$.

Level	var(X)	var(Y)	cov(X, Y)	Components			corr(X, Y)
				X	Y	XY	
1	4·17	0·53	0·33	2·90	0·39	0·09	0·22
2	2·73	0·34	0·28	0·19	0·20	−0·07	0·30
3	1·90	0·14	0·30	0·32	−0·03	0·19	0·58
4	1·34	0·05	0·21	0·38	−0·06	−0·08	0·76
5	0·86	0·05	0·20	0·82	0·06	0·30	0·99

model 3 and equations (28) and (32). The variance component for potatoes (Y) drops away sharply as the level of aggregation increases, and the recorded estimates are negative for levels 3 and 4. For both X and Y, the local component (level 1) is by far the most important.

The above example deals with counties of irregular size and shape, so that application of our approach is not strictly valid. As an example where the assumptions are more closely satisfied, we consider data on land use drawn from the *Atlas of London* (Jones and Sinclair, 1968, sheets 43–45). The data set comprises a 24 × 24 lattice of cells[1], each of which is a square of side 500 metres. For each cell, the area of floor space (in thousands of square feet[2]) in office, commercial and industrial use in 1962 is given. Two empirical features of note, which we would expect from bid-rent theory (Alonso, 1964), are that
(1) high-demand areas, such as the City of London, tend to make intensive use of the land available, so that total floor space in neighbouring cells and the type of use to which it is put, tend to be positively correlated;
(2) there is an element of competition, at least locally, between the different potential uses.

The data were analyzed using model 3 as before. Successive aggregations took in 4, 4, 4, 3, and 3 cells of the previous level; the results are presented in table 7. The features of increasing correlation and decreasing variance, noted for the wheat-yield data, are present again, although the correlation function flattens out from level 3 onwards in the industry and offices table [table 7(b)]. The 'correlation coefficients' for individual levels based upon the corresponding *components* are given in figure 7.

Table 7. Scale analysis of London land-use data. Aggregation steps are 4 × 4 × 4 × 3 × 3; $n = 576$.

Level	var(X)	var(Y)	cov(X, Y)	Components			corr(X, Y)
				X	Y	XY	
(a) $X \equiv$ commerce; $Y \equiv$ offices							
1	116712	886554	60691	60220	248162	−25371	0·19
2	71547	700432	79719	18738	318847	−10598	0·36
3	46203	414767	92425	29746	195119	71664	0·67
4	17557	197011	41853	2969	105111	6910	0·71
5	9211	78547	26008	9644	58588	31090	0·97
(b) $X \equiv$ industry; $Y \equiv$ offices							
1	88954	886554	24856	52025	248162	−7390	0·09
2	49935	700432	30399	10801	318847	−8734	0·16
3	32080	414767	38355	15104	195119	28170	0·33
4	16289	197011	19192	7286	105111	10470	0·34
5	7923	78547	7958	7701	58588	6319	0·32

[1] The northwest corner of the grid was located at grid reference 255860.
[2] Our apologies are offered to readers who are more purist over the matter of units.

These coefficients provide an even sharper contrast. Despite the fact that the components estimates are unrestricted, so that 'variances' and 'correlations' may wander beyond their feasible regions, the general pattern is clear. For smaller cells, the competitive influence dominates (negative correlation between land uses), whereas the common features of neighbourhood effects (positive correlation between land uses) are seen for levels 3 and above. Clearly, any talk of a single correlation coefficient between variables measured on modifiable units is erroneous. The advantage of the method of analysis we have proposed is that the spectral decomposition enables the variations in the process with different spatial scales to be determined, whereas such variations are masked by the simple correlation coefficient. In the next section, we attempt to isolate some of the major influences leading to correlation functions such as those shown in figure 7.

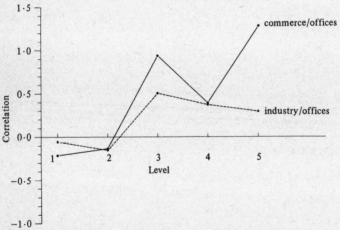

Figure 7. 'Correlation coefficients' based upon scale components for the London land-use data.

A land-use model

We now develop a model which exhibits the same qualitative features as the data, although a formal calibration is not attempted.

First, we assume that the study area is partitioned into a large number of small parcels of land which we term (planning) plots. In this idealised situation, we suppose that a unique use is allocated to each complete plot, independently of all other plots. Then it is evident that the procedure of combining units will lead to steady reductions in the variance. However, neighbouring plots will tend to have similar uses; we incorporate this feature into the model by assuming that the probability function,

$$\Theta_r(i) = P(\text{plot } i \text{ has land use } r) ,$$

takes on related values for plots contiguous to i. We would expect the relation between $\Theta_r(i)$ and $\Theta_r(j)$ to depend upon the distance between plots i and j, but as a first approximation we suppose that:

(a) in a block of n contiguous plots, the number of plots devoted to each use r ($r = 1, ..., k$) follows a multinomial distribution with parameters n and $\{\Theta_r\}$, $r = 1, ..., k$;

(b) the values $\{\Theta_r\}$ are selected from all possible values according to the correlated Dirichlet distribution which has the density function

$$f(\Theta_1, ..., \Theta_k) \propto \int \prod_{r=1}^{k} y_r^{\alpha_r - 1} \prod_{r < s} u_{rs}^{\gamma_{rs} - 1} \, du_{rs} \, , \qquad (35)$$

where

$$\Theta_r = y_r + \tfrac{1}{2} \sum_{r \neq s} u_{rs} \, , \qquad \sum y_r + \sum_{r < s} u_{rs} = 1 \, .$$

In equation (35), the integration is over the $\tfrac{1}{2}k(k-1)$-dimensional space, U, with elements u_{rs}. The exact form is very involved, but we are able to reverse orders of integration when computing the expectations given below.

There is no special reason behind the choice of equation (35), but it is a convenient generalization of the Dirichlet multinomial (Johnson and Kotz, 1972, page 234). As we are concerned primarily with the resulting variance and covariance structure, equation (35) should be viewed as one possibility among many, rather than a form with strong empirical backing. With assumption (a), it follows that the expectations of use, X, for particular cells i and j, given n, $\Theta_r(i)$, and $\Theta_r(j)$, are

$$E[X_i | n, \Theta_r(i)] = n\Theta_r(i), \qquad E[X_i^2 | n, \Theta_r(i)] = n(n-1)[\Theta_r(i)]^2 + n\Theta_r(i) \, ,$$

$$E[X_i X_j | n, \Theta_r(i), \Theta_r(j)] = n(n-1)\Theta_r(i)\Theta_r(j) \, .$$

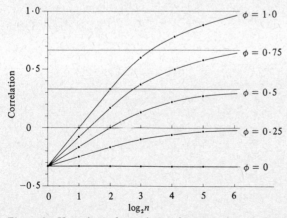

Figure 8. Hypothetical correlation functions for the planning-plots model (see text for model parameters).

The mixing distribution can now be used to obtain unconditional means, variances and covariances for X_i and X_j:

$$E(X_i|n) = \frac{n\beta_r}{\omega}, \qquad \text{var}(X_i|n) = \frac{n\beta_r(\omega - \beta_r)(\omega + n)}{\omega^2(\omega+1)},$$

$$\text{cov}(X_i, X_j|n) = \frac{-n(n+\omega)\beta_r\beta_s + n(n-1)\gamma_{rs}\,\omega}{\omega^2(\omega+1)}.$$

Here

$$\beta_r = \alpha_r + \tfrac{1}{2}\sum_{r \neq s}\gamma_{rs}\,, \qquad r = 1, ..., k\;;$$

$$\omega = \sum_{r=1}^{k}\beta_r\,.$$

Upon inspection, it can be seen that the covariance must be negative when $n = 1$, and its sign thereafter will depend upon the relative magnitudes of the coefficients β and γ, as n increases. Clearly, the exact form of the correlation function will depend upon the values of the β_r and the γ_{rs}. However, the general shape can be discerned from an example. In figure 8 we plot the correlation function for the case

$$k = 4\;; \qquad \beta_r = \beta_s = \tfrac{1}{2}\;; \qquad \omega = 2\;; \quad n = 2^{j-1}\,, \quad j = 1, 2, ..., 6\;;$$

$$\phi = 2\gamma = \frac{m}{4}\,, \quad m = 0, 1, ..., 4\,.$$

The behaviour predicted by the model (figure 8) is in reasonable qualitative accord with the empirical findings (figure 7) up to level 3. A reason for the discrepancy thereafter is not hard to find. As the zone size becomes greater, the plots within any zone will become more diversified, so that assumption (b) needs to be weakened to allow for a neighbourhood effect which decays with increasing zone size. It is noticeable that this decay does not occur for the wheat data, where the grouping scheme is different. The zone-size decay could be built into the model, but the results we have obtained so far are rather messy and do not add to an understanding of the processes involved.

A further point worthy of note is that the present model allows for the proportion of a zone given to a particular land use, rather than the total amount of floorspace available for that use. As different uses attract different levels of rental, it is highly probable that the proportion of usage in a particular category will depend upon the intensity of development in a given plot (for example, skyscraper office blocks).

Summary and conclusions
In this essay, we have defined the term 'spatial autocorrelation', and have presented several coefficients which enable its presence to be detected among geographically located variate values. However, many methods of statistical hypothesis testing assume that the variates being analysed are

independently distributed. Regional concepts in geography mean that positive spatial autocorrelation is likely to be much more commonly encountered than negative autocorrelation. Using Student's t test for differences between means, the linear regression model, and the X^2 goodness-of-fit test as examples, we have shown that if the independence assumption is violated by applying such methods to positively autocorrelated variate values, a substantially increased risk of a type I error is likely to result. Ways of overcoming the difficulty have been discussed, either by designing models which allow for the presence of autocorrelation in the data or by removing the autocorrelation with a variate-differencing filter so that standard models may be applied.

Many statistical models are also scale-dependent, so that it is not only difficult to determine 'the most appropriate scale' for data analysis, but also hard to disentangle scale effects from results of genuine scientific interest. It has been shown how this can hamper attempts to establish the degree of correlation among variate values. A spectral approach for decomposing variate values into variance and covariance components, which helps to overcome the difficulty, has been proposed. Its use has been illustrated by an application to patterns of office, commercial, and industrial land use in London in 1962.

Acknowledgement. The research material in this paper was carried out as part of a four-year project funded by SSRC, entitled "Geographical modelling and forecasting for epidemiological and population data".

References
Alonso W, 1964 *Location and Land Use* (Harvard University Press, Cambridge, Mass)
Bennett R J, 1974 "A review of spatial autocorrelation, by A D Cliff and J K Ord" *Environment and Planning A* **6** 241
Besag J E, 1974 "Spatial interaction and the statistical analysis of lattice systems" (with discussion) *Journal of the Royal Statistical Society, series B* **36** 192–236
Cliff A D, Martin R L, Ord J K, 1975 "A test for spatial autocorrelation in choropleth maps based upon a modified X^2 statistic" *Transactions and Papers, Institute of British Geographers* **65** 109–129
Cliff A D, Ord J K, 1969 "The problem of spatial autocorrelation" in *London Papers in Regional Science 1. Studies in Regional Science* Ed. A J Scott (Pion, London) pp 25–55
Cliff A D, Ord J K, 1973 *Spatial Autocorrelation* (Pion, London)
Cliff A D, Ord J K, 1975a "The choice of a test for spatial autocorrelation" in *Display and Analysis of Spatial Data* Eds J Davis, M McCullagh (John Wiley, New York) pp 54–77
Cliff A D, Ord J K, 1975b "The comparison of means when samples consist of spatially autocorrelated observations" *Environment and Planning A* **7** 725–734
Cliff A D, Ord J K, 1975c "Model building and the analysis of spatial pattern in human geography" *Journal of the Royal Statistical Society, series B* **37** 297–348
Cliff A D, Ord J K, 1977 "Large sample-size distribution of statistics used in testing for spatial correlation, a comment" *Geographical Analysis* **9** 297–299
Cliff A D, Ord J K, 1981 *Spatial Processes: Models and Applications* (Pion, London)

Curry L, 1971 "Applicability of space-time moving-average forecasting" in *Regional Forecasting* Eds M D I Chisholm, A E Frey, P Haggett (Butterworth, Sevenoaks, Kent) pp 11-24

Durbin J, Watson G S, 1950 "Testing for serial correlation in least squares regression, I" *Biometrika* **37** 409-428

Durbin J, Watson G S, 1951 "Testing for serial correlation in least squares regression, II" *Biometrika* **38** 159-178

Durbin J, Watson G S, 1971 "Testing for serial correlation in least squares regression, III" *Biometrika* **58** 1-19

Gould P R, 1970 "Is *Statistix Inferens* the geographical name for a wild goose?" *Economic Geography (Supplement)* **46** 439-448

Greig-Smith P, 1952 "The use of random and contiguous quadrats in the study of the structure of plant communities" *Annals of Botany (London)* **16** 293-316

Greig-Smith P, 1964 *Quantitative Plant Ecology* second edition (Butterworth, Sevenoaks, Kent)

Haggett P, Cliff A D, Frey A E, 1977 *Locational Analysis in Human Geography* (Edward Arnold, London)

Johnson N J, Kotz S, 1972 *Distributions in Statistics: Continuous Multivariate Distributions* (John Wiley, New York)

Johnston J, 1972 *Econometric Methods* second edition (McGraw-Hill, Maidenhead, Berks)

Jones E, Sinclair D J (Eds), 1968 *Atlas of London and the London Region* (Pergamon Press, Oxford)

Kendall M G, 1976 *Time Series* (Charles Griffin, High Wycombe, Bucks)

Kendall M G, Stuart A, 1975 *The Advanced Theory of Statistics, Volume 3* third edition (Charles Griffin, High Wycome, Bucks)

Kendall M G, Stuart A, 1979 *The Advanced Theory of Statistics, Volume 2* fourth edition (Charles Griffin, High Wycombe, Bucks)

Kershaw K A, 1957 "The use of cover and frequency in the detection of pattern in plant communities" *Ecology* **38** 291-299

Kooijman S A L M, 1976 "Some remarks on the statistical analysis of grids especially with respect to ecology" *Annals of Systems Research* **5** 113-132

Lebart L, 1969 "Analyse statistique de la contiguïté" *Publications de l'Université de Paris* **18** 81-112

Martin R L, 1974 "On autocorrelation, bias and the use of first spatial differences in regression analysis" *Area* **6** 185-194

Mead R, 1974 "A test for spatial pattern at several scales using data from a grid of contiguous quadrats" *Biometrics* **30** 295-307

Moellering H, Tobler W R, 1972 "Geographical variances" *Geographical Analysis* **4** 34-50

Moran P A P, 1948 "The interpretation of statistical maps" *Journal of the Royal Statistical Society, series B* **10** 245-251

Robinson A H, 1956 "The necessity of weighting values in correlation analysis of aereal data" *Annals of the Association of American Geographers* **46** 233-236

Sen A, 1976 "Large sample-size distribution of statistics used in testing for spatial correlation" *Geographical Analysis* **8** 175-184

Sen A, 1977 "Large sample-size distribution of statistics used in testing for spatial correlation—a reply" *Geographical Analysis* **9** 300

Siegel S, 1956 *Nonparametric Statistics for the Behavioral Sciences* (McGraw-Hill, New York)

Student, 1914 "The elimination of spurious correlations due to position in time or space" *Biometrika* **10** 179-180

Thomas E N, Anderson D L, 1965 "Additional comment on weighting values in correlation analysis of areal data" *Annals of the Association of American Geographers* **55** 492-505

Tobler W R, 1970 "A computer movie simulating urban growth in the Detroit region" *Economic Geography supplement* **46** 234-240

von Neuman J, 1941 "Distribution of the mean square successive difference of the variance" *Annals of Mathematical Statistics* **12** 367-395

Whittle P, 1954 "On stationary processes in the plane" *Biometrika* **41** 434-449

Yule G U, Kendall M G, 1965 *An Introduction to the Theory of Statistics* fourteenth edition (Charles Griffin, High Wycombe, Bucks)

Zellner A, 1962 "An efficient method of estimating seemingly unrelated regressions and tests for aggregation bias" *Journal of the American Statistical Association* **57** 348-368

Part 2

Deductive methods

8

Gravitational stratification

A B Vistelius

Introduction

Two problems will be studied in this essay. The first involves finding the best description of the sedimentation process. Does each bed reflect some special act of sedimentation, or is a combination of beds generated by each sedimentation event, with a single bed as a special case? The second problem is the investigation of whether there are certain beds which signal the initiation of a sedimentary sequence.

Of course, any one bed type (for example, conglomeratic intercalations) can be used to divide a set of sedimentary units into definite subsets. But such divisions are banal; they do not take into consideration the special features of the whole process. More precise definition of the latter idea requires special mathematical concepts which will be developed later in this essay. The solution of both of the above problems requires the use of the theory of stochastic processes, developed along some special lines typical of geological problems. This development is summarized in the appendices.

Stratification is the most typical feature of sedimentary deposits. It is present in many otherwise dissimilar deposits and its origin is, in many cases, not clear. For the purposes of this essay, we are interested in sedimentary strata expressed in terms of bed sequences. These sequences represent repetitions of the same conditions of sedimentation. Each occurrence of these conditions will generate a set of beds which will be called a *sedimentary unit*. In each sedimentary unit, a number of beds of different composition can appear. Sets of these sedimentary units comprise formations (American Geological Institute, 1957). Thus we are interested in the origin of such sets.

Geologists have shown interest in the types of sequences within formations since early in the nineteenth century (Dawson, 1854). This idea has developed along two lines. The first includes numerous studies of sequences in the coal measures (Weller, 1964). The origin of these sequences is not clear as yet and, as far as the author is aware, there is no indication that this problem will be solved in the foreseeable future. The second line of investigation was initiated by de Geer (1940). He found that sediments of the lakes near the terminus of a glacier occur as pairs of beds. In each pair there is one arenaceous and one argillaceous layer. His conclusion that these layers reflect alternation of summer and winter sedimentation is now widely accepted.

The idea about the generation of beds of different composition introduced by de Geer was applied by Vassoevich (1948) to flysch sediments. He found that flysch deposits contain a definite sequence of beds within each sedimentary unit. Kuenen (1948) explained the origin of this sequence in his turbidite theory. The layering generated in such sedimentation can be referred to as gravitational stratification. In this essay, we will develop a model of this phenomenon as a stochastic process.

In their investigations of the results of gravitational stratification, both de Geer and Vassoevich determined the set of possible beds within a sequence. De Geer (1940) found two types of beds, which showed a definite order. This order leads to a single type of sequence in the sedimentary unit, which de Geer called a 'varve'. Each varve includes an arenaceous and an argillaceous bed. Similarly, the sequence found by Vassoevich (1948) in the flysch deposits contains only three types of beds (a rhythm). Although terrigenous and carbonate flysch are distinguished, the sequences are similar in both types of deposit. Fine pebbles and sand are common in the first two beds. The composition of the third bed determines whether the deposit belongs to a terrigenous, or to a carbonate flysch sequence. If it is composed of argillites, one is dealing with terrigenous flysch; if it is a marl, the deposit is a carbonate flysch. It was the existence of this third bed type which forced Vassoevich to accept the idea of two distinct sequences. This idea of two distinct sequences in space generalizes to that of n distinct sequences in time. We wish to determine how many distinct sequences are necessary to describe a particular observed stratigraphic collection.

A particular sequence may be composed of one bed (say a sandstone), or of a short succession of distinct compositions. We will refer to a sequence of beds within a sedimentary unit as a *packet*, and indicate it by $(\alpha\beta)$, where α and β are beds of different composition, if it is necessary to distinguish between them.

Vassoevich (1948) and de Geer (1940) investigated deposits in which packets could be recognized rather easily in the field. However, there are many cases where it is impossible to distinguish the boundaries of packets. The difficulty is compounded when the identity of the packets is not known. The problem is then to distinguish types of packets, and to determine the boundaries of the packets. This latter is equivalent to the problem of defining the first bed in the sequence, and is a common problem in the geological literature, especially in that dealing with the investigation of coal measures. Some geologists believe that the cyclotheme should be started from the sandstones, but most adhere to the idea that the coals indicate the bottom of the cyclotheme (Matthews, 1974). The problem is not trivial, and can only be solved if we study the properties of *sequences of units*, rather than the composition of individual beds. There are sedimentary environments where deposition of clay is the only possible process. In this case the whole packet will be argillaceous, or the first bed will be argillaceous. It is also possible that an environment of sand deposition will be transformed without interruption to an environment of clay deposition. In this case the bottom of the packet will be a sandstone.

Thus, by knowing the first bed in a packet, we might more readily infer the environment of formation.

The concept of packets gives rise to a new problem, namely how sedimentary units join to make a formation. Indeed, if our ideas are correct, the whole sedimentary process should be regarded as a sequence of sedimentary units. But which sequence will it be? As we are not able to predict which sequence of sedimentary units will occur on the basis of available information, we consider it as a nondeterministic sequence. Thus we are studying a stochastic process. It may be Bernoullian[1], simple Markov, higher-order Markov, or essentially non-Markov. There is little information concerning the stochastic properties of sequences of sedimentary units in the geological literature. However, sediments brought into seas by floods seem to compose sequences which are Markov chains. Similarly, measurements of volume of suspended sediment follow Bernoullian sequences or simple Markov chains. Thus it can be supposed that sequences of sedimentary units are Bernoullian sequences or Markov chains. In other words, we believe that nature acts simply in its elementary operations. Our considerations relevant to sedimentation are summarized in the following model.

Gravitational stratification model
The following axioms describe a sequence of beds according to the gravitational stratification model.
1 Beds are generated in packets. Each packet is the result of one act of sedimentation. The packet is an indecomposable trial of a random, homogeneous process of stratum generation. Any sequence of beds is composed of some definite set of the packets.
2 If the packet is composed of more than one bed, the more coarse-grained bed is lower in the sequence of beds within the packet.
3 The sequence of beds can be divided into stages. After each stage, the process of bed generation starts anew. Each separate stage of sedimentation ends with the appearance of an argillaceous bed (the renewal event) at the top of the packet (including packets composed of a single argillaceous bed).
4 The random sequence of packets may be Bernoullian or simple Markov. We assume that the set of packets to be investigated is generated after sufficient time has elapsed for stabilization of the process.
5 After the packet is generated it is buried by the generation of a new packet in the same place. We assume that after deposition occurs there is no erosion. If erosion occurs, its effect will be included in (and confounded with) the depositional packet sequence.
These axioms will be referred to throughout the rest of this essay.

[1] It should be noted that Vistelius [see equation (2)] uses the term 'Bernoullian' as meaning: generated by a random sequence of independent trials with constant probability of events. Normal usage (see, for example, Feller, 1964) restricts the term to independent trials with only two possible outcomes.

Red beds of the Cheleken Peninsula

We investigate the stratification of sedimentary deposits in two Pliocene sections near the Caspian Sea (figure 1). These deposits will be referred to as sections A and B. Section A is located near the center of the Cheleken Peninsula, on the western edge of the Chokhrak Hills, and is typical of many petroliferous deposits. These deposits are part of the red-bed petroliferous deposits of Western Turkmenia. It is the middle of the upper horizons which are studied here, where they are dominated by argillaceous beds.

The section has been described by Vistelius and Romanova (1962, page 143). The argillaceous beds are brick and brown in color, with various intercalated sands and sandstones. Some ostracoda and (rarely) fossil leaves and tree trunks are found. Occasionally, the sandstones are cross-bedded and erosional surfaces may sometimes be seen between the base of the sandstone and the top of an argillaceous bed. There are also conglomeratic lenses containing shale pebbles. These features indicate deposition in shallow brackish water. In such an environment, it is common to find clastic materials as suspended sediment. Consequently, it is to be expected that sedimentation accompanied by gravitational stratification will occur; therefore it would seem that this is an appropriate site for applying our model. This section also satisfies the requirement that the beds should be clearly distinguishable with minimal numbers of fine intercalations; the deposit is also relatively homogeneous (that is, the

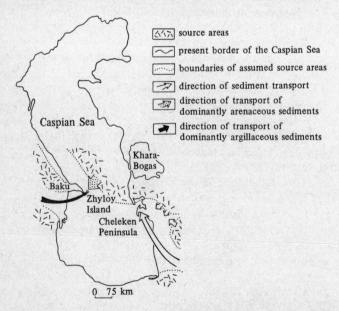

Figure 1. Location of the sequences investigated, and paleographic reconstruction for the time of sedimentation (middle Pliocene).

parameters of the process would appear to be stationary). The sequence of beds is given in table 1, where 's' indicates arenaceous beds and 'a' argillaceous beds. The one-step, two-step, and three-step transition matrices for the sequence of table 1 are provided in table 2.

Section B [2] is bound on Zhyloy Island in the Caspian Sea, off the eastern end of the Apsheron Peninsula (figure 1). The beds are part of the Balakhan suite of the productive series of the Apsheron Peninsula which is the main petroliferous formation of the Baku petroleum region (Vistelius, 1961). The productive series is about 2 126 m thick, dominated by arenaceous beds. The basal portion (lower division) is more argillaceous. It is separated by an erosional surface from the middle division, which consists predominately of conglomerates and coarse sandstones. At the base of the upper division are the beds of the Balakhan suite. These are fine-grained sands with thin argillaceous beds. About 55·8 m of the Balakhan suite were investigated. The observed sequence is given in table 3. The one-step, two-step, and three-step transition matrices of the sequence are given in table 4.

Table 1. The sequence of beds (from bottom to top) in section A, Chokhrak Hills, Cheleken Peninsula. Read table as normal script—left to right, top to bottom.

asasas	aassas	asasas	asasas	aassaa	assasa	sssaas	aasasa	asasaa	ssassa	aasaas	asasas
asssas	asaaaa	sasasa	saasas	asasas	asasas	asassa	ssasas	asasas	asasas	aaasas	asasas
asasas	aaasaa	asasas	asasas	aasasa	sasasa	asasas	asasas	aasasa	sasaas	saasas	sasssa
sasasa	sasasa	sasasa	sasass	asaasa	asasaa	sassaa	ssasas	aasaaa	sasasa	sasasa	ssssas
asasas	aaasaa	ssasas	asasas	asasas	asasas	asaasa	sasaas	asaasa	sasasa	saaasa	sasasa
ssasas	aasaas	asasss	ssasas	asaaas	asasas	asaaas	asasaa	asasas	assasa	sasasa	sasasa
sasasa	ssassa	aasasa	ssaaaa	asassa	saasas	ssasas	asasas	asa			

Table 2. The matrices of the transitional frequencies for the sequence of beds from table 1. The total number of beds, N, in section A is 483; h is an index of the position of the bed within the sequence and n_i is the number of transitions in the ith row of each matrix.

$h-2$	$h-1$	h		n_i		$h-3$	$h-2$	$h-1$	h		n_i
		s	a						s	a	
	s	34	196	230		s	s	s	3	6	9
	a	196	56	252		s	s	a	18	7	25
						s	a	s	18	138	156
s	s	9	25	34		s	a	a	25	14	39
s	a	156	39	195		a	s	s	6	19	25
a	s	25	171	196		a	s	a	138	32	170
a	a	39	17	56		a	a	s	7	32	39
						a	a	a	14	3	17

---

[2] The section was denoted 'O Zh' in the original paper (Vistelius, 1961, page 113).

Table 3. The sequence of beds (from bottom to top) in section B, Zhyloy Island, Caspean Sea. Read table as normal script—left to right, top to bottom.

asssas	asassa	sssssa	sassas	sasass	asasss	asassa	sassas	asasas	sasssa	sasass	sasssa
sasass	ssasas	sassas	sasass	assasa	ssasss	sasasa	sasssa	sassas	assssas	asasss	asasas
assass	assasa	ssasas	asssas	asasas	ssasss	asasas	asasss	assass	asasas	asasas	asasas
asasas	assass	asasas	asasas	sasass	ssasas	assasa	sasasa	sasasa	sasass	asssss	sssssa
sssasa	ssssas	asssas	asssas	sasasa	ssasas	asassa	ssasas	sasass	asasss	sassss	ssssas
asasas	asssas	asasas	asssas	asssss	asasas	asasas	asasas	asassa	sassas	asaass	asassa
asasas	asaasa	aaasss	ssasss	asasas	asasas	asassa	sssasa	sassas	ssasas	asasas	asssas
sasasa	sasass	asasas	asasas	asssas	asasas	asasas	asasas	asasas	assasa	sasass	ssssas
asassa	saassa	ssassa	ssasas	asassa	ssasas	ssasas	sssasa	sasasa	sasasa	sasasa	sassas
asasas	asasas	sssass	asasas	asasas	ssasas	sassas	asasas	asasas	sasass	sasasa	sassas
sasssa	saasaa	sasasa	sasasa	sasas							

Table 4. The matrices of the transitional frequencies for the sequence of beds from table 3. $N = 749$.

$h-2$	$h-1$	h s	h a	n_i		$h-3$	$h-2$	$h-1$	h s	h a	n_i
	s	161	286	447		s	s	s	28	38	66
	a	287	9	296		s	s	a	95	0	95
						s	a	s	91	187	278
s	s	66	95	161		s	a	a	5	1	6
s	a	279	7	286		a	s	s	38	57	95
a	s	95	191	286		a	s	a	184	7	191
a	a	6	2	8		a	a	s	3	3	6
						a	a	a	1	1	2

Mathematical techniques

To establish the precise nature of the models, certain ideas about runs are employed, as follows. We will refer to an uninterrupted sequence of precisely r beds of type α ($\alpha \in \{a, s\}$) as an α run of length r. Let $p_\alpha(r)$ be the probability of occurrence of an α run of length r. The sequence $p_\alpha(r)$, $r = 1, 2, ...$, is a discrete probability distribution with $\sum_{r=1}^{\infty} p_\alpha(r) = 1$. The complimentary event (negation of α) will be written as $\bar{\alpha}$. Each bed type must occur at least once in the sequence, so a run of length zero is not defined.

We consider the form of the distribution in the two cases where the bed transitions are partial-first-order and second-order Markov (see appendix 1). In both cases, it is presumed that the sequence is homogeneous. For a first-order Markov sequence, $p_\alpha(r)$ is a conditional probability, namely

$$p_\alpha(r) = P(\bar{\alpha}_{h+r}\alpha_{h+r-1} \ldots \alpha_{h+1} | \alpha_h \bar{\alpha}_{h-1}) ,$$

and, since the transition through α is first-order Markov, this is equivalent to the expression

$$p_\alpha(r) = P(\bar{\alpha}_{h+r}\alpha_{h+r-1} \ldots \alpha_{h+1}|\alpha_h) \,,$$

which can be restated (by using standard operations with conditional variables):

$$p_\alpha^{(r)} = \frac{p(\alpha_h \alpha_{h+1} \ldots \alpha_{h+r-1}\bar{\alpha}_{h+r})}{p(\alpha_h)} \,.$$

Owing to homogeneity of the chain, we can therefore write

$$\frac{p(\alpha_k)[P(\alpha_k|\alpha_{k-1})]^{r-1}P(\bar{\alpha}_k|\alpha_{k-1})}{p(\alpha_k)} = [P(\alpha_k|\alpha_{k-1})]^{r-1}P(\bar{\alpha}_k|\alpha_{k-1}) \,,$$

thus

$$p_\alpha(r) = [P(\alpha|\alpha)]^{r-1}P(\bar{\alpha}|\alpha) \,.$$

The above formulae show that the probability distribution is a one-parameter distribution, since $P(\bar{\alpha}|\alpha) = 1 - P(\alpha|\alpha)$.

If the sequence is second-order Markov, we have:

$$p_\alpha(r) = P(\bar{\alpha}_{h+r}\alpha_{h+r-1} \ldots \alpha_{h+1}|\alpha_h\bar{\alpha}_{h-1}) = \frac{p(\bar{\alpha}_{h+r}\alpha_{h+r-1} \ldots \alpha_{h+1}\alpha_h\bar{\alpha}_{h-1})}{p(\alpha_h\bar{\alpha}_{h-1})}$$

$$= p(\bar{\alpha}_k)P(\alpha_k|\bar{\alpha}_{k-1})P(\alpha_k|\alpha_{k-1}\bar{\alpha}_{k-2})[P(\alpha_k|\alpha_{k-1}\alpha_{k-2})]^{r-2}$$

$$\times \left[\frac{P(\bar{\alpha}_k|\alpha_{k-1}\alpha_{k-2})}{p(\bar{\alpha}_k)P(\alpha_k|\bar{\alpha}_{k-1})}\right]$$

$$= P(\alpha_k|\alpha_{k-1}\bar{\alpha}_{k-2})[P(\alpha_k|\alpha_{k-1}\alpha_{k-2})]^{r-2}P(\bar{\alpha}_k|\alpha_{k-1}\alpha_{k-2}) \,,$$

and, since the sequence is homogeneous, this simply reduces to

$$p_\alpha(r) = P(\alpha|\alpha\alpha)^{r-2}P(\bar{\alpha}|\alpha\alpha)P(\alpha|\alpha\bar{\alpha}) \,.$$

The formula above applies for $r \geqslant 2$. For $r = 1$, the formula is

$$p_\alpha(1) = P(\bar{\alpha}_{k+1}|\alpha_k\bar{\alpha}_{k-1}) \,,$$

which can be obtained by direct computation. Thus the distribution has two parameters: $P(\alpha|\alpha\alpha)$ and $P(\alpha|\alpha\bar{\alpha})$.

Evaluation of the parameters by minimizing χ^2 uses $(n-2)$ and $(n-3)$ degrees of freedom, where n is the number of classes into which the data are grouped. We use direct estimators of transitional probabilities, namely the observed frequencies; this does not guarantee a minimum in χ^2, thus the fit is worse than optimal. We test the hypotheses kth-order Markov versus $(k+1)$th-order Markov by comparing the observed and expected numbers of runs of length r, under the assumption that the process is ergodic. For this we use the χ^2 test.

We will now investigate how well various sets of packets can account for the observed sequences A and B. For example, if a sequence is generated using only packets (s), (a) and (sa), how closely will this sequence correspond to the observed sequence in section A? The set of packets being considered will be referred to as model {(s), (a), (sa)}, and so on. In accordance with axiom 2 of the gravitational stratification model, we require that any packet with s-beds and a-beds has the s-bed preceding (below) the a-bed. Each model may have one, two, or three packets, and each packet may contain one or two beds. It is apparent that these constraints lead to the simplest models capable of generating the observed sequences.

Within this framework, there are twenty-five [that is, $\binom{5}{1} + \binom{5}{2} + \binom{5}{3} = 25$] models available. In table 5, we list those models which can be rejected on the basis of obvious conflicts with the characteristics of the observed sections (tables 1 and 3).

Six models cannot be rejected in this manner, they are: {(s), (a)}; {(s), (a), (ss)}; {(s), (a), (aa)}; {(s), (a), (sa)}; {(s), (sa), (aa)}; and {(a), (ss), (sa)}. These are checked by comparing predicted and observed properties of the bed sequences (tables 6, 7, and 8). From table 6 we see that both sequences are best explained as second-order Markov chains. We can reject the model {(s), (a)} because it will produce only a first-order chain. We reject the model {(s), (a), (ss)} for section B since it will not

Table 5. Models being direct comparison of models with the observed sections, with reasons for rejection.

Model(s)	Reason for rejection
{(s)}, {(ss)}, {(s), (ss)}	beds of a exist
{(a)}, {(aa)}, {(a), (aa)}	beds of s exist
{(sa)}	runs with $r > 1$ exist
{(s), (sa)}, {(sa), (ss)}, {(s), (sa), (ss)}	runs of bed a with $r > 1$ exist
{(a), (sa)}, {(sa), (aa)}, {(a), (sa), (aa)}	runs of bed s with $r > 1$ exist
{(s), (aa)}, {(ss), (aa)}, {(s), (ss), (aa)}	runs of bed a with $r = 1$ exist
{(a), (ss)}, {(a), (aa), (ss)}	runs of bed s with $r = 1$ exist
{(sa), (aa), (ss)}	{ runs of beds with r even followed by runs of bed a with r odd exist

Table 6. Test of Markov hypotheses for bed sequences in the two sections studied here.

Hypothesis [a]	Degrees of freedom	χ^2 (5% level)	Λ_0 [b] section A	section B
$H_0 : H_1$	1	3·84	192·5	276·3
$H_1 : H_2$	2	5·99	7·14	16·34
$H_2 : H_3$	4	9·49	4·43	5·48

[a] $H_k : H_{k+1}$ represents the test of the null hypothesis kth-order Markov versus the alternative, $(k+1)$th-order.

[b] Λ_0 is the observed value of Λ, which is a likelihood-ratio statistic for the given test.

lead to aa pairs. Similarly, we reject the model {(s), (a), (aa)} for section A since it will not produce ss pairs. These considerations leave only four models to be investigated for each section. These will be studied separately.

Table 7. Test of homogeneity of sequences.

Assumed order of chain	Degrees of freedom	χ^2 (5% level)	χ_0^2 [a]	
			section A	section B
1	1	3·84	0·04	2·86
2	2	5·99	0·36	15·32
3	4	9·49	4·04	13·47

[a] χ_0^2 is the observed value of the χ^2 statistic for the tested hypothesis.

Table 8. Test of order of transitions through bed types s and a.

Alternative hypotheses [a]	Degrees of freedom	χ^2 (5% level)	χ_0^2 [b]			
			section A		section B	
			s	a	s	a
$H_1 : H_2$	1	3·84	4·40	2·74	2·69	13·65
$H_2 : H_3$	2	5·99	1·42	3·01	0·89	4·54

[a] $H_k : H_{k+1}$ represents the test of the null hypothesis kth-order Markov versus the alternative, $(k+1)$th-order.
[b] χ_0^2 is the observed value of the χ^2 statistic for the tested hypothesis.

Models of section A
The model {(s), (a), (sa)}
Mathematical formulation On the basis of axiom 4, we know that all the information about the model is contained in the matrix of transitional probabilities, $\mathbf{P}$, and the stationary distribution vector, p, which are given below.

$$
\begin{array}{c}
\quad\quad (s)\ \ (a)\ \ (sa) \\
\mathbf{P} = \begin{array}{c}(s)\\(a)\\(sa)\end{array} \begin{bmatrix} p_{11} & p_{12} & 1-p_{11}-p_{12} \\ p_{21} & p_{22} & 1-p_{21}-p_{22} \\ p_{21} & p_{22} & 1-p_{21}-p_{22} \end{bmatrix} ; \quad p = \begin{bmatrix} p[(s)] \\ p[(a)] \\ p[(sa)] \end{bmatrix}.
\end{array} \tag{1}
$$

In general, the sequence of packets is not reversible. We assume that the compositions of sediment in the packet (a) and in the a-bed of the (sa) packet are indistinguishable. Thus, these compose the subset of renewal events implied in the axiomatic model. The rows of transitional probabilities for (a) and (sa) in $\mathbf{P}$ are therefore identical.

According to the general model of sedimentation, the sequence of *packets* can be either Bernoullian or simple Markov. If the sequence is Bernoullian, the matrix of transitional probabilities and the stationary vector will be as follows

$$
\begin{array}{cc}
& \begin{array}{ccc} \text{(s)} & \text{(a)} & \text{(sa)} \end{array} \\
\begin{array}{c} \text{(s)} \\ \mathbf{P} = \text{(a)} \\ \text{(sa)} \end{array} &
\left[
\begin{array}{ccc}
p_{11} & p_{12} & 1 - p_{11} - p_{12} \\
p_{11} & p_{12} & 1 - p_{11} - p_{12} \\
p_{11} & p_{12} & 1 - p_{11} - p_{12}
\end{array}
\right]
\end{array}
\quad ; \quad
p = \left[
\begin{array}{c}
p_{11} \\
p_{12} \\
1 - p_{11} - p_{12}
\end{array}
\right] . \tag{2}
$$

In either case the sequence of *beds* will be a homogeneous Markov chain, as follows from the general model. We will study random sequences of states 's' and 'a' which are referred to as *the beds in the packets*. Here, for example, $s^{(1)}$ is the bed from the packet (s); $s^{(2)}$ is from (sa), $a^{(1)}$ is from (a); and $a^{(2)}$ is from (sa). Properties of the random sequences of the beds in the packets are described in lemma 3 (appendix 2). The observed sequences of beds are obtained by combinations

$$ s = s^{(1)} \cup s^{(2)}, \qquad \text{and} \quad a = a^{(1)} \cup a^{(2)} . $$

For the case of a Bernoullian sequence, the matrix of transitional probabilities for beds in the packets, $\mathbf{P}'$, is given by

$$
\begin{array}{cc}
& \begin{array}{cccc} s^{(1)} & s^{(2)} & \quad a^{(1)} & a^{(2)} \end{array} \\
\begin{array}{c} s^{(1)} \\ s^{(2)} \\ \mathbf{P}' = \\ a^{(1)} \\ a^{(2)} \end{array} &
\left[
\begin{array}{cc:cc}
p_{11} & 1 - p_{11} - p_{12} & p_{12} & 0 \\
0 & 0 & 0 & 1 \\
\hdashline
p_{11} & 1 - p_{11} - p_{12} & p_{12} & 0 \\
p_{11} & 1 - p_{11} - p_{12} & p_{12} & 0
\end{array}
\right]
\end{array} . \tag{3}
$$

Inspection of matrix (3) shows that the transition through 'a' will be simple Markov (see appendix 2). The Markov order for transitions through 's' can be obtained from the matrix for the reversed sequence of the beds in the packets. This matrix can be obtained through the vector of stationary probabilities for the states. From matrix (3), and use of the well-known formula

$$ \bar{P}(\beta|\alpha) = P(\alpha|\beta) \frac{p(\beta)}{p(\alpha)} , $$

where the bar ($^-$) indicates transitional probability in the reversed sequence, we obtain:

$$ p(s^{(1)}) = \frac{p_{11}}{2 - p_{11} - p_{12}} , \qquad p(s^{(2)}) = p(a^{(2)}) = \frac{1 - p_{11} - p_{12}}{2 - p_{11} - p_{12}} ; $$

$$ p(a^{(1)}) = \frac{p_{12}}{2 - p_{11} - p_{12}} . $$

Matrix (3) then gives the following matrix for the reversed sequence

$$
\bar{\mathbf{P}}' = \begin{array}{c} \\ s^{(1)} \\ s^{(2)} \\ a^{(1)} \\ a^{(2)} \end{array}
\begin{array}{cccc} s^{(1)} \quad s^{(2)} \quad\; a^{(1)} \quad a^{(2)} \\ \left[\begin{array}{cc:cc} p_{11} & 0 & p_{12} & 1-p_{11}-p_{12} \\ p_{11} & 0 & p_{12} & 1-p_{11}-p_{12} \\ \hdashline p_{11} & 0 & p_{12} & 1-p_{11}-p_{12} \\ 0 & 1 & 0 & 0 \end{array}\right] \end{array} .
\tag{4}
$$

Theorem 1 (appendix 2) shows that matrices (3) and (4) describe a process in which the sequence of beds is a simple Markov chain. However, the analysis summarized in table 6 allows us to conclude that the sequence is a second-order Markov chain. Thus it appears that the matrix given in equation (1) is the appropriate one for this analysis.

The theoretical study of the properties of the sequence of beds in section A, based on the general model, leads us to the following theorem.

If section A is generated according to the model $\{(s), (a), (sa)\}$, then the sequence of beds in this section is a homogeneous reversible Markov chain of the second order, with the partial transition through 'a' being first-order Markov and the partial transitions through 's' being second-order Markov.

Proof: The matrix of transitional probabilities for the beds in the packets is as follows:

$$
\mathbf{P}' = \begin{array}{c} \\ s^{(1)} \\ s^{(2)} \\ a^{(1)} \\ a^{(2)} \end{array}
\begin{array}{cccc} s^{(1)} \quad\; s^{(2)} \qquad\quad a^{(1)} \quad a^{(2)} \\ \left[\begin{array}{cc:cc} p_{11} & 1-p_{11}-p_{12} & p_{12} & 0 \\ 0 & 0 & 0 & 1 \\ p_{21} & 1-p_{21}-p_{22} & p_{22} & 0 \\ p_{21} & 1-p_{21}-p_{22} & p_{22} & 0 \end{array}\right] \end{array} .
\tag{5}
$$

We express the vector of the stationary distribution corresponding to matrix (5) through the component of the vector corresponding to the matrix in equation (1). Thus

$$
p(s^{(1)}) = \frac{p(s)}{p(s) + p(a) + 2p(sa)}; \qquad p(s^{(2)}) = p(a^{(2)}) = \frac{p(sa)}{p(s) + p(a) + 2p(sa)};
$$

$$
p(a^{(1)}) = \frac{p(a)}{p(s) + p(a) + 2p(sa)} .
$$

Lemma 2 (appendix 2) indicates that the transition through 'a' is simple Markov, as is clear from matrix (5).

We now show that transition through 's' in the sequence of section A is not simple Markov. For this purpose, we obtain the matrix of transitional probabilities for the beds in the packets for the reversed sequence:

$$
\bar{\mathbf{P}}' =
\begin{array}{c}
\\
s^{(1)} \\
s^{(2)} \\
a^{(1)} \\
a^{(2)}
\end{array}
\overset{\begin{array}{cccc} s^{(1)} & \quad s^{(2)} & a^{(1)} & \quad\quad a^{(2)} \end{array}}
{\begin{bmatrix}
p_{11} & 0 & p_{21}\dfrac{p[(a)]}{p[(s)]} & p_{21}\dfrac{p[(sa)]}{p[(s)]} \\[2mm]
(1-p_{11}-p_{12})\dfrac{p[(s)]}{p[(sa)]} & 0 & (1-p_{21}-p_{22})\dfrac{p[(a)]}{p[(sa)]} & 1-p_{21}-p_{22} \\[2mm]
p_{12}\dfrac{p[(s)]}{p[(a)]} & 0 & p_{22} & p_{22}\dfrac{p[(sa)]}{p[(a)]} \\[2mm]
0 & 1 & 0 & 0
\end{bmatrix}}
. \tag{6}
$$

If the transition through 's' were simple Markov, lemma 2 (appendix 2) would require that

$$
\frac{p_{11}}{1-p_{11}-p_{12}} = \frac{p_{21}}{1-p_{21}-p_{22}} = \frac{p[(s)]}{p[(sa)]} \, ;
$$

from this we see that

$$
\frac{p_{11}}{1-p_{11}-p_{12}} = \frac{p_{21}}{1-p_{21}-p_{22}} \, , \qquad \text{or} \quad \frac{p_{11}}{p_{21}} = \frac{1-p_{12}}{1-p_{22}} \, . \tag{7}
$$

Thus the transition through 's' is simple Markov if and only if the parameters satisfy equation (7). In all other cases, the transition through 's' is not simple Markov and we must determine the Markov order by other methods.

Let us now prove that the transition through 's' (and so the whole chain) is second-order Markov for arbitrary values of the parameters. If this is true, the matrix of transitional probabilities for the sequence of beds will be

$$
\begin{array}{c}
\\
A_{h-2}, s_{h-1}, s_h \\
A_{h-2}, s_{h-1}, a_h \\
A_{h-2}, a_{h-1}, s_h \\
A_{h-2}, a_{h-1}, a_h
\end{array}
\overset{\begin{array}{cc} s_{h+1} & \quad\quad\quad a_{h+1} \end{array}}
{\begin{bmatrix}
P(s_{h+1}|s_h s_{h-1}) & P(a_{h+1}|s_h s_{h-1}) \\[1mm]
P(s_{h+1}|a_h s_{h-1}) & P(a_{h+1}|a_h s_{h-1}) \\[1mm]
P(s_{h+1}|s_h a_{h-1}) & P(a_{h+1}|s_h a_{h-1}) \\[1mm]
P(s_{h+1}|a_h a_{h-1}) & P(a_{h+1}|a_h a_{h-1})
\end{bmatrix}}
, \tag{8}
$$

where A_{h-2} is an arbitrary event dependent only on the sequence of beds up to $(h-2)$. This has been shown [see matrix (5)] for the rows of the above matrix which end in 'a'.

For the transitions (ss) to (ss) and (ss) to (a) the following two equalities must hold:

$$
P(s_{h+1}|s_h s_{h-1} A_{h-2}) = P(s_{h+1}|s_h s_{h-1}) \, , \tag{9}
$$

and

$$
P(a_{h+1}|s_h s_{h-1} A_{h-2}) = P(a_{h+1}|s_h s_{h-1}) \, . \tag{10}
$$

It is sufficient to prove only one of the above, since these equalities must jointly sum to one. If one equality is independent of A_{h-2}, the second must also be independent of A_{h-2}. Similarly, it is sufficient to prove only one of the following equalities

$$P(s_{h+1}|s_h a_{h-1} A_{h-2}) = P(s_{h+1}|s_h a_{h-1}) , \qquad (11)$$

and

$$P(a_{h+1}|s_h a_{h-1} A_{h-2}) = P(a_{h+1}|s_h a_{h-1}) , \qquad (12)$$

First, we prove equation (10). Expanding the left-hand side, we obtain

$$P(a_{h+1}|s_h s_{h-1} A_{h-2}) = P[a_{h+1}^{(1)} \cup a_{h+1}^{(2)}|(s_h^{(1)} \cup s_h^{(2)})(s_{h-1}^{(1)} \cup s_{h-1}^{(2)})A_{h-2}] . \qquad (13)$$

On the bases of the definition of conditional probability, viz

$$P(B|C) = \frac{p(BC)}{p(C)} , \qquad p(C) > 0 ,$$

and of incompatibility of the events, the right-hand side of equation (12) can be rewritten as

$$[p(A_{h-2} s_{h-1}^{(1)} s_h^{(1)} a_{h+1}^{(1)}) + p(A_{h-2} s_{h-1}^{(2)} s_h^{(1)} a_{h+1}^{(1)}) + p(A_{h-2} s_{h-1}^{(1)} s_h^{(2)} a_{h+1}^{(1)})$$
$$+ p(A_{h-2} s_{h-1}^{(2)} s_h^{(2)} a_{h+1}^{(1)}) + p(A_{h-2} s_{h-1}^{(1)} s_h^{(1)} a_{h+1}^{(2)}) + p(A_{h-2} s_{h-1}^{(2)} s_h^{(1)} a_{h+1}^{(2)})$$
$$+ p(A_{h-2} s_{h-1}^{(1)} s_h^{(2)} a_{h+1}^{(2)}) + p(A_{h-2} s_{h-1}^{(2)} s_h^{(2)} a_{h+1}^{(2)})] / [p(A_{h-2} s_{h-1}^{(1)} s_h^{(1)})$$
$$+ p(A_{h-2} s_{h-1}^{(2)} s_h^{(1)}) + p(A_{h-2} s_{h-1}^{(1)} s_h^{(2)}) + p(A_{h-2} s_{h-1}^{(2)} s_h^{(2)})] .$$

From lemma 3 (appendix 2) we know that the sequence of beds chosen from the set $\{(s^{(1)}, s^{(2)}, a^{(1)}, a^{(2)}\}$ is a simple, homogeneous Markov chain with matrix of transitional probabilities corresponding to matrix (5). This permits us to rewrite the term above as

$$\frac{p(A_{h-2})P(s^{(1)}|A_{h-2})P(s^{(1)}|s^{(1)})P(a^{(1)}|s^{(1)}) + p(A_{h-2})P(s^{(1)}|A_{h-2})P(s^{(2)}|s^{(1)})}{p(A_{h-2})P(s^{(1)}|A_{h-2})P(s^{(1)}|s^{(1)}) + p(A_{h-2})P(s^{(1)}|A_{h-2})P(s^{(2)}|s^{(1)})} \qquad (14)$$

The separate terms in expression (14) are computed as follows. We know that

$$p(A_{h-2})P(s_{h-1}^{(2)}|A_{h-2})P(s_h^{(2)}|s_{h-1}^{(2)})P(a_{h+1}^{(2)}|s_h^{(2)}) = p(A_{h-2} s_{h-1}^{(2)} s_h^{(2)} a_{h+1}^{(2)}) = 0$$

since $P(s_h^{(2)}|s_{h-1}^{(2)}) = 0$ from matrix (5). Similarly, the term

$$p(A_{h-2} s_{h-1}^{(1)} s_h^{(2)} a_{h+1}^{(2)}) = p(A_{h-2})P(s_{h-1}^{(1)}|A_{h-2})P(s_h^{(2)}|s_{h-1}^{(1)})P(a_{h+1}^{(2)}|s_h^{(2)})$$

can be reduced, by matrix (5), to

$$p(A_{h-2} s_{h-1}^{(1)} s_h^{(2)} a_{h+1}^{(2)}) = p(A_{h-1})P(s_{h-1}^{(1)}|A_{h-2})P(s_h^{(2)}|s_{h-1}^{(1)}) ; \qquad (15)$$

since the sequence is homogeneous, the right-hand side of expression (15) simplifies to

$$p(A)P(s^{(1)}|A)P(s^{(2)}|s^{(1)}) .$$

Reducing and factoring term (14), we therefore obtain

$$\frac{p(A_{h-2})P(s^{(1)}|A_{h-2})[P(s^{(1)}|s^{(1)})P(a^{(1)}|s^{(1)}) + P(s^{(2)}|s^{(1)})]}{p(A_{h-2})P(s^{(1)}|A_{h-2})[P(s^{(1)}|s^{(1)}) + P(s^{(2)}|s^{(1)})]}$$

$$= \frac{P(s^{(1)}|s^{(1)})P(a^{(1)}|s^{(1)}) + P(s^{(2)}|s^{(1)})}{P(s^{(1)}|s^{(1)}) + P(s^{(2)}|s^{(1)})} . \tag{16}$$

Finally, we insert the appropriate values from matrix (5) into term (16), so that equation (10) now reduces to

$$P(a_{h+1}|s_h s_{h-1}) = P(a_{h+1}|s_h s_{h-1}) = \frac{p_{11}p_{12} + 1 - p_{11} - p_{12}}{p_{11} + 1 - p_{11} - p_{12}} = 1 - p_{11} \tag{17}$$

and similarly

$$P(a_h|s_{h-1} s_{h-2}) = 1 - p_{11} .$$

Thus we have shown that the transition through $s_h s_{h-1}$ is second-order Markov. Similar computations for equation (12) show that

$$P(a_h|s_{h-1} a_{h-2}) = P(a_h|s_{h-1} a_{h-2} A_{h-3})$$

$$= \frac{p_{21}p_{12} + 1 - p_{21} - p_{22}}{p_{21} + 1 - p_{21} - p_{22}} = 1 - \frac{p_{21}(1 - p_{12})}{1 - p_{22}} , \tag{18}$$

so the transition through $s_h a_{h-1}$ is also second-order Markov.

It will now be proved that the sequence of beds is reversible. Since (axiomatically) the initial and stationary distributions coincide, the reversibility of the chain (if we deal with the second-order Markov chain) depends only on the matrices of transitional frequencies, $N_{\alpha\beta}$ and $N_{\alpha\beta\gamma}$, as expressed by the numbers of observations per cell. By the definition of reversibility it follows that

$$p(\alpha_{h+1}\beta_h) = p(\alpha_h\beta_{h+1}) ; \qquad p(\alpha_h\beta_{h+1}\gamma_{h+2}) = p(\gamma_h\beta_{h+1}\alpha_{h+2}) ;$$

for all $\alpha, \beta, \gamma \in \{s, a\}$, and for arbitrary integers h. If these two equalities hold for a stationary, homogeneous, second-order Markov chain with two states, then it is reversible. These equalities should now be proved correct.

The following matrix indicates the probabilities of the various combinations of three events:

$$
\begin{array}{c c}
 & \begin{array}{cc} s_{h+1} & a_{h+1} \end{array} \\
\begin{array}{c} s_{h-1}s_h \\ s_{h-1}a_h \\ a_{h-1}s_h \\ a_{h-1}a_h \end{array} &
\left[\begin{array}{cc} p(sss) & p(ssa) \\ p(sas) & p(saa) \\ p(ass) & p(asa) \\ p(aas) & p(aaa) \end{array}\right] .
\end{array}
\tag{19}
$$

For each element of matrix (19), it is necessary to prove that

$$p(\alpha_{h-1}\beta_h\gamma_{h+1}) = p(\gamma_{h-1}\beta_h\alpha_{h+1}) .$$

This is trivial for the probabilities: p(sss), p(sas), p(asa), and p(aaa). We shall prove that

$$p(s_{h-1} s_h a_{h+1}) = p(a_{h-1} s_h s_{h+1}) .$$

We can write

$$p(s_{h-1} s_h) = p(s_{h-1} s_h s_{h+1}) + p(s_{h-1} s_h a_{h+1}) ,$$

and

$$p(s_h s_{h+1}) = p(s_{h-1} s_h s_{h+1}) + p(a_{h-1} s_h s_{h+1}) ;$$

since the chain is homogeneous,

$$p(s_{h-1} s_h) = p(s_h s_{h+1}) ; \quad \text{thus} \quad p(s_{h-1} s_h a_{h+1}) = p(a_{h-1} s_h s_{h+1}) .$$

The proof is similar for the remaining pair of cells. As the sequence with probability matrix (19) is reversible, then the sequence with matrix of transitional probabilities:

$$
\begin{array}{c|cc}
 & s_{h+1} & a_{h+1} \\
\hline
s_{h-1} s_h & P(s|ss) & P(a|ss) \\
s_{h-1} a_h & P(s|as) & P(a|as) \\
s_{h-1} s_h & P(s|sa) & P(a|sa) \\
a_{h-1} a_h & P(s|aa) & P(a|aa)
\end{array} ,
$$

is also a reversible sequence.

The reversibility of a 2×2 matrix can be proved in a similar way. Thus a homogeneous second-order Markov chain with two states and with stationary initial probability vector is always reversible, and the theorem stated on page 151 is proved. We will now compare the properties of the observed sequence of beds in section A with the properties of the theoretically obtained sequence.

Comparison of the model with data for section A It is clear from table 7 that the sequence of beds in section A does not contradict the assumption of a homogeneous sequence. Inspection of table 6 shows that it is a second-order Markov chain[3]. Table 8 shows that the transition through 'a' is simple Markov, and that the transition through 's' is second-order Markov. Summarizing the results of our investigation of the model {(s), (a), (sa)}, we see that this model does indeed correspond to the observed sequence of beds in section A. Let us check some additional properties. Table 9 provides statistical proof of the second-order dependence structure for each of the two possible transitions through 's'. Tables 8 and 9 show that in section A there are no transitions of higher than second order. It is also possible to check the order of dependence

[3] It must be emphasized again that it is only possible to check the existence of Markov *characteristics* (see appendix 1).

by using the ideas about runs developed earlier. Table 10 lists the observed and expected numbers of runs as a function of run length. The goodness of fit is tested by a χ^2 procedure.

We have shown that the model $\{(s), (a), (sa)\}$ explains the properties of the bed intercalations in section A. It is now necessary to estimate the parameters of the model. The parameters are the transitional probabilities in the matrix:

$$
\begin{array}{c}
\quad\ (s)\ \ (a)\ \ (sa) \\
\begin{array}{c}(s)\\(a)\\(sa)\end{array}
\begin{bmatrix}
p_{11} & p_{12} & 1-p_{11}-p_{12} \\
p_{21} & p_{22} & 1-p_{21}-p_{22} \\
p_{21} & p_{22} & 1-p_{21}-p_{22}
\end{bmatrix},
\end{array}
\tag{20}
$$

along with the corresponding vector of stationary probabilities. This vector is completely determined by matrix (20).

We estimate the parameters by computing the transitional probabilities in the matrices $\mathbf{P}_{\alpha\beta}$ and $\mathbf{P}_{\alpha\beta\gamma}$ for the beds in the section, and equating these with the elements of the observed matrix. Lumping the states in the matrix

Table 9. Test of the second-order Markov property[a] of the possible transitions through 's' in section A.

Transition	Observed value[b] of χ^2
$s_{h-1}s_h$	0·29
$a_{h-1}s_h$	1·15

[a] Alternative hypothesis is third-order Markov.
[b] One degree of freedom; $\chi^2_{0.05} = 3\cdot84$.

Table 10. Distributions of lengths of 's' and 'a' runs in section A.

Runs of a-beds				Runs of s-beds			
length	number observed (O)	number expected (E)	$\dfrac{(O-E)^2}{E}$	length	number observed (O)	number expected (E)	$\dfrac{(O-E)^2}{E}$
1	155	150·4	0·14	1	171	170·7	0·00
2	25	33·8	2·29	2	19	18·6	0·01
3	12	7·6	2·55	3	4	4·9	0·17
4	1	1·7	0·29	4	1	1·3	0·07
⩾5	1	0·6	0·27	⩾5	1	0·47	0·60
Total	194	194·1	5·54[a]	Total	196	195·97	0·85[b]

[a] $0\cdot10 < p_a\,(\chi^2 > 5\cdot54) < 0\cdot25$; three degrees of freedom.
[b] $0\cdot50 < p_s\,(\chi^2 > 0\cdot85) < 0\cdot75$; two degrees of freedom.

for the beds in the packets, as shown by the dotted lines in the matrix and vector.

$$\mathbf{P}' = \begin{array}{c} \\ s^{(1)} \\ s^{(2)} \\ a^{(1)} \\ a^{(2)} \end{array} \overset{\displaystyle s^{(1)} \quad s^{(2)} \qquad\qquad a^{(1)} \quad a^{(2)}}{\left[\begin{array}{cc:cc} p_{11} & 1-p_{11}-p_{12} & p_{12} & 0 \\ 0 & 0 & 0 & 1 \\ \hdashline p_{21} & 1-p_{21}-p_{22} & p_{22} & 0 \\ p_{21} & 1-p_{21}-p_{22} & p_{22} & 0 \end{array}\right]} , \qquad p' = \begin{bmatrix} p(s^{(1)}) \\ p(s^{(2)}) \\ \hdashline p(a^{(1)}) \\ p(a^{(2)}) \end{bmatrix} , \qquad (21)$$

we obtain the matrices below

$$\mathbf{P}_{\alpha\beta} = \begin{array}{c} \\ s \\ a \end{array} \overset{\displaystyle \quad s \qquad\qquad\qquad a}{\left[\begin{array}{cc} \dfrac{p(s^{(1)})(1-p_{12})}{p(s)} & 1-\dfrac{p(s^{(1)})(1-p_{12})}{p(s)} \\ 1-p_{22} & p_{22} \end{array}\right]} ; \qquad (22)$$

$$\mathbf{P}_{\alpha\beta\gamma} = \begin{array}{c} \\ s_{h-1}\,s_h \\ s_{h-1}\,a_h \\ a_{h-1}\,s_h \\ a_{h-1}\,a_h \end{array} \overset{\displaystyle \quad s_{h+1} \qquad\qquad a_{h+1}}{\left[\begin{array}{cc} p_{11} & 1-p_{11} \\ 1-p_{22} & p_{22} \\ \dfrac{p_{21}(1-p_{12})}{1-p_{22}} & 1-\dfrac{p_{21}(1-p_{12})}{1-p_{22}} \\ 1-p_{22} & p_{22} \end{array}\right]} . \qquad (23)$$

All of the information about the probabilistic behavior of the beds in the section is contained, according to the model, in matrices (22) and (23). The estimators for the components of the vector of stationary probabilities in equation (21) can be found from the parameters p_{ij}, by use of the equations below

$$\left.\begin{aligned} p(s^{(1)}) &= \frac{p_{21}}{(2-p_{22})(1-p_{11})+p_{21}(1-p_{12})} ; \\[2mm] p(s^{(2)}) &= \frac{1-p_{11}p_{22}-p_{12}p_{21}-p_{11}-p_{22}}{(2-p_{22})(1-p_{11})+p_{21}(1-p_{12})} ; \\[2mm] p(a^{(1)}) &= \frac{p_{22}+p_{12}-p_{21}-p_{11}p_{22}}{(2-p_{22})(1-p_{11})+p_{21}(1-p_{12})} ; \\[2mm] p(a^{(2)}) &= p(s^{(2)}) . \end{aligned}\right\} \qquad (24)$$

In this case, the parameters are p_{11}, p_{12}, p_{21}, and p_{22}. We estimate these parameters from the frequencies in the observed section. Thus $\bar{P}(a|a)$ is the observed random value which we accept as the estimator[4] of the parameter p_{22}.

[4] Estimators will be marked ($\check{}$).

Equating the components of matrices (22) and (23) to the observed values gives the equalities

$$\tilde{p}_{22} = \tilde{P}(a|a) , \qquad \tilde{p}_{11} = \tilde{P}(s|sa) , \tag{25}$$

and the system

$$\left. \begin{array}{l} \dfrac{\tilde{p}_{21}(1-\tilde{p}_{12})}{1-\tilde{p}_{22}} = \tilde{P}(s_{h+1}|s_h a_{h-1}) , \\[3mm] \dfrac{\tilde{p}(s^{(1)})(1-\tilde{p}_{12})}{\tilde{p}(s)} = \tilde{P}(s|s) . \end{array} \right\} \tag{26}$$

By combining equations (25) with the value for $p(s^{(1)})$ from equations (24) in equations (26), we obtain

$$\left. \begin{array}{l} \tilde{p}_{21}(1-\tilde{p}_{12}) = \tilde{P}(s|a)\tilde{P}(s_{h+1}|s_h a_{h-1}) , \\[3mm] \dfrac{\tilde{p}_{21}(1-\tilde{p}_{12})}{[2-\tilde{P}(a|a)][1-\tilde{P}(s_{h+1}|s_h s_{h-1})] + \tilde{p}_{21}(1-\tilde{p}_{12})} = \tilde{p}(s)\tilde{P}(s|s) . \end{array} \right\} \tag{27}$$

Substitution of the right-hand side of the first of the above equations for the numerator on the left-hand side of the second results, after transformations, in the equation:

$$\frac{\tilde{P}(s|a)\tilde{P}(s_h|s_{h-1}a_{h-2})}{[2-\tilde{P}(a|a)][1-\tilde{P}(s_h|s_{h-1}s_{h-2})] + \tilde{P}(s|a)\tilde{P}(s_h|s_{h-1}a_{h-2})} = \tilde{p}(s)\tilde{P}(s|s) .$$

We replace the term $2-\tilde{P}(a|a)$ by $1+\tilde{P}(s|a)$, and $1-\tilde{P}(s_h|s_{h-1}s_{h-2})$ by $\tilde{P}(a_h|s_{h-1}s_{h-2})$, to give:

$$\frac{\tilde{P}(s|a)\tilde{P}(s_h|s_{h-1}a_{h-2})}{[1+\tilde{P}(s|a)]\tilde{P}(a_h|s_{h-1}s_{h-2}) + \tilde{P}(s|a)\tilde{P}(s_h|s_{h-1}a_{h-2})} = \tilde{p}(s)\tilde{P}(s|s) ,$$

or equivalently,

$$\tilde{P}(s|a)\tilde{P}(s_h|s_{h-1}a_{h-2}) = \tilde{P}(a_h|s_{h-1}s_{h-2})\tilde{p}(s)\tilde{P}(s|s)$$
$$+ \tilde{P}(s|a)\tilde{P}(a_h|s_{h-1}s_{h-2})\tilde{p}(s)\tilde{P}(s|s) + \tilde{p}(s)\tilde{P}(s|s)\tilde{P}(s|a)\tilde{P}(s_h|s_{h-1}a_{h-2}) \tag{28}$$

The homogeneity and reversibility of the sequence of beds in the section permit us to write down the probability of joint appearance of the beds not indicated, for example:

$$\tilde{p}(s_{h-1}s_h a_{h+1}) = \tilde{p}(a_{h+k-1}s_{h+k}s_{h+k+1}) = \tilde{p}(ass) = \tilde{p}(ssa) ,$$

although this expression does not hold for $\tilde{p}(sas)$.

Upon substituting the probabilities of joint realization of events for the conditional probabilities in equation (28), we obtain

$$\frac{\tilde{p}(ssa)}{\tilde{p}(a)} = \tilde{p}(ssa) + \frac{\tilde{p}(ssa)\tilde{p}(as)}{\tilde{p}(a)} + \frac{\tilde{p}(ss)\tilde{p}(ass)}{\tilde{p}(a)} ,$$

or

$$\frac{1}{\tilde{p}(a)} = 1 + \frac{\tilde{p}(as)}{\tilde{p}(a)} + \frac{\tilde{p}(ss)}{\tilde{p}(a)} \ ,$$

or

$$\tilde{p}(a) + \tilde{p}(as) + \tilde{p}(ss) = 1 \ ;$$

and noting that $\tilde{p}(as) + \tilde{p}(ss) = \tilde{p}(s)$, we obtain the obvious identity

$$\tilde{p}(a) + \tilde{p}(s) = 1 \ .$$

Thus we have only one equation connecting two estimators. Let us accept $\tilde{p}_{12}$ as a free parameter; thus,

$$\tilde{p}_{21} = \frac{\tilde{P}(s|a)\tilde{P}(s_h|s_{h-1}a_{h-2})}{1 - \tilde{p}_{12}} \ .$$

Application of the data from table 2 yields the transitional probability matrix, $\tilde{P}$, and initial probability vector, $\tilde{p}_0$, as follows

$$\tilde{P} = \begin{array}{c} \\ (s) \\ (a) \\ (sa) \end{array} \begin{array}{ccc} (s) & (a) & (sa) \\ \left[\begin{array}{ccc} 0\cdot265 & \tilde{p}_{12} & 0\cdot735 - \tilde{p}_{12} \\ \dfrac{0\cdot096}{1-\tilde{p}_{12}} & 0\cdot223 & 0\cdot777 - \dfrac{0\cdot096}{1-\tilde{p}_{12}} \\ \dfrac{0\cdot096}{1-\tilde{p}_{12}} & 0\cdot223 & 0\cdot777 - \dfrac{0\cdot096}{1-\tilde{p}_{12}} \end{array} \right] \end{array} \ . \tag{29}$$

$$\tilde{p}_0 = \begin{bmatrix} \tilde{p}[(s)] \\ \tilde{p}[(a)] \\ \tilde{p}[(sa)] \end{bmatrix} = \begin{bmatrix} \dfrac{0\cdot1307}{1\cdot1307 - \tilde{p}_{12}} \\ \dfrac{0\cdot2230 - 0\cdot0924\tilde{p}_{12}}{1\cdot1307 - \tilde{p}_{12}} \\ \dfrac{0\cdot7770 - 0\cdot9076\tilde{p}_{12}}{1\cdot1307 - \tilde{p}_{12}} \end{bmatrix} \ . \tag{30}$$

Since $\tilde{p}_{12}$ is a free parameter, the components of $\tilde{p}_0$ are not determined. The values of $\tilde{p}_{12}$ are determined by those of the stationary probabilities and, it is easy to show, are included in the segment $0 \leqslant \tilde{p}_{12} \leqslant 0\cdot735$. Inserting the end values of this range into equation (30) gives the following possible values of the various probabilities

$$0\cdot12 \leqslant p[(s)] \leqslant 0\cdot33 \ ; \qquad 0\cdot20 \leqslant p[(a)] \leqslant 0\cdot39 \ ;$$

and

$$0\cdot29 \leqslant p[(sa)] \leqslant 0\cdot69 \ .$$

The general picture of the interrelations among the probabilities of the packets and the values of $\tilde{p}_{12}$ is illustrated in table 11.

The analysis shows that the most common component of section A should be the packet (sa), with the (s) and (a) packets much more rare. The (a) packet should be more common than the (s) packet. Moreover, the transition from (a) and (sa) is most frequently to (sa). Finally, the transition from (s) to (s) is more frequent than that from (a) to (a).

The other four models mentioned at the beginning of this section (page 149) must now be examined. We will show that each of these models fails to predict adequately the observed sequence in section A.

The model {(s), (a), (aa)} should be rejected because the transition through 's' in the lumped sequence in this model would be simple Markov, which contradicts the results presented in tables 6, 8, and 10.

Table 11. Values of the initial probability vector, P_0, for various values of $\tilde{p}_{12}$.

$\tilde{p}_{12}$								
0·00	0·10	0·20	0·30	0·40	0·50	0·60	0·70	0·73
$\tilde{p}[(s)]$ 0·12	0·12	0·14	0·16	0·18	0·21	0·25	0·30	0·32
$\tilde{p}[(a)]$ 0·19	0·22	0·23	0·23	0·25	0·28	0·32	0·37	0·39
$\tilde{p}[(sa)]$ 0·69	0·66	0·63	0·61	0·57	0·51	0·43	0·33	0·29

The model {(s), (sa), (aa)}

This model requires special study. In this case the matrix of transitional probabilities for the packets will be

$$
\begin{array}{c}
\quad\;\; \text{(s)} \quad\; \text{(sa)} \quad\; \text{(aa)} \\
\mathbf{P} = \begin{array}{c} \text{(s)} \\ \text{(sa)} \\ \text{(aa)} \end{array}
\begin{bmatrix}
p_{11} & p_{12} & 1-p_{11}-p_{12} \\
p_{21} & p_{22} & 1-p_{21}-p_{22} \\
p_{21} & p_{22} & 1-p_{21}-p_{22}
\end{bmatrix} ;
\end{array}
$$

and for the beds in the packets, the matrix and corresponding stationary probability vector will be

$$
\begin{array}{c}
\quad\;\; s^{(1)} \;\; s^{(2)} \;\; a^{(1)}\, a^{(2)} \qquad\quad a^{(3)} \\
\mathbf{P}' = \begin{array}{c} s^{(1)} \\ s^{(2)} \\ a^{(1)} \\ a^{(2)} \\ a^{(3)} \end{array}
\begin{bmatrix}
p_{11} & p_{12} & 0 & 1-p_{11}-p_{12} & 0 \\
0 & 0 & 1 & 0 & \\
p_{21} & p_{22} & 0 & 1-p_{21}-p_{12} & 0 \\
0 & 0 & 0 & 0 & 1 \\
p_{21} & p_{22} & 0 & 1-p_{21}-p_{22} & 0
\end{bmatrix} ; \qquad
p' = \begin{bmatrix}
p(s^{(1)}) \\
p(s^{(2)}) \\
p(a^{(1)}) \\
p(a^{(2)}) \\
p(a^{(3)})
\end{bmatrix} .
\end{array}
$$

After the lumping process (which does not violate simple Markovity): $a'_h = a_h^{(1)} \cup a_h^{(3)}$, we have a sequence defined on the set of states $\{s^{(1)}, s^{(2)}, a', a^{(2)}\}$. Since $p(a^{(1)}) + p(a^{(3)}) = p(s^{(2)}) + p(a^{(2)})$, the matrix of transitional probabilities and stationary probability vector are given by

$$\mathbf{P''} = \begin{array}{c} \\ s^{(1)} \\ s^{(2)} \\ a' \\ a^{(2)} \end{array} \begin{array}{cccc} s^{(1)} & s^{(2)} & a' & a^{(2)} \\ \begin{bmatrix} p_{11} & p_{12} & 0 & 1-p_{11}-p_{12} \\ 0 & 0 & 1 & 0 \\ p_{21} & p_{22} & 0 & 1-p_{11}-p_{12} \\ 0 & 0 & 1 & 0 \end{bmatrix} \end{array} \quad ; \quad \mathbf{p''} = \begin{bmatrix} p(s^{(1)}) \\ p(s^{(2)}) \\ p(a^{(1)})+p(a^{(3)}) \\ p(a^{(2)}) \end{bmatrix}.$$

For the reversed sequence, the matrix of transitional probabilities will be:

$$\bar{\mathbf{P}}'' = \begin{array}{c} \\ s^{(1)} \\ s^{(2)} \\ a' \\ a^{(2)} \end{array} \begin{array}{cccc} \quad s^{(1)} & \quad s^{(2)} & \quad a' & \quad a^{(2)} \\ \begin{bmatrix} p_{11} & 0 & p_{21}\dfrac{p(s^{(2)})+p(a^{(2)})}{p(s^{(1)})} & 0 \\[2ex] p_{12}\dfrac{p(s^{(1)})}{p(s^{(2)})} & 0 & p_{22}\dfrac{p(s^{(2)})+p(a^{(2)})}{p(s^{(2)})} & 0 \\[2ex] 0 & \dfrac{p(s^{(2)})}{p(s^{(2)})+p(a^{(2)})} & 0 & \dfrac{p(a^{(2)})}{p(s^{(2)})+p(a^{(2)})} \\[2ex] (1-p_{11}-p_{12})\dfrac{p(s^{(1)})}{p(a^{(2)})} & 0 & (1-p_{21}-p_{22})\dfrac{p(s^{(2)})+p(a^{(2)})}{p(a^{(2)})} & 0 \end{bmatrix} \end{array}$$

Theorem 1 (appendix 2) shows that the simple Markov structure will be retained after repeated lumping, viz $a_h'' = a_h' \cup a_h^{(2)}$, if and only if

$$\left. \begin{aligned} \frac{p(s^{(2)})}{p(s^{(2)})+p(a^{(2)})} &= (1-p_{11}-p_{12})\frac{p(s^{(1)})}{p(a^{(2)})} \; ; \\ \frac{p(a^{(2)})}{p(s^{(2)})+p(a^{(2)})} &= (1-p_{21}-p_{22})\frac{p(s^{(2)})+p(a^{(2)})}{p(a^{(2)})} \; . \end{aligned} \right\} \tag{31}$$

Solving for $[p(s^{(2)})+p(a^{(2)})]$ in the first of equations (31), and substituting in the second, we obtain

$$[p(s^{(1)})]^2(1-p_{11}-p_{12})^2 = [p(s^{(2)})]^2(1-p_{21}-p_{22})^2 \; ;$$

thus,

$$\frac{p(s^{(1)})}{p(s^{(2)})} = \frac{1-p_{21}-p_{22}}{1-p_{11}-p_{12}} \; . \tag{32}$$

Since the stationary probabilities of the beds in the packets have the same ratio as that of the probabilities of the packets, we substitute these probabilities in equation (32) to obtain

$$\frac{p[(s)]}{p[(sa)]} = \frac{P[(aa)|(sa)]}{P[(aa)|(s)]} \; . \tag{33}$$

We will show from equation (33) that the probability of an a-run of length two is equal to the probability of an a-run of length three. However, the *observed* sequence displays 12 a-runs of length three and 25 of length two.

In terms of conditional probabilities, we have:

$$P(s_{h+3}\,a_{h+2}\,a_{h+1}|a_h\,s_{h-1}) = P(s_{k+2}\,a_{k+1}|a_k\,s_{k-1})\;,$$

or, by using operations with conditional variables,

$$\frac{p(s_{h+3}\,a_{h+2}\,a_{h+1}\,a_h\,s_{h-1})}{p(a_h\,s_{h-1})} = \frac{p(s_{k+2}\,a_{k+1}\,a_k\,s_{k-1})}{p(a_k\,s_{k-1})}\;.$$

This is equivalent to showing that

$$p(s_{h+3}\,a_{h+2}\,a_{h+1}\,a_h\,s_{h-1}) = p(s_{k+2}\,a_{k+1}\,a_k\,s_{k-1})\;;$$

but, for the model $\{(s), (sa), (aa)\}$, the event $(s_{h+3}\,a_{h+2}\,a_{h+1}\,a_h\,s_{h-1})$ can be realized if and only if the event $[(sa)_r(aa)_{r+1}((s)_{r+2} \cup (sa)_{r+2})]$ is realized. The probability of this latter event is

$$p[(sa)]P[(aa)|(sa)]P[(a) \cup (sa)|(aa)]\;.$$

Similarly, the event $(s_{k+2}\,a_{k+1}\,a_k\,s_{k-1})$ is realized if and only if the event $[(s)_{r-1}(aa)_r((s)_{r+1} \cup (sa)_{r+1})]$ is realized. The probability of this event is

$$p[(s)]P[(aa)|(s)]P[(s) \cup (sa)|(aa)]\;.$$

So, it suffices to show that

$$p[(sa)]P[(aa)|(sa)] = p[(s)]P[(aa)|(s)]\;.$$

But this corresponds exactly to equation (33). Thus the probability of an a-run of length two should equal the probability of an a-run of length three. Since these values are not equal in the observed sequence, we must reject the model.

The model $\{(a), (ss), (sa)\}$

In this model, the matrix of transitional probabilities for the packets for the Markov case is:

$$
\mathbf{P} = \begin{array}{c} \\ (a) \\ (ss) \\ (sa) \end{array}
\begin{array}{ccc} (a) & (ss) & (sa) \end{array} \atop
\begin{bmatrix}
p_{11} & p_{12} & 1-p_{11}-p_{12} \\
p_{21} & p_{22} & 1-p_{21}-p_{22} \\
p_{11} & p_{12} & 1-p_{11}-p_{12}
\end{bmatrix}\;.
$$

The corresponding matrix for the beds in the packets, $\mathbf{P}'$, will be given by

$$
\mathbf{P}' = \begin{array}{c} \\ s^{(1)} \\ s^{(2)} \\ s^{(3)} \\ a^{(1)} \\ a^{(2)} \end{array}
\begin{array}{ccccc} s^{(1)} & s^{(2)} & s^{(3)} & a^{(1)} & a^{(2)} \end{array} \atop
\begin{bmatrix}
0 & 1 & 0 & 0 & 0 \\
p_{22} & 0 & 1-p_{21}-p_{22} & p_{21} & 0 \\
0 & 0 & 0 & 0 & 1 \\
p_{12} & 0 & 1-p_{11}-p_{12} & p_{11} & 0 \\
p_{12} & 0 & 1-p_{11}-p_{12} & p_{11} & 0
\end{bmatrix}\;,
$$

where the superscripts correspond to the scheme $\{(a^{(1)}), (s^{(1)}s^{(2)}), (s^{(3)}a^{(2)})\}$. Lumping the states, viz $a_h' = a_h^{(1)} \cup a_h^{(2)}$, we obtain the matrix and stationary probability vector

$$\mathbf{P''} = \begin{array}{c} \\ s^{(1)} \\ s^{(2)} \\ s^{(3)} \\ a' \end{array} \begin{array}{cccc} s^{(1)} & s^{(2)} & s^{(3)} & a' \\ \begin{bmatrix} 0 & 1 & 0 & 0 \\ p_{22} & 0 & 1-p_{21}-p_{22} & p_{21} \\ 0 & 0 & 0 & 1 \\ p_{12} & 0 & 1-p_{11}-p_{12} & p_{11} \end{bmatrix} \end{array} ; \qquad \mathbf{p''} = \begin{bmatrix} p(s^{(1)}) \\ p(s^{(2)}) \\ p(s^{(3)}) \\ p(a') \end{bmatrix} . \qquad (34)$$

We now compute the three transitional probabilities $P(a_h|s_{h-1}s_{h-2})$, $P(a_h|s_{h-1}s_{h-2}s_{h-3})$, and $P(a_h|s_{h-1}s_{h-2}a_{h-3})$. Inspection of tables 6-9 will show that the bed sequence of section A can be considered as a second-order Markov chain. Thus, the transitional probabilities stated above should equal one another. Calculations based on equations (34) give

$$\left. \begin{aligned} P(a_h|s_{h-1}s_{h-2}) &= \frac{1-p_{22}}{2-p_{21}} ; \\ P(a_h|s_{h-1}s_{h-2}a_{h-3}) &= p_{21} ; \\ \text{and} \\ P(a_h|s_{h-1}s_{h-2}s_{h-3}) &= \frac{(1-p_{21})(1-p_{22})}{1-p_{21}+p_{22}} . \end{aligned} \right\} \qquad (35)$$

Equations (35) show that the model is correct only if:

$$p_{21} = 1-(p_{22})^{\frac{1}{2}} . \qquad (36)$$

If the p parameters are taken to be random variables with a continuous density (de Groot, 1970), observations show that relation (36) has a zero probability; the model should therefore be rejected.

The model $\{(s), (a), (ss)\}$

In this case the matrix of transitional probabilities for the packet is

$$\begin{array}{c} \\ (s) \\ (a) \\ (ss) \end{array} \begin{array}{ccc} (s) & (a) & (ss) \\ \begin{bmatrix} p_{11} & p_{12} & 1-p_{11}-p_{12} \\ p_{21} & p_{22} & 1-p_{21}-p_{22} \\ p_{31} & p_{32} & 1-p_{31}-p_{32} \end{bmatrix} \end{array} .$$

This matrix contains six independent parameters which determine the behavior of the beds in the packets, and of the beds in the observed sequence. The transition through 'a' is simple Markov, since there is no lumping of the 'a' states. The transition through 's' should be second-order Markov, which implies certain special conditions of the type given by equation (36), as shown in theorem 1 (appendix 2). Thus, for this case also, the probability of fulfilling these conditions is zero. Fitting the observations by the matrices $P_{\alpha\beta}$ and $P_{\alpha\beta\gamma}$ is impossible because there are six independent parameters which should be estimated; this does not leave the necessary number of degrees of freedom.

We conclude that the model {(s), (a), (sa)} is the only model available which easily explains the properties of the intercalations of beds in section A.

Models of section B

It is clear from table 3 that the sedimentological environment during the generation of this section differed greatly from the conditions for formation of section A. Section B was deposited near the source of arenaceous material. From table 8 we see that the transition through 's' is simple Markov, whereas the transition through 'a' is second-order Markov. From table 12 we see that the order of transition through 'a' is no higher than two, for either combination of beds. It is obvious, both from the observed characteristics and from the axiomatic framework, that the models {(s), (a), (sa)} and {(s), (a), (ss)} are not appropriate because the transition through 'a' for these models is necessarily simple Markov.

Table 12. Test of the second-order Markov property[a] of the possible transitions through 'a' in section B.

Transition	Observed value[b] of χ^2
$s_{h-1}a_h$	3·64
$a_{h-1}a_h$	0·89

[a] Alternative hypothesis is third-order Markov.
[b] One degree of freedom; $\chi^2_{0\cdot05} = 3\cdot84$.

The model {(a), (ss), (sa)}
The model {(a), (ss), (sa)} requires special investigation. We use the standard notation for the various beds, viz $(a^{(1)})$, $(s^{(1)}s^{(2)})$, $(s^{(3)}a^{(2)})$. The matrix of transitional probabilities for the packets is given by

$$
\begin{array}{c}
 \text{(ss)} \quad \text{(sa)} \quad \text{(a)} \\
P = \begin{array}{c} \text{(ss)} \\ \text{(sa)} \\ \text{(a)} \end{array}
\left[
\begin{array}{ccc}
p_{11} & p_{12} & 1-p_{11}-p_{12} \\
p_{21} & p_{22} & 1-p_{21}-p_{22} \\
p_{21} & p_{22} & 1-p_{21}-p_{22}
\end{array}
\right],
\end{array}
$$

where the equality of the lower two rows follows from the axioms of the general model of sedimentation (pages 143 and 149). The matrix of transitional probabilities for the beds in the packets is given by

$$
\mathbf{P'} = \begin{array}{c} \\ s^{(1)} \\ s^{(2)} \\ s^{(3)} \\ a^{(1)} \\ a^{(2)} \end{array}
\begin{array}{cccccc}
s^{(1)} & s^{(2)} & s^{(3)} & a^{(1)} & & a^{(2)} \\
\left[\begin{array}{cccccc}
0 & 1 & 0 & 0 & & 0 \\
p_{11} & 0 & p_{12} & 1-p_{11}-p_{12} & & 0 \\
0 & 0 & 0 & 0 & & 1 \\
p_{21} & 0 & p_{22} & 1-p_{21}-p_{22} & & 0 \\
p_{21} & 0 & p_{22} & 1-p_{21}-p_{22} & & 0
\end{array}\right]
\end{array}.
$$

Inspection of this matrix indicates that after lumping, the transition through 'a' is simple Markov. Thus, this model should be rejected.

The model {(s), (sa), (aa)}
For this model, the matrix of transitional probabilities for the packets is

$$
\begin{array}{c} (s) \\ (sa) \\ (aa) \end{array}
\begin{array}{c}
\begin{array}{ccc} (s) & (sa) & (aa) \end{array} \\
\left[\begin{array}{ccc}
p_{11} & p_{12} & 1-p_{11}-p_{12} \\
p_{21} & p_{22} & 1-p_{21}-p_{22} \\
p_{21} & p_{22} & 1-p_{21}-p_{22}
\end{array}\right]
\end{array};
$$

and for the beds in the packets it is

$$
\begin{array}{c} s^{(1)} \\ s^{(2)} \\ a^{(1)} \\ a^{(2)} \\ a^{(3)} \end{array}
\begin{array}{c}
\begin{array}{ccccc} s^{(1)} & s^{(2)} & a^{(1)} & a^{(2)} & a_3 \end{array} \\
\left[\begin{array}{ccccc}
p_{11} & p_{12} & 0 & 1-p_{11}-p_{12} & 0 \\
0 & 0 & 1 & 0 & 0 \\
p_{21} & p_{22} & 0 & 1-p_{21}-p_{22} & 0 \\
0 & 0 & 0 & 0 & 1 \\
p_{21} & p_{22} & 0 & 1-p_{21}-p_{22} & 0
\end{array}\right]
\end{array}.
$$

The beds $s^{(1)}$, $s^{(2)}$ should be lumped; furthermore the transitions through 'a' are unimportant and may be neglected. The matrix of transitional probabilities for the reversed sequence, $\bar{\mathbf{P}}'$, is given by

$$
\bar{\mathbf{P}}' = \begin{array}{c} s^{(1)} \\ \\ s^{(2)} \\ \\ a^{(1)} \\ a^{(2)} \\ a^{(3)} \end{array}
\begin{array}{c}
\begin{array}{ccccc} s^{(1)} & \quad s^{(2)} & a^{(1)} & \quad a^{(2)} & a^{(3)} \end{array} \\
\left[\begin{array}{ccc:ccc}
p_{11} & 0 & p_{21}\dfrac{p(a^{(1)})}{p(s^{(1)})} & 0 & p_{21}\dfrac{p(a^{(3)})}{p(s^{(1)})} \\[2ex]
p_{12}\dfrac{p(s^{(1)})}{p(s^{(2)})} & 0 & p_{22}\dfrac{p(a^{(1)})}{p(s^{(2)})} & 0 & p_{22}\dfrac{p(a^{(3)})}{p(s^{(2)})} \\[2ex]
\hdashline
& & 0 & & \\
& & & 0 & \\
& & & & 0
\end{array}\right]
\end{array}.
$$

Thus, to retain simple Markovity after the lumping process $s_h' = s_h^{(1)} \cup s_h^{(2)}$, we must either forbid the 'ss' transition, which contradicts geological observations, or require that the following relations exist between the parameters:

$$p_{11} = p_{12}\frac{p(s^{(1)})}{p(s^{(2)})}\,, \qquad\qquad \frac{p_{21}}{p(s^{(1)})} = \frac{p_{22}}{p(s^{(2)})}\,;$$

or

$$\frac{p_{11}}{p_{12}} = \frac{p(s^{(1)})}{p(s^{(2)})}\,, \qquad\qquad \frac{p_{21}}{p_{22}} = \frac{p(s^{(1)})}{p(s^{(2)})}\,.$$

This implies that

$$\frac{p_{11}}{p_{21}} = \frac{p_{12}}{p_{22}} = C\,,$$

where C is a constant. This is a necessary and sufficient condition for a simple Markov transition through 's', as shown in lemma 2 (appendix 2). It is clear that this restriction of parameters will occur only with probability zero (if we assume that the parameters are random variables with continuous density, and so the model $\{(s), (sa), (aa)\}$ is rejected. We have now investigated twenty-four models, twenty-three of which should be rejected on geological or mathematical grounds. Thus we have only one model that remains uninvestigated.

The model $\{(s), (a), (aa)\}$

We have mentioned earlier that section B is highly arenaceous. Additionally, paleogeographic reconstruction (Vistelius, 1961) suggests that the source area for the argillaceous material is the Kura river valley, far to the west (figure 1). On the basis of this evidence we may introduce an additional restriction into the model $\{(s), (a), (aa)\}$: that there are no two succeeding sedimentary acts generating (aa). In other words, because the source area is far from Zhyloy Island, the probability of argillaceous sedimentation on the island with repeated splitting into two beds is practically zero. This restriction seems natural, and can be expressed in mathematical form as

$$P[(aa)|(aa)] = 0\,.$$

The matrix of transitional probabilities for the packets is therefore given by

$$\mathbf{P} = \begin{array}{c} \\ (s) \\ (a) \\ (aa) \end{array} \begin{array}{ccc} (s) & (a) & (aa) \\ \begin{bmatrix} p_{11} & p_{12} & 1-p_{11}-p_{12} \\ p_{21} & 1-p_{21} & 0 \\ p_{21} & 1-p_{21} & 0 \end{bmatrix} \end{array}. \qquad (37)$$

From matrix (37), we can see that the sequence of packets must be simple Markov. Otherwise, if the sequence is Bernoullian, we must stipulate that

$1 - p_{11} - p_{12} = 0$ in the column corresponding to (aa). But this is impossible, as we have already accepted the full model including (aa) packets. Since the sequence of packets is simple Markov it follows that the transition through 's' is simple Markov, and that there is no lumping for 's'.

To investigate the behavior of the transition through 'a', we first compute the matrix for the beds in the packets which (with the usual numbering for the beds) is as follows

$$
\begin{array}{c}
\qquad\qquad\qquad \overbrace{\phantom{a^{(1)} \qquad a^{(3)}}}^{a'} \\
\begin{array}{ccccc}
 & s^{(1)} & a^{(1)} & a^{(3)} & a^{(2)}
\end{array} \\
\mathbf{P}' =
\begin{array}{c}
s^{(1)} \\
a^{(1)} \\
a^{(3)} \\
a^{(2)}
\end{array}
\left[
\begin{array}{cc:c:c}
p_{11} & p_{12} & 0 & 1 - p_{11} - p_{12} \\ \hdashline
p_{21} & 1 - p_{21} & 0 & 0 \\ \hdashline
p_{21} & 1 - p_{21} & 0 & 0 \\ \hdashline
0 & 0 & 1 & 0
\end{array}
\right] .
\end{array}
$$

Lumping the states is performed in two steps; first we obtain

$$a'_h = a_h^{(1)} \cup a_h^{(3)} ,$$

and the simple Markovity of the sequence of states $s^{(1)}$, a' and $a^{(2)}$ is retained. This new sequence has the matrix of transitional probabilities, $\mathbf{P}''$, given by

$$
\mathbf{P}'' =
\begin{array}{c}
 \\
s^{(1)} \\
a' \\
a^{(2)}
\end{array}
\begin{array}{c}
\begin{array}{ccc}
s^{(1)} & a' & a^{(2)}
\end{array} \\
\left[
\begin{array}{ccc}
p_{11} & p_{12} & 1 - p_{11} - p_{12} \\
p_{21} & 1 - p_{21} & 0 \\
0 & 1 & 0
\end{array}
\right] .
\end{array}
\qquad (38)
$$

The observed sequence in section B would be obtained from a sequence with matrix (38) after further lumping:

$$a_h = a'_h \cup a_h^{(2)} .$$

Let us now prove that the transition through 'a' is second-order Markov. For this, it is necessary and sufficient to prove that

$$
\left.
\begin{array}{l}
P(s_{h+1} | a_h s_{h-1} A_{h-2}) = P(s_{h+1} | a_h s_{h-1}) , \\
P(s_{h+1} | a_h a_{h-1} A_{h-2}) = P(s_{h+1} | a_h a_{h-1}) ,
\end{array}
\right\}
\qquad (39)
$$

and that

$$P(s_{h+1} | a_h) \neq P(s_{h+1} | a_h a_{h-1}) , \qquad (40)$$

or

$$P(s_{h+1} | a_h) \neq P(s_{h+1} | a_h s_{h-1}) .$$

There are two other sets of equations, but they combine with those given to sum to one. We prove the first equality in equations (39) as follows:

$$P(s_{h+1}|a_h s_{h-1} A_{h-2}) = P[s_{h+1}|(a'_h \cup a_h^{(2)})s_{h-1}A_{h-2}]$$

$$= \frac{p[(s_{h+1}a'_h s_{h-1}A_{h-2}) \cup (s_{h+1}a_h^{(2)}s_{h-1}A_{h-2})]}{p[(a'_h s_{h-1}A_{h-2}) \cup (a_h^{(2)}s_{h-1}A_{h-2}]}$$

$$= \frac{p(s_{h+1}a'_h s_{h-1}A_{h-2}) + p(s_{h+1}a_h^{(2)}s_{h-1}A_{h-2})}{p(a'_h s_{h-1}A_{h-2}) + p(a_h^{(2)}s_{h-1}A_{h-2})} \, ,$$

or, after transformation using matrix (38) and the homogeneity of the sequence,

$$P(s_{h+1}|a_h s_{h-1}A_{h-2})$$

$$= \frac{p(A_{h-2})P(s|A_{h-2})P(a'|s)P(s|a') + p(A_{h-2})P(s|A_{h-2})P(a^{(2)}|s)P(s|a^{(2)})}{p(A_{h-2})P(s|A_{h-2})P(a'|s) + p(A_{h-2})P(s|A_{h-2})P(a^{(2)}|s)}$$

$$= \frac{p(A_{h-2})P(s|A_{h-2})p_{12}p_{21}}{p(A_{h-2})P(s|A_{h-2})(1-p_{11})} = \frac{p_{12}p_{21}}{1-p_{11}} \, . \tag{41}$$

Computations for $P(s_{h+1}|a_h s_{h-1})$ give the same expressions. Similarly, for the second equality in equations (39), we obtain

$$P(s_{h+1}|a_h a_{h-1}A_{h-2}) = P(s_{h+1}|a_h a_{h-1}) = p_{21} \, . \tag{42}$$

Thus the Markov order of the transition through 'a' is no higher than two.

Now we will estimate the parameters of the model represented by matrix (37) for the observed sequence, and use these to prove the inequality of equation (40). As we have proved that the chain is at most second-order Markov, and since there are only two states, the reversibility of the chain is trivially true, as shown in the discussion of section A.

It is obvious that in the matrix of transitional probabilities, $P_{\alpha\beta}$,

$$\bar{P}(s|s) = \bar{p}_{11} \, , \tag{43}$$

as there is no lumping of 's'.

From equation (42), we can write

$$\bar{p}_{21} = \bar{P}(s_h|a_{h-1}a_{h-2}) \, . \tag{44}$$

Finally, insertion of the estimators $\bar{p}_{11}$ and $\bar{p}_{21}$ into equation (41) gives the estimator $\bar{p}_{12}$:

$$\bar{p}_{12} = \frac{\bar{P}(s_h)|a_{h-1}s_{h-2})\bar{P}(a|s)}{\bar{P}(s_h|a_{h-1}a_{h-2})} \, . \tag{45}$$

It must be noted that this method does not necessarily yield statistically optimal estimators.

We now have all of the estimators needed in order to determine the theoretical matrices $P_{\alpha\beta}$, $P_{\alpha\beta\gamma}$, and $P_{\alpha\beta\gamma\delta}$ for the model, except that for $P(s|a)$. To find the matrix $P_{\alpha\beta}$, it is necessary to obtain the components of the stationary distribution vector corresponding to matrix (38). To do this, we first compute the eigenvector of the transposed matrix that corresponds to the eigenvalue λ_1 ($= 1$) in nonstochastic form. We will mark the components of this vector by an asterisk, viz $p(a')^*$. In this way, we obtain the equations:

$$p_{11}p(s^{(1)})^* + p_{21}p(a')^* = p(s^{(1)})^* \ ;$$

$$(1 - p_{11} - p_{12})p(s^{(1)})^* = p(a^{(2)})^* \ ;$$

so that

$$p(s^{(1)})^* = \frac{p_{21}}{1 - p_{11}} \ ;$$

$$p(a^{(2)})^* = \frac{(1 - p_{11} - p_{12})p_{21}}{1 - p_{11}} \ ;$$

$$p(a')^* = 1 \ ;$$

or

$$p(a')^* + p(a^{(2)})^* + p(s^{(1)})^* = \frac{(p_{21} + 1)(1 - p_{11}) + p_{21}(1 - p_{12})}{(1 - p_{11})} \ .$$

After transformation of the vector to stochastic form, we have

$$p(s^{(1)}) = \frac{p_{21}}{(1 - p_{11})(1 + p_{21}) + p_{21}(1 - p_{12})} \ ;$$

$$p(a') = \frac{1 - p_{11}}{(1 - p_{11})(1 + p_{21}) + p_{21}(1 - p_{12})} \ ;$$

$$p(a^{(2)}) = \frac{p_{21}(1 - p_{11} - p_{12})}{(1 - p_{11})(1 + p_{21}) + p_{21}(1 - p_{12})} \ .$$

We are now able to compute the estimator for the probability $P(s|a)$:

$$\tilde{P}(s|a) = \frac{\tilde{p}(a'_{h-1}s_h^{(1)}) + \tilde{p}(a_{h-1}^{(2)}s_h^{(1)})}{\tilde{p}(a') + \tilde{p}(a^{(2)})}$$

$$= \frac{\tilde{p}(a')\tilde{P}(s^{(1)}|a') + \tilde{p}(a^{(2)})\tilde{P}(s^{(1)}|a^{(2)})}{\tilde{p}(a') + \tilde{p}(a^{(2)})}$$

$$= \frac{\tilde{p}_{21}(1 - \tilde{p}_{11})}{(1 - \tilde{p}_{11})(1 + \tilde{p}_{21}) - \tilde{p}_{12}\tilde{p}_{21}} \ . \tag{46}$$

Comparison of equations (42) and (46) shows that inequality (40) is proved. Simultaneously, we have proved that the transition through 'a' is second-order Markov. These results provide us with the following matrices of transitional probabilities

$$
\mathbf{P}_{\alpha\beta} = \begin{array}{c} \\ s_h \\ a_h \end{array}
\begin{array}{c} s_{h+1} \quad\; a_{h+1} \\ \left[\begin{array}{cc} p_{11} & 1-p_{11} \\ x & 1-x \end{array}\right] \end{array} ;
$$

$$
\mathbf{P}_{\alpha\beta\gamma} = \begin{array}{c} \\ s_{h-1}s_h \\ s_{h-1}a_h \\ a_{h-1}s_h \\ a_{h-1}a_h \end{array}
\begin{array}{c} s_{h+1} \qquad\quad a_{h+1} \\ \left[\begin{array}{cc} p_{11} & 1-p_{11} \\ \dfrac{p_{12}p_{21}}{1-p_{11}} & 1-\dfrac{p_{12}p_{21}}{1-p_{11}} \\ p_{11} & 1-p_{11} \\ p_{21} & 1-p_{21} \end{array}\right] \end{array} ;
$$

$$
\mathbf{P}_{\alpha\beta\gamma\delta} = \begin{array}{c} \\ s_{h-2}s_{h-1}s_h \\ s_{h-2}s_{h-1}a_h \\ s_{h-2}a_{h-1}s_h \\ s_{h-2}a_{h-1}a_h \\ a_{h-2}s_{h-1}s_h \\ a_{h-2}s_{h-1}a_h \\ a_{h-2}a_{h-1}s_h \\ a_{h-2}a_{h-1}a_h \end{array}
\begin{array}{c} s_{h+1} \qquad\quad a_{h+1} \\ \left[\begin{array}{cc} p_{11} & 1-p_{11} \\ \dfrac{p_{12}p_{21}}{1-p_{11}} & 1-\dfrac{p_{12}p_{21}}{1-p_{11}} \\ p_{11} & 1-p_{11} \\ p_{21} & 1-p_{21} \\ p_{11} & 1-p_{11} \\ \dfrac{p_{12}p_{21}}{1-p_{11}} & 1-\dfrac{p_{12}p_{21}}{1-p_{11}} \\ p_{11} & 1-p_{11} \\ p_{21} & 1-p_{21} \end{array}\right] \end{array} .
$$

In the first of these matrices, the quantity x is given by

$$
x = \frac{p_{21}(1-p_{11})}{(1-p_{11})(1+p_{21}) - p_{12}p_{21}} .
$$

Computation of the matrix $\mathbf{P}_{\alpha\beta\gamma\delta}$ is trivial, since we have a second-order Markov chain; however, this matrix is needed for statistical tests. We can compare the matrices $\mathbf{P}_{\alpha\beta}$, $\mathbf{P}_{\alpha\beta\gamma}$, and $\mathbf{P}_{\alpha\beta\gamma\delta}$ with the observed matrices by substituting the estimators given by equations (43)–(46) for the parameters and using the χ^2 test. The comparison is trivial for the matrix $\mathbf{P}_{\alpha\beta}$; there is one degree of freedom for comparison with the matrix $\mathbf{P}_{\alpha\beta\gamma}$; there are five degrees of freedom for comparison with matrix $\mathbf{P}_{\alpha\beta\gamma\delta}$. In general, the number of degrees of freedom, ν, is equal to the difference between the number of independently estimated parameters, ν_1, for the hypothesis H_1, and the corresponding number, ν_0, for the hypothesis H_0. Here, H_0 is the hypothesis that the sequence is a Markov chain with a simple Markov transition through 's' and a second-order Markov transition through 'a'; H_1 is the hypothesis corresponding to all other Markov chains of the second order. The computation of values of ν for the three matrices is summarized in table 13.

The estimated values of the parameters are given in the matrices below.

$$
\tilde{\mathbf{P}}_{\alpha\beta} = \begin{array}{c} \\ s \\ a \end{array}
\begin{array}{c} s \qquad\quad a \\ \left[\begin{array}{cc} 0\cdot3602 & 0\cdot6398 \\ 0\cdot9685 & 0\cdot0315 \end{array}\right] \end{array} ;
$$

$$
\tilde{\mathbf{P}}_{\alpha\beta\gamma} = \begin{array}{c} \\ s_{h-1}s_h \\ s_{h-1}a_h \\ a_{h-1}s_h \\ a_{h-1}a_h \end{array}
\begin{array}{c} s_{h+1} \qquad\quad a_{h+1} \\ \left[\begin{array}{cc} 0\cdot3602 & 0\cdot6398 \\ 0\cdot9756 & 0\cdot0244 \\ 0\cdot3602 & 0\cdot6398 \\ 0\cdot7500 & 0\cdot2500 \end{array}\right] \end{array} ;
$$

$$
\tilde{\mathbf{P}}_{\alpha\beta\gamma\delta} = \begin{array}{c} \\ s_{h-2}s_{h-1}s_h \\ s_{h-2}s_{h-1}a_h \\ s_{h-2}a_{h-1}s_h \\ s_{h-2}a_{h-1}a_h \\ a_{h-2}s_{h-1}s_h \\ a_{h-2}s_{h-1}a_h \\ a_{h-2}a_{h-1}s_h \\ a_{h-2}a_{h-1}a_h \end{array}
\begin{array}{c} s_{h+1} \qquad\quad a_{h+1} \\ \left[\begin{array}{cc} 0\cdot3602 & 0\cdot6398 \\ 0\cdot9756 & 0\cdot0244 \\ 0\cdot3602 & 0\cdot6398 \\ 0\cdot7500 & 0\cdot2500 \\ 0\cdot3602 & 0\cdot6398 \\ 0\cdot9756 & 0\cdot0244 \\ 0\cdot3602 & 0\cdot6398 \\ 0\cdot7500 & 0\cdot2500 \end{array}\right] \end{array} .
$$

If the χ^2 test is used to compare $\bar{\mathbf{P}}_{\alpha\beta\gamma}$ and $\bar{\mathbf{P}}_{\alpha\beta\gamma\delta}$ with the observed matrices in table 4, we have

$$0\cdot10 \leqslant p(\chi^2 \geqslant 2\cdot69) \leqslant 0\cdot25 , \qquad \nu = 1 ;$$

$$0\cdot10 \leqslant p(\chi^2 \geqslant 7\cdot99) \leqslant 0\cdot25 , \qquad \nu = 5 .$$

The goodness of fit is quite sufficient.

The order of the Markov chain for section B was also checked by the runs test, as described for section A(table 10). The results of the test for section B are given in table 14. We conclude that the model {(s), (a), (aa)} accounts for the properties of the sequence of beds in section B if we assume that $\mathbf{P}[(aa)|(aa)] = 0$, which is quite natural as far as is indicated from the probable paleogeographic conditions.

Table 13. Degrees of freedom, ν, in test of significance of parameters of the model {(s), (a), (aa)} in section B.

Matrix	ν_0	ν_1	ν
$\mathbf{P}_{\alpha\beta}$	2	2	0
$\mathbf{P}_{\alpha\beta\gamma}$	3	4	1
$\mathbf{P}_{\alpha\beta\gamma\delta}$	3	8	5

Table 14. Distributions of lengths of 's' and 'a' runs in section B.

length	number observed (O)	number expected (E)	$\frac{(O-E)^2}{E}$	length	number observed (O)	number expected (E)	$\frac{(O-E)^2}{E}$
Runs of a-beds				Runs of s-beds			
1	281	280·94	0·00	1	192	182·98	0·44
2	6	5·30	0·09	2	56	65·92	1·48
3	0	1·32	1·32	3	25	23·74	0·07
$\geqslant$4	1	0·44	0·71	4	7	8·55	0·28
Total	288	288	2·12 [a]	5	3	3·09	0·00
				$\geqslant$6	3	1·72	0·95
				Total	286	286	3·22 [b]

[a] $0\cdot10 < p_a \,(\chi^2 \geqslant 2\cdot12) < 0\cdot25$; one degree of freedom.
[b] $0\cdot50 < p_s \,(\chi^2 \geqslant 3\cdot22) < 0\cdot75$; four degrees of freedom.

Summary and conclusions

Investigations have been made of different models of stratification in sections of Pliocene deposits in the Cheleken Peninsula and Zhyloy Island. For all models it was assumed that the sections are composed of packets of beds which follow Bernoullian[5] or simple Markov sequences, and that

[5] See footnote (1), page 143.

some beds in the packets are renewal events. From these conditions, we can predict the observed matrices of transitional probabilities between beds.

Two special lines of investigation were used. Firstly, it was assumed that the sequences of packets are simple Markov chains. This assumption is based upon general considerations, and does not contradict the observations. However, it is necessary in the future to investigate the sequences of packets directly. This could be done by investigation of flysch deposits. Secondly, good agreement between observation and theory was obtained when some renewal event was included in the axiomatic structure of the models. This is a very important result. It would seem that the entire theory of cyclic sedimentation should be reworked on the basis of the theory of renewal events[6]. The boundaries of cycles should be distinguished by beds which constitute renewal events. It would seem that deposits such as flysch are not cyclical deposits, but sequences with renewal events. The deposits in the interval between these events are an arbitrary set under the influence of the gravitational field. Thus in many cases it is necessary, or at least more convenient, to talk about 'packet sedimentation', and to separate these cases from the 'cyclical' sediments. These problems await solution in the future.

Acknowledgements. During my work on stratification problems, my correspondence and personal discussions with Professor J C Griffiths were stimulating. The constant assistance of A V Faas has been exceedingly useful; in particular, all proofs of theorems in this essay and the following appendices are his work.

[6] The models introduced by Vistelius conform to the concept of a hierarchical stratigraphy, as developed by Dacey and Krumbein (1970). It is interesting to note that sequences containing packets which are structured either by an independent Markov process ('Bernoullian' in Vistelius's terminology), or by a Markov chain, can be modelled effectively by semi-Markov processes which have been used in stratigraphical analysis by Krumbein and Dacey (1969) and Schwarzacher (1972). The importance of renewal theory and its connection with the semi-Markov process is discussed at some length by Schwarzacher (1975). (Editors.)

Some additional concepts related to Markov sequences

In the text of this essay, the concepts 'renewal event' and 'partial Markovity', and a special definition of 'Markov characteristic' are used. These are discussed below.

Renewal event   It is well known that in a simple Markov chain, the equality:

$$P(\alpha_h|\beta_{h-1}A_{h-2}) = P(\alpha_h|\beta_{h-1}) , \qquad (A1.1)$$

holds for any $\alpha, \beta \in S$, where S is the set of states $\{\alpha\}$, and for any A_{h-2}, that is, for any event whose outcome is dependent only upon trials before the $(h-1)$th. From equation (A1.1), we obtain the equation for a simple Markov chain:

$$P(A_h|\beta_{h-1}A_{h-2}) = P(A_h|\beta_{h-1}) , \qquad (A1.2)$$

where A_h is an event dependent only upon trials after the $(h-1)$th. For a simple Markov chain, equation (A1.2) can be rewritten (symmetrically) with respect to the future as well as to the past, as shown below

$$P(A_h|\beta_{h-1})P(A_{h-2}|\beta_{h-1}) = P(A_h, A_{h-2}|\beta_{h-1}) . \qquad (A1.3)$$

Thus, for a simple Markov chain, the future and the past are conditionally independent, provided that the realization at the present moment, $(h-1)$, is known. If the event A_0 may be inserted in equation (A1.2) in the place of trial β_{h-1} it is referred to as a renewal event; A_0 is found as the event which satisfies the equation

$$P(A_+|A_0A_-) = P(A_+|A_0) , \qquad (A1.4)$$

where A_0 is the renewal event, A_+ (future) is any event dependent only upon trials appearing after those composing A_0 (present), and A_- (past) is any event dependent only upon trials preceding A_0.

The appearance of any state in a simple Markov chain is a renewal event, as is the appearance of any combination $\alpha_{h-1}\beta_{h-2} \ldots \delta_{h-r}$ for a Markov chain of rth order. The converse is also true if S is finite. If the appearance of any state from S is a renewal event, then the generated sequence is a simple Markov chain. If the appearance of any combination $\alpha_{h-1}\beta_{h-2} \ldots \delta_{h-r}$ with states $\alpha, \beta, \ldots, \delta \in S$ is a renewal event, then the sequence is a Markov chain of rth order.

Partial Markovity   We say that there is partial Markovity with a simple Markov transition through the fixed state α if and only if the sequence satisfies the equation

$$P(\beta_h|\alpha_{h-1}A_{h-2}) = P(\beta_h|\alpha_{h-1}) , \qquad (A1.5)$$

for all $\beta \in S$. It should be emphasized here that α is a fixed state and β can be any state from S. If a random sequence possesses the property that for a fixed ordered combination of trials of length r, $\alpha_{h-1} \ldots \delta_{h-r}$, there is a Markov transition of order r such that the equation

$$P(\beta_h|\alpha_{h-1} \ldots \delta_{h-r} A_{h-r-1}) = P(\beta_h|\alpha_{h-1} \ldots \delta_{h-r}) \tag{A1.6}$$

is satisfied for any integer h, then this sequence is said to have partial Markovity, with Markov transition of order r through the fixed combination $\alpha_{h-1} \ldots \delta_{h-r}$. Again we note that β can be any state of S. Finally, we say that the sequence has partial Markovity in the transition through $\alpha_{h-1} \ldots \delta_{h-r}$ of order k $(k > r)$, if and only if the equality

$$P(\beta_h|\alpha_{h-1} \ldots \delta_{h-r} A'_{h-r-1} A_{h-k-1}) = P(\beta_h|\alpha_{h-1} \ldots \delta_{h-r} A'_{h-r-1}) \tag{A1.7}$$

is satisfied. Here, A$'$ is any event dependent only upon events from the $(h-k)$th to $(h-r-1)$th trials, inclusive. The event A_{h-k-1}, as always, is dependent only on the results of trials before $(h-k)$.

The interrelations of Markov chains, sequences with partial Markovity, and Markov transitions are as follows. If the transition through any combination of r trials is Markov of order k $(k \geq r)$, the sequence is a Markov chain of order k. If the transition through a *fixed* combination of r trials is Markov of order r, the sequence may or may not be a Markov chain of order r. If the transitions through other combinations are not Markov of order r, the sequence will not be Markov of order r. If the sequence is a simple Markov chain, all transitions through any state are simple Markov. For a Markov chain of order k, all transitions through any combination of trials of length r $(r \leq k)$ are Markov of order k.

If the appearance of some fixed combination of trials of length r is a renewal event, the transition through this fixed combination is Markov of order r. This comes directly from the definition of a renewal event [equation (A1.4)]: insert the above combination into equation (A1.4) in place of A_0, and for A_+ insert β_h, $\beta \in S$, for any integer h. The converse is not true, as will now be demonstrated. Assume that there is a sequence of arenaceous (s) and argillaceous (a) beds. Let the transition through 'a' be simple Markov, and the type of transition through 's' be unknown. Consider the transitional probabilities $P(s_h|a_{h-1}s_{h-2})$ and $P(\alpha_{h+1}s_h|a_{h-1}s_{h-2})$, where $\alpha \in \{s, a\}$. Since the transition through 'a' is simple Markov, equation (A1.5) yields

$$P(s_h|a_{h-1}s_{h-2}) = P(s_h|a_{h-1}) . \tag{A1.8}$$

If the transition through 'a' is also a renewal event, then

$$P(\alpha_{h+1}s_h|a_{h-1}s_{h-2}) = P(\alpha_{h+1}s_h|a_{h-1}) . \tag{A1.9}$$

We can check equation (A1.9) by computations. We assume that the transition through a_{h-1} is simple Markov, but that it is not known whether 'a' is a renewal event. Using the formula for conditional probabilities, we obtain from the left-hand side of equation (A1.9):

$$\frac{p(\alpha_{h+1}s_h a_{h-1}s_{h-2})}{p(a_{h-1}s_{h-2})} = P(s_h|a_{h-1}s_{h-2})P(\alpha_{h+1}|s_h a_{h-1}s_{h-2}) . \tag{A1.10}$$

As the transition through 'a' is simple Markov, the right-hand side of equation (A1.10) becomes $p(s_h|a_{h-1})P(\alpha_{h+1}|s_h a_{h-1} s_{h-2})$. Transforming the right-hand side of equation (A1.9) we obtain

$$\frac{p(a_{h-1} s_h \alpha_{h+1})}{p(a_{h-1})} = P(s_h|a_{h-1})P(\alpha_{h+1}|s_h a_{h-1}) . \qquad (A1.11)$$

So to verify equation (A1.9) is equivalent to testing the equality

$$P(\alpha_{h+1}|s_h a_{h-1} s_{h-2}) = P(\alpha_{h+1}|s_h a_{h-1}) ; \qquad (A1.12)$$

and equation (A1.12) is true if the transition through 's' is first-order or second-order Markov. If the sequence has a Markov structure of order greater than two the correctness of equation (A1.12) is indeterminate. Thus we can expand the action of the Markov transition to a future event composed of two trials, if the transition through 's' is first-order or second-order Markov.

This example has shown that in sequences with partial Markovity, the concept of a Markov transition differs from that of a renewal event. For a renewal event, sequences in both past and future can be of any length. For a Markov transition in sequences with partial Markovity, the past can be of any length, but the future should contain only one trial. In a Markov sequence, the probability of any random event in the future is determined if the present is fixed. With partial Markov sequences this is only possible if some special assumptions are made concerning the properties of the sequence.

Markov characteristics The presence of Markovity, partial Markovity, or a renewal event in the sequence cannot be verified by observations, because exhaustive investigation requires an infinite number of observations. Well-investigated statistical procedures are used to compare the transitional probabilities for the equality

$$P(\beta_h|\alpha_{h-1} \ldots \gamma_{h-k}) = P(\beta_h|\alpha_{h-1} \ldots \gamma_{h-k} \delta_{h-k-1}) . \qquad (A1.13)$$

This is the only necessary condition for Markovity of order k, or a partial Markov transition of order k, to exist. However, this equality is not sufficient to prove either Markovity or a partial Markov transition in the sequence. The random sequences satisfying equation (A1.13) for all $\alpha, \beta, \gamma, \delta \in S$ and for any integer h are referred to as sequences with *Markov characteristic* of order k. If equation (A1.13) is satisfied for a random sequence for any $\beta, \delta \in S$ and for a fixed combination $\alpha_{h-1} \ldots \gamma_{h-k}$, the sequence is taken to have a partial Markov characteristic for the transition through this fixed combination.

In numerical applications, we always deal with a certain type of Markov characteristic. The extension of conclusions based on this characteristic to cases of true Markovity (the Markov property) can lead to misunderstandings.

Statement of the main theorems[7] used in the paper

Here we assume that we are dealing with a finite Markov chain which is ergodic, with an initial probability distribution coinciding with the stationary distribution [if such a chain can be lumped it is said that there is 'weak lumpability' (Burke and Rosenblatt, 1958)]. The terminology used here is based on that of Kemeny and Snell (1960). Thus the chain before lumping is referred to as *fine*; after lumping, it is referred to as a *lumped sequence*.

Lemma 1: Let the sequence before lumping be the simple Markov chain described in appendix 1. The partial transition through state α would obtain a simple Markov characteristic after lumping within all classes α, β, ..., μ, independently of the Markov properties of other transitions, if at least one of two conditions

$$\sum_j \Gamma(\beta^{(j)}|\alpha^{(i)}) = C_{\alpha\beta} \,;$$

$$\sum_j \bar{P}(\beta^{(j)}|\alpha^{(i)}) = K_{\alpha\beta} \,. \tag{A2.1}$$

is fulfilled, where $C_{\alpha\beta}$ and $K_{\alpha\beta}$ are constants independent of i, α is fixed, and $\beta \in S$ is arbitrary; $\beta^{(j)}$, $\alpha^{(i)}$ are states from the fine chain being lumped; $\bar{P}$ is a probability for the reversed sequence.

It should be pointed out that lemma 1 deals only with the Markov characteristic and contains no information on true Markovity (see appendix 1). The Markov characteristic of the sequence may be checked by statistical computations. Such calculations cannot be used to check Markovity; Markovity is a purely theoretical concept.

A reversible lumped sequence is a more general case than a reversible fine sequence. If a fine sequence is reversible then the lumped sequence is reversible too. However, a lumped sequence may be reversible even if the fine sequence is not reversible. When the lumped sequence is reversible, the following lemma can be used.

Lemma 2: If a fine sequence is a homogeneous simple Markov chain and the lumped sequence is reversible, then to retain simple Markovity in the transition through state α in the lumped sequence it is sufficient that one of the conditions (A2.1) hold.

Theorem 1: Let a fine sequence be a homogeneous simple Markov chain and let the lumped sequence be reversible. Furthermore, let the lumped subset, S, of S consist of not more than two states. Then the transition through state α in the lumped sequence is simple Markov if and only if at least one of the conditions (A2.1) hold.

From the proof of theorem 1, the following corollary can be shown to hold.

---

[7] The proofs of all theorems and lemmas in this appendix are given in Vistelius (1980).

Corollary: Under the above assumptions, the sufficiency of conditions (A2.1) holds for lumping in the subset S by an arbitrary number of states, not just two. Necessity follows without assuming reversibility of the lumped sequence.

Finally, the following lemma is useful for determination of the properties of the model sequences introduced in this essay.

Lemma 3: If a sequence of packets (as defined in this essay) is a simple Markov chain, then the sequence of beds in the packets, obtained from the sequence of packets, is also a simple Markov chain.

References

American Geological Institute, 1957 *Glossary of Geology* (American Geological Institute, Washington, DC)

Burke C K, Rosenblatt M, 1958 "Markovian function of a Markov chain" *Annals of Mathematical Statistics* **29** 1112-1122

Dacey M F, Krumbein W C, 1970 "Markovian models in stratigraphy" *Journal of the International Association of Mathematical Geologists* **2** 175-191

Dawson J W, 1854 *Acadian Geology* second edition (Macmillan, London)

de Geer G, 1940 "Geochronologia Suecica Principles" *Kungliga Svenska Vetenskapsakademiens Handlingar, series 3* **18**(6)

de Groot M H, 1970 *Optimal Statistical Decisions* (McGraw-Hill, New York)

Feller W, 1964 *An Introduction to Probability Theory and its Applications* volume 1 (John Wiley, New York)

Kemeny J G, Snell J L, 1960 *Finite Markov Chains* (Van Nostrand, New York)

Krumbein W C, Dacey M F, 1969 "Markov chains and embedded Markov chains in geology" *Journal of the International Association of Mathematical Geologists* **1** 79-96

Kuenen Ph H, 1948 "Turbidity currents of high density" in *International Geological Congress Report, 18th Session, London* part 3, pp 44-52

Matthews R K, 1974 *Dynamic Stratigraphy* (Prentice-Hall, Englewood Cliffs, NJ)

Schwarzacher W, 1972 "The semi-Markov processes a general sedimentation model" in *Mathematical Models of Sedimentary Processes* Ed. D F Merriam (Plenum, New York) pp

Schwarzacher W, 1975 *Sedimentation Models and Quantitative Stratigraphy* (Elsevier, Amsterdam)

Vassoevich N B, 1948 *Flysch and Methods of its Investigation* (Gostoptekhizdat, Leningrad)

Vistelius A B, 1961 *Materials for Lithostratigraphy of Azerbaidzhan Productive Series* (USSR Academy of Sciences, Moscow)

Vistelius A B, 1980 *Principles of Mathematical Geology* (Nauka, Moscow)

Vistelius A B, Romanova M A, 1962 *Red-Beds of Cheleken Peninsula, Geology* (USSR Academy of Sciences, Moscow)

Weller M, 1964 "Development of the concept and interpretation of cyclic sedimentation" *University of Kansas Science Bulletin* **2** 607-621

Similarities between models for particle-size distributions and stream-channel networks

M F Dacey, W C Krumbein

Introduction

This essay arises as a by-product of two earlier studies by Dacey and Krumbein (1976; 1979) on stochastic processes for the growth of stream-channel networks and for the development of particle-size distributions. Network growth was formulated in terms of channel bifurcations and tributary developments, whereas particle-size distributions were the product of a succession of selection and breakage events operating on an initial particle. The net results were three network-growth models and two models for breakage and selection.

The network-growth models were conveniently shown as types of planted plane trees, known to geologists as channel networks. The possible outcomes of the network-growth models were described at each stage of development by the probability of occurrence of each topologically distinct channel network. For one network-growth model, all channel networks of the same magnitude have the same probability of occurrence, which is in accord with Shreve's (1966; 1967) model of topologically random channel networks. The outcomes of the breakage–selection models for particles can also be represented by the probability of occurrence of topologically distinct channel networks. These probabilities are identical with those for two of the three network-growth models, but a breakage–selection model that has probabilities conforming with the Shreve model of topological randomness was not developed. The question accordingly arises whether a third breakage–selection model, that does have this property of topological randomness, can be formulated.

In seeking an answer to this question, we reviewed the allowable growth processes for channel networks to see whether a criterion for branching could be developed that could also be applied to the selection and breakage rules for particles. If such a criterion exists, it would serve two important purposes. The first would be the specification of a set of rules for a particular breakage–selection model satisfying topological randomness; the second might lead to a set of more general rules for transferring models developed in one field to their counterparts in entirely different fields. The main objective of this essay is to explore the first purpose, although some implications of the more general approach are also considered.

The b criterion in network growth

Three-channel network-growth processes introduced in an earlier paper (Dacey and Krumbein, 1976) were based on certain allowable events for network growth from magnitude n to $(n+1)$. These events involved the bifurcation of exterior links, and the development of tributaries either on exterior or on interior links. However, the results of tributary development on an exterior link and those of bifurcation of the same link are topologically indistinguishable, whereas tributary developments on the right and left sides of interior links are topologically distinguishable. 'Branching' is used as a general term that refers to bifurcations and tributary developments on exterior links and to tributary developments on interior links.

The change in network magnitude from n to $(n+1)$ can be characterized by contrasting the probability that growth occurs by branching of exterior links against that for branching of interior links. The criterion b is defined as the ratio:

$$\frac{p(\text{exterior link branches})/(\text{number of exterior links})}{p(\text{interior link branches})/(\text{number of interior links})} . \tag{1}$$

Three values of b are of particular interest in terms of stream-channel network growth. If growth is restricted to interior-link branching, $b = 0$; if growth occurs only along exterior links, $b = \infty$. A particularly interesting case arises when $b = 1$, which involves growth both along exterior and along interior links, and generates channel networks that satisfy the condition of topological randomness.

It will be shown that the two major breakage–selection models dealt with in our 1979 study (Dacey and Krumbein, 1979) may be represented as channel networks that are identical with those for the growth models with $b = 0$, and $b = \infty$. As stated in the introduction, a main objective here is to develop a new breakage–selection model that may be represented as channel networks that are identical with those for the growth model with $b = 1$; for this breakage–selection model, the particle-size distribution has been obtained. But, in turn, this model for particle sizes suggests a new model for growth of stream-channel networks that obeys the condition of topological randomness. This new growth model has $b = \infty$, so it and the growth model with $b = 1$ have quite different branching probabilities, but they both generate stream-channel networks satisfying topological randomness. The last part of this essay examines some of the implications that competing and contradictory growth models with the same pattern of growth may have for geomorphological theory.

Statement of network-growth and particle-size models

The similarities between the models of network growth and of particle breakage are formulated in the language of channel networks. Because the pertinent terminology is rather widely used in the geological literature, basic concepts and definitions are not repeated here, but Shreve (1966; 1967), and Dacey and Krumbein (1976) may be consulted for explicit statements. It is, however, convenient to restate here the two basic models of network growth and particle sizes.

Definition 1   A network-growth model is a sequence of channel networks. At each time, t_n $(n = 1, 2, ...)$, there is a channel network of magnitude n that, for $n \geqslant 2$, is the result of branching of one link of the channel network at time t_{n-1}.

Assumption 1   One link of the channel network at time t_n branches, and the probabilities of branching are as follows:

p(tributary develops on the right of each interior link) $= \alpha_n$,

p(tributary develops on the left of each interior link) $= \alpha_n$,

p(each exterior link branches) $= \beta_n$,

with

$$2(n-1)\alpha_n + n\beta_n = 1 . \tag{2}$$

Additionally, either all α_n are zero, or there is a nonnegative real number, b, such that

$$b = \frac{\beta_n}{2\alpha_n} , \qquad n \geqslant 2 . \tag{3}$$

If all $\alpha_n = 0$, set $b = \infty$.

Notice that equations (1) and (3) are consistent. Also, for a given value of b, the values of α_n and β_n are determined by equation (2) for each value of n.

The model for particle sizes is stated next in a slightly different language to that used by Dacey and Krumbein (1978), but its structure is not altered.

Definition 2   A breakage–selection model is defined by three conditions: *initial state, nth breakage rule, and nth selection rule.* For the *initial state*, there is at time t_0 a single particle of unit size. The particles and their sizes at t_n $(n = 1, 2, ...)$ are obtained from the particle sizes at t_{n-1} by selecting one particle and breaking it into two smaller particles. The *nth selection rule* describes the manner of selecting this particle from the particles at t_{n-1}, this is called the *nth selection*. The *nth breakage rule* describes the manner in which this particle is divided.

Definition 3   A particle of size x is *evenly divided* when a breakage produces two particles, each having size $\frac{1}{2}x$; it is *uniformly divided* when a breakage produces two particles, one of size xX and the other of size $x(1-X)$, where X is a random variable uniformly distributed on the open interval $(0, 1)$. A random variable is *uniformly distributed* on the interval 0 to 1 when it has the density function

$$g(x) = 1 , \qquad 0 < x < 1 .$$

Assumption 2   The *n*th breakage rule for all n is either that particles are evenly divided, or that they are uniformly divided. The first selection is the initial particle, and for $n \geqslant 2$, the nth selection rule satisfies:

p[particle produced by $(n-1)$th breakage is selected] $= p_n$,

p[particle not produced by $(n-1)$th breakage is selected] $= q_n$,

with

$$2p_n + (n-1)q_n = 1 \ .$$

Our study of the network-growth model considered three sets of branching probabilities, and thus three different values of b, which were called models (a), (b), and (c). Our study of particle sizes considered two sets of selection probabilities, which were called models 1 and 2; for each set, particles were evenly and uniformly divided, but for the present purposes only the selection probabilities are relevant. These models, their b criteria, and their probability sets are summarized in table 1. There are also several additional models listed in this table. Model (d) is not new, although it was not explicitly considered by Dacey and Krumbein (1976); model (c) was considered, but is not used in this study. Models (e) and 3 are newly developed in this study.

First, it is shown that models (a) and 2 have identical graphic structures, as do models (d) and 1. We then construct the new particle-size model 3, which has the graphic structure of model (b), but a different value of b. This, in turn, prompts the development of the new network-growth model (e), which also has the same graphic structure as model (b) and a different value of b.

Table 1. Definitions of network-growth and particle-size models.

Network-growth model			Particle-size model		
b [a]	selection probabilities		b [a]	selection probabilities	
(a) ∞	$\alpha_n = 0$	$\beta_n = \dfrac{1}{n}$	1 ∞	$p_n = \tfrac{1}{2}$	$q_n = 0$
(b) 1	$\alpha_n = \dfrac{1}{2(2n-1)}$	$\beta_n = \dfrac{1}{2n-1}$	2 ∞	$p_n = \dfrac{1}{n+1}$	$q_n = \dfrac{1}{n+1}$
(c) $\tfrac{1}{2}$	$\alpha_n = \dfrac{1}{3n-2}$	$\beta_n = \dfrac{1}{3n-2}$	3 ∞	illustrated in figure 5	
(d) 0	$\alpha_n = \dfrac{1}{2(n-1)}$	$\beta_n = 0$			
(e) ∞	illustrated in figure 5				

[a] If $b = 0$, only interior links branch and all channel networks are 'fishbones'.
If $b = \infty$, only exterior links branch and channel networks have all allowable shapes.
If $0 < b < \infty$, interior and exterior links both branch; channel networks have all allowable shapes.

Similarities between models

The statement of relations between models is facilitated by use of Smart's (1969) concept of ambilateral channel networks: two channel networks belong to the same *ambilateral class*, and are called *ambilaterally equivalent*, if they can be made topologically identical by interchange of the left and right branches at one or more forks of one of the channel networks. Two links belong to the same *ambilateral* (link) *class*, and are called *ambilaterally equivalent*, if they belong either to the same channel network or to ambilaterally equivalent channel networks, and if they can be made topologically identical by interchange of the branches at one or more forks of one of the channel networks. Figure 1 illustrates these concepts. Clearly, channel networks can be ambilaterally equivalent only if they are of the same magnitude.

As another preliminary, it is well known (for example, from Shreve, 1966) that the number, $N(n)$, of topologically distinct channel networks of magnitude n is given by

$$N(n) = \frac{(2n-2)!}{(n-1)!\,n!} ,$$

and Shreve illustrates them for small n. The network-growth model implies that at each t_n, there is a channel network of magnitude n that is one of the $N(n)$ topologically distinct forms. One link of this channel network branches, in accordance with the probabilities given in assumption 1, to produce the magnitude-$(n+1)$ channel network for time t_{n+1}. For some sets of branching probabilities, Dacey and Krumbein (1976) gave the probability for the occurrence of each of the topologically distinct channel networks, and these are the same for ambilaterally equivalent channel networks.

For the particle-size models, at t_n there are $(n+1)$ particles. Dacey and Krumbein (1979) showed that the sequence of selections that precede breakage of particles may be represented by diagrams of the type illustrated on the left-hand side of figures 2(a) and 2(b). By making obvious changes these diagrams may be shown as channel networks, as illustrated on the right-hand side of these two figures. On the channel network for particles at t_n, the outlet corresponds to the initial particle, the n forks correspond

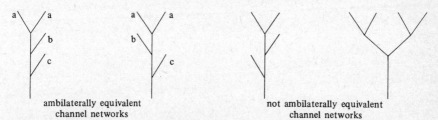

ambilaterally equivalent
channel networks

not ambilaterally equivalent
channel networks

Figure 1. Examples of use of the terms *ambilaterally equivalent* channel networks and *ambilaterally equivalent* links. Ambilaterally equivalent links are indicated by the same letter.

to the n selections, and the $(n+1)$ exterior links correspond to the $(n+1)$ particles produced by the n breakages. The $(n+1)$th selection identifies one particle, which is broken by the $(n+1)$th breakage into two particles to produce the $(n+2)$ particles for t_{n+1}. The selected particle corresponds to an exterior link; bifurcation of this link produces the channel network corresponding to the particles at t_{n+1}. When the selection rules for particles are probabilistic, each sequence of particle selections has a probability of occurrence that is assigned to the channel network corresponding to this sequence. Because the probabilities assigned to channel networks depend only upon the selection probabilities, and not on the manner in which the particles are broken, models with the same selection rules, but with different breakage rules, are represented by channel networks having the same probabilities.

In representing a breakage–selection model by channel networks, only the sequence of selections is taken into account. Each sequence of n selections and breakages is represented by a unique channel network of magnitude $(n+1)$. However, some channel networks result from more than one sequence, and figure 3 shows a simple example.

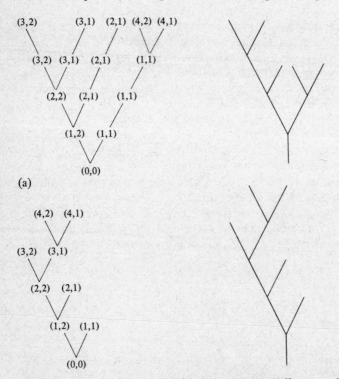

Figure 2. Channel networks (right) representing two different models for particle selection and breakage (left). (a) Model 2; (b) model 1, channel network is a fishbone. (Source: Dacey and Krumbein, 1979.)

Notice that the growth model starts at time t_1 and generates a magnitude-n channel network for time t_n, whereas the breakage–selection model starts at time t_0 and generates a magnitude-$(n+1)$ channel network for time t_n. The different starting times were selected to simplify the notation for the derivations in Dacey and Krumbein (1976; 1979). The unfortunate implication for this essay is that the identification of identical channel networks involves comparing the breakage–selection model at t_n with the network-growth model at t_{n+1}, but this should cause less confusion than shifting the starting time for one of the models.

Models (a) and 2 are related in the following way. For the network-growth model with $b = \infty$, $\alpha_{n+1} = 0$, and $\beta_{n+1} = 1/(n+1)$, all branching is by bifurcation of exterior links; at t_{n+1} there are $(n+1)$ such links, and each bifurcates with equal probability. For breakage–selection model 2 with $p_n = q_n = 1/(n+1)$, the nth selection is, with equal probability, one of the $(n+1)$ particles existing at t_n, and the channel network that

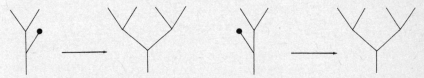

Figure 3. Bifurcation of the link marked by a dot produces the channel network on the right from the one on the left. Bifurcation of other links of these magnitude-3 channel networks produces a fishbone (see definition 4).

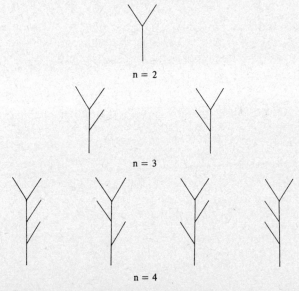

n = 2

n = 3

n = 4

Figure 4. Examples of the possible fishbones of magnitudes two, three, and four. There are eight fishbones of magnitude five.

represents the resulting breakage is obtained by bifurcation of the exterior link associated with this particle. As the initial channel network for both sequences has magnitude one, this breakage–selection model at t_n and this growth model at t_{n+1} are represented by the same channel networks, with the same probabilities of generation.

Models (d) and 2 are related through the *fishbones*, which are defined below and illustrated in figure 4.

Definition 4 A *fishbone* is a channel network for which at least one link of each fork is an exterior link. By convention, a channel network of magnitude 1 is a fishbone.

Notice that, since $(n-2)$ of the n exterior links of a magnitude-n channel network may have a left or right orientation, the number of topologically distinct fishbones of magnitude n $(n \geqslant 2)$ is 2^{n-2}.

For breakage–selection model 1, the nth selection is, with equal probability, one of the two particles produced by the $(n-1)$th breakage. It is readily verified (by induction, if a proof is needed) that the channel networks that represent this model at t_n are restricted to the 2^{n-1} fishbones of magnitude $(n+1)$, and that each occurs with equal probablity. For network-growth model (d), all β are equal to zero, so that all branching is (for $n \geqslant 2$) by tributary developments on interior links. It can be verified (again, by induction) that the resulting channel networks at t_n are the fishbones with magnitude n, and that each occurs with equal probability. Thus, model (d) at t_{n+1} and model 1 at t_n are represented by the same channel networks, with the same probabilities of generation.

The preceding discussion leads to the following results.

Property 1 Every network-growth model may be represented at each time t_n by a channel network of magnitude n, and the particles produced for time t_n by every breakage–selection model may be represented by a channel network of magnitude $(n+1)$.

Property 2 The channel networks that represent network-growth model (d) at time t_{n+1} and breakage–selection model 1 at time t_n are limited to the 2^{n-1} fishbones of magnitude $(n+1)$, each of which occurs with equal probability.

Property 3 The channel networks that represent network-growth model (a) at time t_{n+1} and breakage–selection model 2 at time t_n include all of the $N(n+1)$ topologically distinct channel networks of magnitude $(n+1)$; each of these channel networks occurs with the same probability.

These observations establish the identities between the channel networks representing models 1 and 2 and those for the two extreme conditions, $b = 0$, $b = \infty$, for the growth of channel networks. The next section develops the particle-size distributions for a breakage–selection model with the same channel-network representation as model (b).

Breakage–selection model 3

This particle breakage–selection model can be represented by channel networks, as illustrated by figure 2. It is of the type delimited by definition 2, but the selection probabilities do not satisfy assumption 2 and, instead, have the following attributes.

Assumption 3 The particles for each t_n are represented by the exterior links of a channel network of magnitude $(n+1)$; particles represented by ambilaterally equivalent links have the same probability of being the nth selection.

Definition 5 If the particles at each t_n can be represented with equal probability by the $N(n+1)$ topologically distinct channel networks of magnitude $(n+1)$, then the particles are said to *conform to topological randomness* (although the topological randomness is not an attribute of the particles, but only of the channel networks that represent them).

The selection probabilities required by assumption 3 are stated as conditional probabilities which give the probability that each of the n particles at t_{n-1} is the nth selection, given the channel network that represents these particles at t_{n-1}. Accordingly, the selection probabilities for each channel network sum to unity. When these probabilities are the same for ambilaterally equivalent links, they are completely known when they are known for a single channel network in each ambilateral class. For small numbers of particles, it is convenient to enumerate the selection probabilities by actually displaying them on diagrams of channel networks, and the probability given for an exterior link is the selection probability for the corresponding particle. Figure 5 is of this type, and gives the probability associated with each exterior link of one channel network in each ambilateral class, of magnitudes 1, 2, ..., 6.

The analysis of model 3 has two distinct parts. One is to obtain the size distribution of particles that conform to topological randomness at t_n. This is obtained here for breakage rules that evenly or uniformly divide all selected particles. The other part of the analysis is to demonstrate that there are selection probabilities which satisfy assumption 3 and are represented by channel networks that conform to topological randomness. We are not at present able to show that such selection probabilities exist for each t_n, although we are able to specify them for small n. This evidence is presented before the size distributions are obtained.

Property 4 For selection rules defined by the probabilities displayed in figure 5, the particles at time t_n, for $n \leqslant 6$, conform to topological randomness.

To show this property of the particles at t_n, it is necessary to show that they are represented with equal probability by each of the $N(n+1)$ topologically distinct channel networks of magnitude $(n+1)$. Our method of proof is numerical, and involves actually calculating the probability of occurrence of each topologically distinct channel network. In addition to this proof, it is also instructive to indicate how selection probabilities that satisfy property 4 are obtained, and this task is completed first.

To obtain selection probabilities for magnitude-n channel networks, these channel networks, and those of magnitude $(n+1)$, are arbitrarily numbered 1, ..., $N(n)$, and 1, ..., $N(n+1)$, respectively. The exterior links

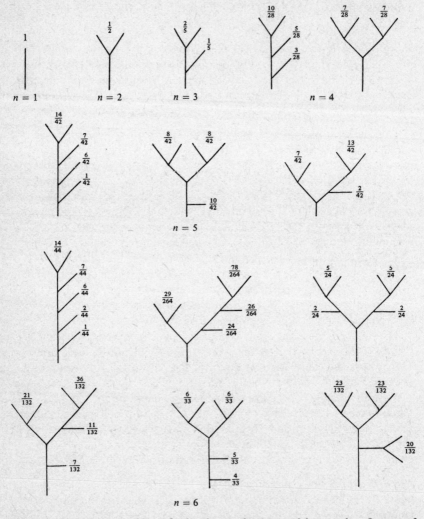

Figure 5. Selection probabilities for breakage-selection model networks. One set of selection probabilities is given for a channel network in each ambilateral class of magnitudes 1, ..., 6. In some places, one probability is given for a pair of ambilaterally equivalent links which have the same probability. For $n = 6$, the probabilities for each ambilateral class are expressed independently, to avoid unwieldy fractions.

of each magnitude-n channel network are also arbitrarily numbered 1, ..., n. Let p_{ij} be the selection probability for the ith link of the jth, magnitude-n channel network, let q_{ijk} equal one or zero as bifurcation of this link produces, or does not produce, the kth magnitude-$(n+1)$ channel network, and let λ_k be the probability of occurrence of the kth magnitude-$(n+1)$ channel network. If the magnitude-n channel networks occur with equal probability, given by $1/N(n)$, thereby satisfying the hypothesis of topological randomness, then

$$\lambda_k = \sum_{i=1}^{n} \sum_{j=1}^{N(n)} \frac{p_{ij} q_{ijk}}{N(n)}, \qquad k = 1, 2, ..., N(n+1) .$$

The magnitude-$(n+1)$ channel networks obey the hypothesis of topological randomness for selection probabilities, p_{ij}, such that $\lambda_k = 1/N(n+1)$ for each k. A trial-and-error process can be used to identify selection probabilities that satisfy the $N(n+1)$ simultaneous equations, with the constraint, required by assumption 3, that $p_{ij} = p_{hj}$ when the ith and hth links are ambilaterally equivalent. Because the selection probabilities are the same for all ambilaterally equivalent channel networks, the number of equations can be reduced to one for each ambilateral class. Although the number of equations remains large even for small n, we have found it easy to identify appropriate selection probabilities, and figure 5 shows one set of solutions for $n \leqslant 6$.

Probabilities for the 7th selection have not been obtained because they are needed for eleven ambilateral classes, and the number of classes escalates rapidly for later selection rules. However, we have no evidence to indicate that suitable probabilities do not exist, and it is, accordingly, conjectured that there are selection probabilities which produce particles conforming to topological randomness at each t_n.

Particle sizes for topological randomness
There is evidence that there are selection probabilities which produce particles conforming to topological randomness, but these probabilities are not needed to obtain the size distribution of these particles; it suffices to accept that these probabilities exist. A key step in the derivation of the particle-size distribution is the establishment of the relation between the numbers of breakages that produce the particles at t_n, and the lengths of the *paths* in the collection of topologically distinct channel networks of magnitude $(n+1)$.

Definition 6 A *path* for an exterior link is a sequence of links, in which no link occurs more than once, that connects the exterior link and the outlet link; the (topological) *length* of a path is the number of links in the sequence.

Property 5 For particles that conform to topological randomness at time t_n, let X_n represent the number of breakages that produced a particle selected with equal probability from the particles existing at t_n. Then for each n $(n > 1)$,

$$p(X_n = k) = 2^k k \frac{(2n-k-1)!\,n!}{(n-k)!\,(2n)!}, \qquad k = 1, 2, ..., n .$$

Proof Let Y_n represent the length of a path which is selected with equal probability from all paths of the $N(n)$ topologically distinct channel networks of magnitude n. From equation (13a) of Werner and Smart (1973), it may be deduced that $p(Y_1 = 1) = 1$, and that for $n \geqslant 2$, each $Y_n \geqslant 2$, and

$$p(Y_n = k) = 2^{k-1}(k-1) \frac{(2n-k-2)!\,(n-1)!}{(2n-2)!\,(n-k)!}, \qquad k = 2, 3, ..., n .$$

It can now be shown that

$$p(X_n = k) = p(Y_{n+1} = k+1) . \tag{4}$$

The condition that particles conform to topological randomness requires that these particles at t_n are represented with equal probability by each of the $N(n+1)$ topologically distinct channel networks of magnitude $(n+1)$. As each such network has $(n+1)$ exterior links, X_n is also the number of breakages that produce a particle selected with equal probability from the particles represented by all exterior links of all topologically distinct channel networks of magnitude $(n+1)$. The number of breakages that produce a particle is one less than the length of the path for the exterior link that corresponds to that particle. Thus, X_n is also one less than the length of the path for an exterior link selected with equal probability from all exterior links of all topologically distinct channel networks of magnitude $(n+1)$. As the length for this exterior link is $(Y_{n+1} - 1)$, $p(X_n = k) = p(Y_{n+1} - 1 = k)$. This implies relation (4), which, in turn, implies the probability terms for X_n.

Equation (4) identifies a fundamental relation between models of particle-size distributions and channel-network growth that obey the hypothesis of topological randomness.

The following notation for particle sizes is the same as that used by Dacey and Krumbein (1979), which facilitates use of the properties they gave in appendix A for all breakage–selection models which satisfy the conditions of definition 2.

Definition 7 For particles conforming to topological randomness with even division of particles, let U_n denote the size (relative to that of the original particle at t_0) of a particle selected with equal probability from the $(n+1)$ particles at t_n, and let V_n denote the size when each is selected with probability proportional to its size. When there is uniform division of particles, the corresponding random variables are U_n^* and V_n^*, which have density functions u_n^* and v_n^*, respectively. Expectation is denoted by E.

The random variables U_n and U_n^* correspond to sampling of particles by *number frequency*, whereas V_n and V_n^* correspond to sampling by *size frequency*. The following properties of these four random variables are verified in the appendix to this chapter. Some of these properties use the hypergeometric series function, $_{m+1}F_l$, given by

$$_{m+1}F_l(-n, a_1, a_2, ..., a_m; b_1, b_2, ..., b_l; t)$$

$$= \sum_{i=0}^{n} \frac{\Gamma(i-n)\Gamma(i+a_1)\Gamma(i+a_2) ... \Gamma(i+a_m)}{\Gamma(-n)\Gamma(a_1)\Gamma(a_2) ... \Gamma(a_m)}$$

$$\times \frac{\Gamma(b_1)\Gamma(b_2) ... \Gamma(b_l)}{\Gamma(i+b_1)\Gamma(i+b_2) ... \Gamma(i+b_l)} \frac{t_i}{i!} ,$$

where $\Gamma(z)$ is the gamma function.

Property 6 For each n, $n \geqslant 1$

$$p(U_n = 2^{-k}) = 2^k k \frac{(2n-k-1)! n!}{(n-k)!(2n)!} , \qquad k = 1, 2, ..., n ;$$

$$E(t^{U_n}) = \frac{t^{-\frac{1}{2}}}{2n-1} \, _2F_1(-n+1, 2; -2n+2; 2t^{-\frac{1}{2}}) ;$$

$$p(V_n = 2^{-k}) = k \frac{(2n-k-1)!(n+1)!}{(n-k)!(2n)!} , \qquad k = 1, 2, ..., n ;$$

$$E(t^{V_n}) = \frac{n+1}{2(2n-1)} \, _2F_1(-n+1, 2; -2n+2; t^{\frac{1}{2}}) .$$

Property 7 For each n, $n \geqslant 1$

$$u_n^*(x) = \frac{1}{2n-1} \, _2F_1(-n+1, 2; -2n+2; 1; -2\ln x) , \qquad 0 < x < 1 ;$$

$$E[(U_n^*)^j] = \frac{1}{(2n-1)(j+1)} \, _2F_1\left[-n+1, 2; -2n+2; \frac{2}{(j+1)}\right] ,$$
$$j = 1, 2, ... ;$$

$$v_n^*(x) = \frac{x(n+1)}{2n-1} \, _2F_2(-n+1, 2; -2n+2, 1; -2\ln x) , \qquad 0 < x < 1 ;$$

$$E[(V_n^*)^j] = \frac{n+1}{(2n-1)(j+2)} \, _2F_1\left[-n+1, 2; -2n+2; \frac{2}{(j+2)}\right] ,$$
$$j = 1, 2, ...$$

The hypergeometric series functions occurring in the expressions for moments are evaluated only by numerical methods, except for the case $E(U_n) = E(U_n^*) = 1/(n+1)$.

As n increases, there is a corresponding increase in the number of particles which divide the unit size of the initial particle. Thus the mean size of particle goes to zero as n increases:

$$\lim_{n \to \infty} E(U_n) = \lim_{n \to \infty} E(U_n^*) = \lim_{n \to \infty} \frac{1}{n+1} = 0 \ .$$

It follows that

$$\lim_{n \to \infty} p(U_n > 0) = \lim_{n \to \infty} p(U_n^* > 0) = 0 \ .$$

A sufficiently large number of random breakages will therefore produce sizeless particles, in the sense that

$$\lim_{n \to \infty} p(U_n = 2^{-k}) = 0 \ , \qquad \text{for all finite } k \ , \tag{5}$$

and

$$\lim_{n \to \infty} u_n^*(x) = 0 \ , \qquad \text{for all } x > 0 \ . \tag{6}$$

Nevertheless, it can now be established that for sampling by size frequency, the particles having nonzero size account for the entire unit size of the initial particle, in the sense that the limits of V_n and V_n^* are nondegenerate, positive-valued random variables.

Property 8 If the limits as n goes to infinity of V_n, V_n^*, and v_n^* are denoted by V, V^*, and v^*, respectively, then V satisfies

$$p(V = 2^{-k}) = k2^{-k-1} \ , \qquad k = 1, 2, \dots ,$$

and

$$E(V^j) = \frac{2^j}{(2^{j+1}-1)^2} \ , \qquad j = 1, 2, \dots ,$$

with

$$p(V > 0) = 1 \ .$$

Also, V^* satisfies

$$v^*(x) = \tfrac{1}{2}x \, {}_1F_1(2; \ 1; \ -\ln x) \ , \qquad 0 < x < 1 \ ,$$

$$E[(V^*)^j] = \frac{j+2}{2(j+1)^2} \ , \qquad j = 1, 2, \dots ,$$

with

$$p(V^* > 0) = 1.$$

Property 9 Let assumption 3 define model 3, and let $\mu_i(n)$ and $\sigma_i^2(n)$ be the mean and variance of the particle size U_n^* at time t_n for model i ($i = 1, 2,$ or 3). For all n,

$$\mu_1(n) = \mu_2(n) = \mu_3(n) \ ,$$

and for sufficiently large n,

$$\sigma_2(n) < \sigma_1(n) < \sigma_3(n) .$$

Property 10 Let $\rho_i(n)$ be the ratio variance/mean for the particle size V_n^* for model i. Then, for sufficiently large n,

$$\rho_2(n) < \rho_1(n) < \rho_3(n) .$$

It also follows from comments in Dacey and Krumbein (1978) that the variance–mean ratio for these particles is larger than for particles with sizes described by the lognormal probability law.

Property 11 Let $Y = -\ln V^*$. The density function, $f_Y(y)$, and moments of Y are then given by

$$f_Y(y) = \tfrac{1}{2} \exp(-2y)\,{}_1F_1(2; 1; y) , \qquad y > 0 ,$$

and

$$E(Y^j) = 2^{-j+1}(j-1)!\,{}_2F_1[j, 2; 1; \tfrac{1}{2}] .$$

The function ${}_1F_1$ is a confluent hypergeometric series function, and Y is considered to obey a confluent-hypergeometric-series probability law.

Extension of random breakage process to a growth model for channel networks

Selection rules of the type illustrated in figure 5 for particles also have implications for growth models of stream-channel networks. Dacey and Krumbein (1976) introduced three growth models based on bifurcation and tributary development, and showed that both processes have to occur with essentially equal probabilities to generate networks that satisfy both Shreve's (1966; 1967) model of topological randomness, and the conditions of assumption 1. We have also pointed out that these probabilities for link branching are independent of their position within the network.

Geomorphological consensus favors network growth mainly by exterior-link branching at the upstream end of the network, and very likely it would support the view that the probability of branching diminishes as one moves downstream. By adopting the probability trees of the random particle-breakage model for channel networks, it is possible to construct a growth model that simultaneously satisfies the probabilities of Shreve's model, and the consensus that network growth is most rapid in its upstream end. In this model all growth is by bifurcation of exterior links only, with no provision for later stages of growth by tributary development. We first examine some implications of the bifurcation model, and then discuss tributary development.

For the bifurcation growth sequence to satisfy the condition of topological randomness at each growth stage, it is required that at each t_n, each of the $N(n)$ topologically distinct channel networks of magnitude n occurs with equal probability. Under these conditions, the growth model (in the sense of definition 1, but not satisfying assumption 1) has probabilities of exterior-link bifurcation identical with those specified in figure 5. For each channel network, these probabilities sum to unity. It is an immediate consequence of property 4 that each of the $N(n)$ topologically distinct channel networks of magnitude n occurs with equal probability for $1 \leqslant n \leqslant 7$. For $n > 7$, we proceed more tentatively.

Assumption 4   Only exterior links branch, and all ambilaterally equivalent exterior links bifurcate with the same probability.

Conjecture   There is a network-growth model obeying assumption 4 which, at each t_n, satisfies the condition of topologically random channel networks.

If this conjecture is true, and if the evidence suggesting the validity of the generalization of property 4 is applicable, there is conflict with conclusions derived from assumption 1 on the importance of tributary development in the growth of channel networks which obey the model of topological randomness.

There seems to be no *a priori* basis for judging the relative merits of these two conflicting assumptions, mainly because criteria are not available for unambiguous interpretation of field and map observations. The weight of evidence does suggest that channel networks tend to approach an optimum drainage density by developing tributaries from interior links as the network grows; this intuitive judgment suggests that growth by bifurcation alone may not be sufficient, in the real world, to develop topological randomness in the sense of Shreve's (1966; 1967) model.

Despite these present difficulties, it is possible to express the two choices set out above as special cases of a more general formulation which allows all ambilaterally equivalent links to branch with the same probability. Assumption 1 is a special case which, in addition, combines the exterior and interior links into different classes, and assigns separate probabilities (one of which may be zero) to each class. Assumption 4 combines the interior links into a single class which is assigned probability zero, whereas the classes of ambilaterally equivalent exterior links are assigned separate probabilities. Thus, both models impose equal probabilities on ambilaterally equivalent links. This means that the postulated growth processes do not depend upon right and left distinctions but, as stated previously, they do reflect, in varying ways, other positional differences of links in ambilaterally equivalent channel networks.

Concluding remarks
It was stated in the introduction that one purpose of this essay is to discuss some general rules for transferring models developed in one field to their counterparts in entirely different fields. Model transfer is by no means a new departure in the sciences; there are many examples in

geology that involve the direct transfer of models from physics or chemistry to geological processes. In almost all of these, however, the criterion for transfer is the demonstration that the geological process is isomorphous with the physical or chemical process. In such cases model transfer is direct, and mainly involves changes in initial and boundary conditions.

In this essay, the model-transfer process is distinctly different, in that it is not contended that the physical process of selection and breakage of particles is also the physical process involved in channel-network growth. On the contrary, the processes may be entirely different, and the motive for transfer may arise from similarity in graphic output (such as the tree diagrams in this example); or because the probabilities of events occurring in one process may seem reasonable for another (such as the diminishing likelihood of selection in succeeding particle generations); or for other reasons. Model transfer is obviously more difficult here and, indeed, may be impossible under some conditions.

Our approach has been to develop a symbolic language applicable to both models, and to use these symbols to seek elements held in common by both. Table 1 includes the new models, (e) and 3. Our treatment is entirely mathematical, in terms of allowable network processes (bifurcation and tributary development) translated to their corresponding breakage-selection processes as revealed by tree diagrams. This demonstrates that the transferred model can exist; however, this is only the first step in a sequence that then tests the model experimentally, or by appropriate observational data.

Acknowledgements. The support of the National Science Foundation Grant SOC-75-16103 is gratefully acknowledged. We are indebted to Ronald L Shreve for calling our attention to some ambiguities in an earlier draft.

References
Dacey M F, 1971 "Probability distribution of number of networks in topologically random network patterns" *Water Resources Research* 7 (6) 1652–1657
Dacey M F, Krumbein W C, 1976 "Three growth models for stream channel networks" *Journal of Geology* 84 153–163
Dacey M F, Krumbein W C, 1979 "Models of breakage and selection for particle size distributions" *Mathematical Geology* 11 (2) 193–222
Shreve R L, 1966 "Statistical law of stream numbers" *Journal of Geology* 74 17–37
Shreve R L, 1967 "Infinite topologically random channel networks" *Journal of Geology* 75 179–186
Smart J S, 1969 "Topological properties of channel networks" *Geological Society of America Bulletin* 80 1757–1774
Werner C, Smart J S, 1973 "Some new methods of topologic classification of channel networks" *Geographical Analysis* 5 271–295

Appendix

Proofs of properties assumed for the models of network growth

Proof of property 6 When there is even division of particles, a particle that is produced by k breakages is the result of k halvings of an initial particle of unit size, and therefore has size 2^{-k}. Since X_n is the number of breakages that produced a particle, $p(X_n = k) = p(U_n = 2^{-k})$, and the probability terms for X_n are given by property 5. The probability generating function of U_n is

$$
\begin{aligned}
E(t^{U_n}) &= \sum_{k=1}^{n} t^{2^{-k}} p(U^n = 2^{-k}) \\
&= \sum_{k=0}^{n-1} (2t)^{\frac{1}{2}(k+1)} \frac{(k+1)!}{k!} \frac{(2n-k-2)!\,n!}{(n-k-1)!\,(2n)!} \\
&= \sum_{k=0}^{n-1} (2t)^{\frac{1}{2}(k+1)} \frac{\Gamma(k+2)}{k!} \frac{\Gamma(2n-k-1)\Gamma(n)}{\Gamma(2n-1)\Gamma(n-k)} \frac{1}{2(2n-1)} \\
&= \frac{t^{\frac{1}{2}}}{(2n-1)} \sum_{k=0}^{n-1} (2t)^{\frac{1}{2}k} \frac{\Gamma(k+2)}{k!} \frac{\Gamma(k-n+1)}{\Gamma(-n+1)} \frac{\Gamma(-2n+2)}{\Gamma(k-2n+2)} \qquad \text{(A1)} \\
&\equiv \frac{t^{\frac{1}{2}}}{(2n-1)} \, {}_2\Gamma_1(-n+1, 2; -2n+2; 2t^{\frac{1}{2}}) \ .
\end{aligned}
$$

Equation (A1) is obtained from the preceding line by two uses of the duplication formula for the gamma function.

Dacey and Krumbein (1978) showed that every breakage–selection process satisfies the relation

$$
p(V_n = 2^{-k}) = (n+1)2^{-k} p(U_n = 2^{-k}) \ ,
$$

which implies the probability terms for V_n. The probability generating function for V_n is obtained by calculations similar to those used for U_n.

Let $p(V_n = 2^{-k}) = p(D_n = k)$; D_n was obtained by Dacey (1971, page 1656), in a study of network patterns composed of a sequence of channel networks.

Proof of property 7 Dacey and Krumbein (1978) showed that U_n and U_n^* are related by

$$
u_n^*(x) = \sum_{k=1}^{n} p(U_n = 2^{-k}) f_k(x) \ , \qquad \text{(A2)}
$$

where

$$
f_k(x) = \frac{(-\ln x)^{k-1}}{(k-1)!}, \qquad 0 < x < 1 \ ; \qquad \text{(A3)}
$$

$f_k(x)$ is the density function of the loggamma probability law. They also showed that

$$
\int_0^1 x^j f_k(x)\,dx = (j+1)^{-k} \ . \qquad \text{(A4)}
$$

So, from equations (A2) and (A3),

$$u_n^*(x) = \frac{1}{2n-1} \sum_{k=0}^{n-1} \frac{(k+1)!}{(k!)^2} \frac{(2n-k-2)!(n-1)!}{(n-k-1)!(2n-2)!} (-2\ln x)^k$$

$$= \frac{1}{2n-1} \sum_{k=0}^{n-1} \frac{\Gamma(k+2)}{k!\,\Gamma(k+1)} \frac{\Gamma(k-n+1)\Gamma(-2n+2)}{\Gamma(-n+1)\Gamma(k-2n+2)} (-2\ln x)^k \qquad \text{(A5)}$$

$$= \frac{1}{2n-1} \, _2F_2(-n+1, 2; -2n+2, 1; -2\ln x)$$

[by definition] .

Hence,

$$E[(U_n^*)^j] = \int_0^1 x^j u_n^*(x)\,dx$$

$$= \frac{1}{2n-1} \int_0^1 \sum_{k=0}^{n-1} \frac{\Gamma(k+2)}{k!} \frac{\Gamma(k-n+1)\Gamma(-2n+2)}{\Gamma(-n+1)\Gamma(k-2n+2)} 2^k x^j f_{k+1}(x)\,dx$$

[by equation (A5)]

$$= \frac{1}{(2n-1)(j+1)} \sum_{k=0}^{n-1} \frac{\Gamma(k+2)}{k!} \frac{\Gamma(k-n+1)\Gamma(-2n+2)}{\Gamma(-n+1)\Gamma(k-2n+2)} \left(\frac{2}{j+1}\right)^k$$

[by equation (A4)]   (A6)

$$= \frac{1}{(2n-1)(j+1)} \, _2F_1\left[-n+1, 2; -2n+2; \frac{2}{j+1}\right] .$$

[by definition] .

Dacey and Krumbein (1978) also showed that

$$v_n^*(x) = (n+1)x\,u_n^*(x) , \qquad\qquad\qquad\qquad\qquad\qquad \text{(A7)}$$

which implies that

$$E[(V_n^*)^j] = (n+1)E[(U_n^*)^{j+1}] .$$

These two relations yield directly the density function and moments of V_n^* from those for U_n^*.

Proof of property 8 Let

$$R(k, n) = k\frac{(2n-k-1)!\,n!}{(n-k)!\,(2n)!} .$$

The approximation

$$\frac{(z+a)!}{(z+b)!} \simeq z^{a-b} , \qquad z \to \infty , \qquad\qquad\qquad\qquad \text{(A8)}$$

yields, for large n,

$$R(k, n) \simeq \frac{2^{-k-1}k}{n} .$$

Since

$$p(U_n = 2^{-k}) = 2^k R(k, n) ,$$

then, for each finite value of k,

$$\lim_{n \to \infty} p(U_n = 2^{-k}) = 2^k \lim_{n \to \infty} R(k, n) = \lim_{n \to \infty} \left(\frac{k}{2n}\right) = 0 ,$$

which indicates that the entire mass of the limit of U_n is concentrated at size zero, and thereby verifies equation (5). Furthermore,

$$p(V_n = 2^{-k}) = (n+1)R(k, n) ,$$

so that

$$p(V = 2^{-k}) = k2^{-k-1} \lim_{n \to \infty} \left(\frac{n+1}{n}\right) = k2^{-k-1} .$$

Thus,

$$E(V^j) = \sum_{k=1}^{\infty} 2^{-jk} p(V = 2^{-k}) = 2^{-j-2} \sum_{k=0}^{\infty} (2^{-j-1})^k (k+1)$$

$$= 2^{-j-2} \sum_{k=0}^{\infty} (2^{-j-1})^k \frac{\Gamma(k+2)}{k!} = \frac{2^{-j-2}}{(1-2^{-j-1})^2} = \frac{2^j}{(2^{j+1}-1)^2} .$$

Next, consider

$$v^*(x) = \lim_{n \to \infty} v_n^*(x)$$

$$= \lim_{n \to \infty} (n+1)x u_n^*(x) \qquad \text{[by equation (A7)]} \qquad \text{(A9)}$$

$$= x \lim_{n \to \infty} \left[-\frac{n+1}{2n-1} \sum_{k=0}^{n-1} \frac{\Gamma(k+2)}{k!\Gamma(k+1)} \frac{\Gamma(k-n+1)\Gamma(-2n+2)}{\Gamma(-n+1)\Gamma(k-2n+2)}(-2\ln x)^k\right]$$

$$\text{[by equation (A5)]}$$

$$= \frac{x}{2} \sum_{k=0}^{\infty} \frac{\Gamma(k+2)}{k!\Gamma(k+1)} (-\ln x)^k \qquad \text{[by equation (A8)]} \qquad \text{(A10)}$$

$$\equiv \frac{x}{2} {}_1F_1(2; 1; -\ln x) .$$

Hence,

$$E[(V^*)^j] = \int_0^1 x^j v_n^*(x)\,dx$$

$$= \frac{1}{2} \int_0^1 \sum_{k=0}^{\infty} \frac{\Gamma(k+2)}{k!} x^{j+1} f_{k+1}(x)\,dx \qquad \text{[by equation (A10)]}$$

$$= \frac{1}{2(j+2)} \sum_{k=0}^{\infty} \frac{\Gamma(k+2)}{k!} \left(\frac{1}{j+2}\right)^k \qquad \text{[by equation (A4)]}$$

$$= \frac{j+2}{2(j+1)^2} .$$

Moreover, since the limit in equation (A9) exists for $0 < x \leqslant 1$, the limit of $u_n^*(x)$ does not, which implies that $p(U_n^* > 0) \to 0$ as $n \to \infty$, and thereby verifies equation (6).

Proof of property 9 Let $\lambda_i(n)$ be the second moment of particle size at t_n for model i, so that

$$\sigma_i^2(n) = \lambda_i(n) - \mu_i^2(n) . \tag{A11}$$

From property 9 in Dacey and Krumbein (1978),

$$\mu_1(n) = \frac{1}{n+1} ,$$

and

$$\lambda_1(n) = \frac{3^n + 1}{3^n\, 2(n+1)} \simeq \frac{1}{2(n+1)} .$$

Also, from property 10 of Dacey and Krumbein (1979),

$$\mu_2(n) = \frac{1}{(n+1)} ,$$

and

$$\lambda_2(n) = \frac{(n-\frac{1}{3})!}{(n+1)!\,(-\frac{1}{3})!} \simeq \frac{1}{(-\frac{1}{3})!\, n^{4/3}} ,$$

where $(n+x)!/(n+y)!$ is approximated by n^{x-y}. From property 7 of this essay,

$$\mu_3(n) = \frac{1}{(n+1)} .$$

This verifies that the expected values are equal for the three models. From equation (A6)

$$
\begin{aligned}
\lambda_3(n) &= \frac{1}{3(2n-1)} \sum_{k=0}^{n-1} \frac{\Gamma(k+2)}{k!} \frac{\Gamma(k-n+1)}{\Gamma(-n+1)} \frac{\Gamma(-2n+2)}{\Gamma(k-2n+2)} \left(\frac{2}{3}\right)^k \\
&= -\frac{2}{3} \sum_{k=0}^{n-1} \frac{\Gamma(k+2)}{k!} \frac{\Gamma(k-n+1)}{\Gamma(-n)} \frac{\Gamma(-2n)}{\Gamma(k-2n+2)} \left(\frac{2}{3}\right)^k \\
&\simeq -\frac{2}{3} \sum_{k=0}^{n-1} \frac{\Gamma(k+2)}{k!} (-n)^{k+1}(-2n)^{-k-2} \left(\frac{2}{3}\right)^k \\
&\simeq \frac{1}{6n} \sum_{k=0}^{n-1} \frac{\Gamma(k+2)}{k!} \left(\frac{1}{3}\right)^k = \frac{1}{6n} \sum_{k=0}^{\infty} \frac{\Gamma(k+2)}{k!\,\Gamma(2)} \left(\frac{1}{3}\right)^k \\
&\simeq \frac{1}{6n(1-\frac{1}{3})^2} = \frac{3}{8n} .
\end{aligned}
$$

Thus,

$$\lambda_2(n) < \lambda_1(n) < \lambda_3(n) .$$

In view of equation (A11) and the equality of expected values, this implies the stated inequalities for the variances.

Proof of property 10 For models 1 and 3, V^* exists, and is used as an approximation to V_n^* for large n. From property 8, the expected value, variance, and variance/mean ratio for model 3 are

$$\mu_3 = \frac{3}{8}, \qquad \sigma_3^2 = \frac{47}{576}, \qquad \rho_3 = \frac{47}{216};$$

and from property 8 of Dacey and Krumbein (1979),

$$\mu_1 = \frac{1}{2}, \qquad \sigma_1^2 = \frac{1}{12}, \qquad \rho_1 = \frac{1}{6}.$$

For model 2, property 10 of the same paper gives

$$E[(V_n^*)^j] = \binom{j'}{n},$$

where, in the binomial coefficient $\binom{j'}{n}$, $j' = n + 2/(j+2) - 1$. Using approximation (A8), we obtain

$$\mu_2(n) \simeq \frac{1}{n^{1/3}(-\frac{1}{3})!},$$

and

$$\sigma_2^2(n) = \frac{(n-\frac{1}{2})!}{n!(-\frac{1}{2})!} - \left[\frac{(n-\frac{1}{3})!}{n!(-\frac{1}{3})!}\right]^2 = \frac{(n-\frac{1}{2})!\,n!(-\frac{1}{3})!^2 - (n-\frac{1}{3})!^2(-\frac{1}{2})!}{n!^2(-\frac{1}{2})!(-\frac{1}{3})!^2}$$

$$\simeq \frac{(-\frac{1}{3})!}{n^{1/2}(-\frac{1}{2})!}.$$

So, from these expressions for $\sigma_2^2(n)$ and $\mu_2^2(n)$, we obtain

$$\rho_2(n) \simeq \frac{(-\frac{1}{3})!(-\frac{1}{3})!}{(-\frac{1}{2})!\,n^{1/6}}.$$

Thus, $\rho_2(n) \to 0$ as $n \to \infty$, whereas the corresponding limits for models 1 and 3 are $\frac{1}{6}$ and $\frac{47}{216}$ ($> \frac{1}{6}$), respectively.

Part 3

Resources

A review of the practical gains from applications of geostatistics to South African ore valuation

D G Krige

Introduction

Geostatistics was the name given by Matheron some twenty years ago to the theory and techniques developed mainly for the valuation of ores, and which take account of the spatial relationships between, or structure of, the regionalised variable(s) concerned. Geostatistics can and should, however, be accepted in a somewhat wider context, so as to cover applications of a more classical statistical nature whose development preceded and led to the establishment of geostatistics proper.

The technical details of the theories, models, and techniques developed since the late 1940s have already been covered adequately in various publications (Krige, 1962; 1978; de Wijs, 1951; 1953; Matheron, 1962; 1971; David, 1977), and will also be dealt with by other authors in this publication. This essay will cover only the main practical advantages gained in the South African gold mines from these developments.

Up to the late 1940s ore valuation in the South African gold mines followed orthodox procedures. The seemingly haphazard variation in values did not appear to follow any logic, and some arbitrary adjustments, such as the cutting of high values, were prevalent. As in other branches of science, no significant advances were made until some appropriate pattern or model was discovered for these values.

One of the most important steps forward was therefore the recognition and demonstration by Sichel (1947) that statistical methods could be applied in this field, and that the lognormal-frequency-distribution model was appropriate for gold-ore values.

The introduction of the two-parameter lognormal model—later extended to the three-parameter model (Krige, 1960)—and of the de Wijsian model for the spatial structure of gold-ore values (de Wijs, 1951), led directly or indirectly to the following main advances in South African ore-valuation technology:

1 improved ore-grade estimates, that is, estimates with smaller limits of error, particularly for the valuation of a new gold mine based on a small set of borehole values which can include an exceptionally high value(s);

2 the recognition of skew limits of error for such estimates and the development of the necessary small sampling theory for the lognormal model;

3 a logical model for the distributions of ore-block grades, and hence for estimating the tonnage and grade to be obtained when the ore body is eventually exploited on a selective basis;

4 the application of risk-analysis techniques to the problem of capital investment in new mining ventures;

5 a valid explanation of the conditional biases inherent in orthodox ore-reserve valuations, and a simple statistical regression model for eliminating these biases;

6 as a consequence of point 5 above, the development of all the modern sophisticated models and techniques of geostatistics proper, and the consequent financial gains associated with the more efficient assessment of ore reserves.

Improved ore-grade estimates, limits of error, and risk analysis
The superior efficiency of the likelihood estimator of the mean of a lognormally distributed population (corresponding to the statistical grade estimate of an ore body), relative to the orthodox arithmetic mean, can conveniently be measured by the ratio of the relevant error variances as reflected in table 1 (Sichel, 1952).

From table 1 it is evident that the lognormally based t' estimator provides a substantial advantage when the variability of the ore values is high. However, a practical follow-up demonstration of any such claim is always advisable and, to this end, the next example will be self-evident. The original seven surface boreholes on the Western Holdings mine in the Orange Free State goldfield gave the following basal-reef data (Krige, 1961).

Borehole values (in dwt)	182, 237, 280, 327, 383, 418, 6399
Arithmetic mean	1175 in dwt
Statistical t' estimator	832 in dwt (3rd parameter = 60)
5% lower limit	396 in dwt
5% upper limit	2050 in dwt (σ^2 limited to 1·3)
Average of underground development to 1977	698 in dwt over 163·473 km

(1 in dwt = $4·3543$ cm g t^{-1}, where 1 t $\equiv 10^3$ kg)

The skew error distribution of the lognormal t' estimator, evident from the above limits of error, can be compared with that of Student's distribution for the normal case, but its solution is far more complicated. The distribution is very skew; it approaches the lognormal shape only for very low error variances, and tends towards the normal distribution only

Table 1. Efficiency of arithmetic mean relative to that of t' estimator (accepted as 1).

Population variance (logarithmic), σ	Sample size, n				
	5	10	20	50	∞
0·5	0·98	0·97	0·97	0·97	0·96
1·0	0·94	0·91	0·89	0·88	0·87
1·5	0·87	0·82	0·79	0·77	0·75
2·0	0·81	0·72	0·67	0·65	0·63
3·0	0·68	0·53	0·46	0·42	0·39

in the limiting case (Sichel, 1952; 1966; Wainstein, 1975), as will be evident from table 2.

In the above example of the Western Holdings mine, the 5% lower limit of 396 in dwt, if not determined on the correct error distribution model, would have been as follows.

Based on a lognormal error distribution: 372 in dwt;

based on a normal error distribution: 12 in dwt.

We can now examine the position where, under present economic conditions, the risk of a mine such as the above proving not to be a viable investment would be accepted as equivalent to, say, a 5% chance of the actual average grade being lower than a critical level of, say, 300 in dwt. This criterion would be well met by the correct 5% lower limit of 396 in dwt, as well as by that based on the lognormal error distribution pattern, that is, 372 in dwt. However, on the assumption of symmetrical normal limits of error, the 5% lower limit of 12 in dwt would apply; the chances of the actual grade being below 300 in dwt, and hence the chances of failure, would be assessed at about 15%, and the mine would not be exploited. The advantages to be gained from a decision to proceed with the exploitation of a potentially profitable mine, based on a correct assessment of the limits of error, are obvious.

For *any* grade estimate based on lognormally distributed values, whether for a whole mine or a single ore block, it is therefore essential to allow for the appropriate skew limits of error; this skewness can only be ignored where the logarithmic error variances are substantially below 0·005. The same principle will naturally apply even where the basic value distributions are significantly skew, but not necessarily lognormal.

The above example can be extended to cover the more sophisticated procedures of a detailed risk analysis for such a new mining project. This was done in the case of the Prieska copper–zinc mine (Krige, 1973), and more recently for a large new gold mine (Munro, 1977). As pointed out in these analyses, the major risk variables are ore grade and future metal prices. Both of these factors affect the gross revenue of the project, but only the former can be estimated scientifically and objectively. For this purpose, the use of the correct skew error distribution(s) for the grade

Table 2. Upper and lower (5%) limits of error (relative to estimated mean), for the t' and lognormal distributions.

Distribution	Limit	Logarithmic error variance (%)				
		0·005	0·01	0·05	0·1	0·5
t' estimator ($n = 10$)	upper	19	27	72	119	571
	lower	−11	−14	−26	−34	−55
Lognormal	upper	12	19	48	76	313
	lower	−11	−15	−29	−37	−60

estimate(s) is essential and provides an advantage which can be measured by the improved chances of taking the right decision to go ahead with a potentially profitable proposition, and of not losing very large capital sums [up to R400 million (R1 ≈ $1.30, 1981) for a new large gold mine] in investing in potential failures.

Ore-block distributions for selective mining

The effect of the size of the basic ore unit (that is, a single ore sample or borehole core) versus a large ore block (the population member) on the variance of a gold-ore value distribution (the population), was demonstrated by Ross (1950) and Krige (1951; 1952; 1962); the extension of the application of the lognormal pattern from sample values to ore-block distributions was also made.

Where, as is invariably the case in practice, selection of the ore blocks with payable gold grades (that is, grades above the cutoff grade) has to be effected on the basis of block-grade estimates (the actual grades being unknown), the lognormal-distribution model is also applicable; the appropriate variance can then be estimated by use of the lognormal and de Wijsian models (Krige, 1978). This approach provides realistic estimates of the likely tonnages and grades for a new mining property where only a limited number of borehole values are available.

In 1952, at the inception of gold production in the main sector of the Orange Free State (eventually covering nine large producers), the author published the following statistical estimate of the likely average recovery grade and mill tonnage for this sector, based on the lognormal and de Wijsian models, and using the 91 available borehole values and a simulated ore-block distribution (Krige, 1952).

Likely average recovery grade $\qquad$ 9·8 dwt (short ton)$^{-1}$ (16·8 g t^{-1})
Total milled $\qquad$ 6·1 × 10^8 t
Comparable orthodox estimate (based on
  arithmetic mean of 42 payable boreholes) $\quad$ 4·20 × 10^8 t
$\qquad\qquad\qquad\qquad\qquad\qquad\qquad\qquad$ at 17·8 dwt (short ton)$^{-1}$ (30·5 g t^{-1})

[1 dwt (short ton)$^{-1}$ = 1·7143 g t^{-1}]

As pointed out subsequently by Sichel (discussion in Krige, 1961), this 1952 estimate caused raised eyebrows among the mining fraternity, many of whom had expectations based on Witwatersrand experience of a grade of nothing more than 6 dwt (short ton)$^{-1}$ (10·3 g t^{-1}) and a much lower total mill tonnage. At the time, an orthodox estimate based directly on the payable borehole values would have given unrealistically high grade estimates and low tonnage estimates, as is shown above. Up to the end of 1976, nearly 3 × 10^8 t had already been milled in this sector at an average recovery grade of 16·0 g t^{-1}, and the total tonnage eventually milled on depletion of this sector should not be far from the original statistical estimate. However, the good agreement does not entirely result from the statistical approach, but also from the interaction of various compensating factors in an assessment of this nature.

The general statistical approach as introduced in 1952 is now accepted practice in South Africa for dealing with new gold mining properties and any other mining proposition to which the lognormal model applies. Furthermore, all the sophisticated geostatistical techniques now available (David, 1977; Guarascio et al, 1976) enable similar valuations to be effected for all types of mineral deposits where the lognormal distribution pattern does not apply.

Ore-reserve estimates in producing mines

A feature well known from practical observation from the early days of gold mining in South Africa, but only satisfactorily explained on the basis of lognormal regression theory by the author in 1951 (Krige, 1951), was that of the serious under-valuation of ore blocks in the low-grade category and over-valuation of those in the high-grade category. These biases resulted in inefficiencies in the selective mining process and in the grade control of the mines; they were eliminated by the application of regression adjustments and this led in turn to the now well-known geostatistical Kriging techniques.

It is not possible to demonstrate in real figures the financial advantages gained from the development and implementation of these techniques by specific follow-up studies because the comparable position if the orthodox procedures had been used throughout can only be inferred. However, if the models used in the geostatistical techniques are accepted as representative of any specific ore body, comparisons can be drawn which will be valid and indicative of the relative order of the advantages to be gained.

The comparison below is based on the following basic assumptions for a typical marginal gold mine under 1977 conditions.

Recovery grade with no selective mining		$5 \cdot 0 \text{ g t}^{-1}$
Operating costs per tonne milled		R23
Recovery grade required to break even, at a gold price of \$150 per ounce (R4.193 per gram)		$5 \cdot 5 \text{ g t}^{-1}$
Total remaining mill tonnage (no selective mining)		$50 \times 10^6 \text{ t}$
Logarithmic population variances	actual block values	$0 \cdot 15$
	actual peripheral values	$0 \cdot 16$
	regressed block values	$0 \cdot 062$
	Kriged block values	$0 \cdot 10$
Error variances of estimators	orthodox block values	$0 \cdot 15$
	regressed block values	$0 \cdot 088$
	Kriged block values	$0 \cdot 05$

The extent of the improvement in the error variances of the ore-block estimates from the use of Kriging procedures as compared with regression or orthodox methods ($0 \cdot 05$ versus $0 \cdot 088$ and $0 \cdot 15$), will depend mainly on the spatial structure of the ore-grade values in the ore body. Even more marked improvements can be obtained when this structure is strongly anisotropic.

The financial effects of the different valuation options for the above example are given in table 3, based on the logarithmic–de Wijsian model.

In practice, the real position as to the remaining life of the mine would be complicated by inflation and variations in costs and gold price, but the relative advantages to be gained by employing geostatistical techniques are nevertheless evident. Apart from these overall financial advantages, considerable practical gains arise from the closer and unbiased valuations of individual ore blocks. These in turn ensure improved levels of grade and financial control, particularly where, at any time, ore is drawn only from a limited number of localized ore blocks within the ore body.

Table 3. Projected financial effects of different valuation options, under the assumptions set out above (Krige, 1978, section 1.20).

Valuation option	Total milled (10^6 t)	Recovery grade (g t^{-1})	Total profits [a] (10^6 R)
No selective mining	50·0	5·0	−105
Selection on			
orthodox valuation	16·0	6·45	64
regressed values	15·3	6·35	66
Kriged values	16·1	6·86	92
actual values [b]	16·5	7·26	122

[a] Since 1977 costs have escalated substantially and the gold price even more so. The profit range for a 1981 marginal proposition will be much wider for the relevant valuation procedures. [b] Theoretical optimum with perfect block valuations.

Conclusions

Ore valuation cannot, and never will, be an exact science, because of the variability of ore values and the nature and extent of the ore sampling data available. For practical mining and economic reasons, these data will always be limited. Within these limitations, improvements can only be gained from better interpretations of the data via appropriate models and techniques. The limited information used for a specific grade estimate can, of course, also cover sampling data for other metals and minerals in the ore body, related geological and mineralogical observations, and so on. The optimum use of these limited data is the field in which geostatistics has proved itself over the last three decades and is continuing to develop improved models and techniques. Apart from possible improvements in ore sampling procedures, such as those offered by the measurement of grades *in situ* by, for example, x-ray fluorescence detectors, further progress in this field can only be achieved by the continued development and sophistication of geostatistical models and techniques.

Acknowledgement. Permission for publication was granted by Anglo-Transvaal Consolidated Investment Company Limited, and is appreciated.

References
David M L R, 1977 *Geostatistical Ore Reserve Estimation* (Elsevier, Amsterdam)
de Wijs H J, 1951 "Statistics of ore distribution. Part 1" *Geologie en Mijnbouw* **30** 365-375
de Wijs H J, 1953 "Statistics of ore distribution. Part 2" *Geologie en Mijnbouw* **32** 12-24
Guarascio M, Huijbregts C J, David M L R (Eds), 1976 *Advanced Geostatistics in the Mining Industry* (Reidel, Dordrecht, The Netherlands)
Krige D G, 1951 "A statistical approach to some basic mine valuation problems on the Witwatersrand" *Journal of the Chemical, Metallurgical, and Mining Society of South Africa* **52** 119-139
Krige D G, 1952 "A statistical analysis of some of the borehole values in the Orange Free State goldfield" *Journal of the Chemical, Metallurgical, and Mining Society of South Africa* **53** 47-64
Krige D G, 1960 "On the departure of ore value distributions from the lognormal model in South African gold mines. Part 1" *Journal of the South African Institute of Mining and Metallurgy* **61** 231-244
Krige D G, 1961 "Developments in the valuation of gold mining properties from borehole results" in *Proceedings, Seventh Commonwealth Mining and Metallurgical Congress* (South African Institute of Mining and Metallurgy, Johannesburg) pp 537-561
Krige D G, 1962 "Statistical applications in mine valuation. Parts 1 and 2" *Journal of the Institute of Mine Surveyors of South Africa* **12** 45-84, 95-136
Krige D G, 1973 "Computer applications in investment analysis ore valuation and planning for the Prieska copper mine" in *Proceedings of the Eleventh Symposium on Computer Applications in the Mineral Industries* Ed J R Sturgul (University of Arizona Press, Tuscon, Ariz.) pp G31-G47
Krige D G, 1978 "Lognormal de Wijsian geostatistics for ore valuation" *South African Institute of Mining and Metallurgy Monograph Series* number 1 (South African Institute of Mining and Metallurgy, Johannesburg)
Matheron G, 1962 *Traité de Géostatistique Appliquée, Tome 1* (Technip, Paris) [English translation (unpublished): *Treatise on Applied Geostatistics* Kennecott Copper Corporation, Salt Lake City, Utah]
Matheron G, 1971 "The theory of regionalized variables and its applications" *Cahiers du Centre de Morphologie Mathématique de Fontainebleau, France* number 5 (Centre de Morphologie Mathématique, Fontainebleau, France)
Munro A H, 1977 "Application of risk analysis to a new gold mine" in *APCOM 77* (proceedings of the 15th APCOM symposium) (Australian Institute of Mining and Metallurgy, Parkville, Australia) pp 471-480
Ross F W J, 1950 "The development and some practical applications of statistical value distribution theory for the Witwatersrand auriferous deposits" unpublished masters thesis, University of Witwatersrand, South Africa
Sichel H S, 1947 "An experimental and theoretical investigation of bias error in mine sampling with special reference to narrow gold reefs" *Transactions of the Institute of Mining and Metallurgy, London* **56** 403
Sichel H S, 1952 "New methods in the statistical evaluation of mine sampling data" *Bulletin of the Institute of Mining and Metallurgy, London* **61** 261-288
Sichel H S, 1966 "The estimation of means and associated confidence limits for small samples from lognormal populations" *Journal of the South African Institute of Mining and Metallurgy, Special Symposium Volume on Ore Valuation* 106-123
Wainstein B M, 1975 "An extension of lognormal theory and its application to risk analysis for new mining ventures" *Journal of the South African Institute of Mining and Metallurgy* **75** 221-238

Bridging the gap between small-scale and large-scale petroleum discovery models

L J Drew, D H Root

Introduction

Petroleum exploration has a modern history of more than a hundred years and is worldwide in extent, although most drilling has been done in the USA. As of the end of 1975, approximately $3 \cdot 3 \times 10^6$ holes had been drilled worldwide and $2 \cdot 4 \times 10^6$ of these holes were in the USA (Grossling, 1976). This enormous exploration and production activity can be studied from several different points of view or scales of observation. Large-scale studies are those which include the drilling and discovery record over a large geographical area containing many productive basins. The smallest-scale studies which will be considered here are those focusing on the exploration of a single productive formation. We wish to use both

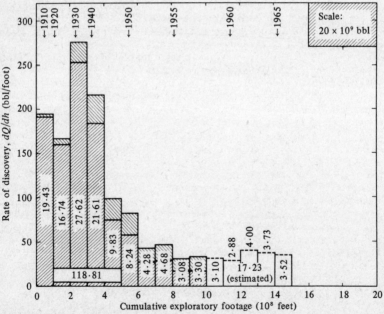

Figure 1. Crude oil discoveries per foot of exploratory hole (averaged for each 10^3 feet) versus cumulative exploratory drilling in the USA. For the first ten columns, lower shaded area represents the National Petroleum Council estimate as of 1st January 1964; upper shaded area represents oil added by application of a correction factor. The last five columns are based on annual American Petroleum Institute estimates of oil added by new discoveries, increased by a factor of $5 \cdot 8$ (Hubbert, 1967).

large-scale and small-scale studies to investigate statistical properties of the petroleum discovery process.

Hubbert (1967) observed that the amount of petroleum being discovered per foot of exploratory drilling in the conterminous United States had declined through the years. His analysis of petroleum exploration incorporated drilling data from 1859 to 1966 over the lower forty-eight states, in the Gulf of Mexico, and off the coast of California. In addition to Hubbert's 1967 publication, there have been other statistical studies of petroleum exploration in smaller areas and covering shorter time spans (see, for example, Arps and Roberts, 1958). These smaller-scale studies provide a better understanding of the decline in the discovery rate for the USA as a whole, shown in figure 1.

Small-scale models

Small-scale analysis is at the level of the exploration play. At this scale, the analysis of the discovery process is usually focused on exploration for petroleum within a group of genetically related structures within a relatively small region, or on the exploration for stratigraphic traps within a single stratigraphic unit. A method of analysis of the exploration process at this scale was first proposed by Arps and Roberts (1958); it was used to predict future rates of discovery within the Cretaceous D–J sandstone unit in the Denver basin. Arps and Roberts proposed a model to describe the drilling and discovery history up to 1958, which would then permit a forecast of the number and sizes of fields which would be discovered by a given number of future exploratory wells. From the data generated during exploration, Arps and Roberts used only the number of exploratory wells which had been drilled, the recoverable volume of oil in each of the fields discovered, the number of oil fields discovered, and the area of the surface projection of the discovered oil fields. Their basic assumption in relating cumulative number of exploratory wells to cumulative number of discoveries was that the probability that the next exploratory well would discover a given field is proportional to the ratio of the area of the surface projection of that field to the area within the basin which is considered worthy of exploration. The constant of proportionality in this relationship is called the efficiency of exploration. To estimate the efficiency of exploration in the Denver basin, Arps and Roberts began with the American Association of Petroleum Geologists (AAPG) statistics on exploratory drilling; these showed that, over the USA as a whole, wildcatting based upon technical advice was 2·75 times more likely to be successful than wildcats which were drilled for nontechnical reasons. They then reasoned that because the traps in the Denver basin were stratigraphic, the comparative advantage of the geologist over the random driller—that is, the efficiency of exploration— would be less than the national average of 2·75, and they chose a value of 2·0 based on their experience and knowledge of the region for the efficiency of exploration in the Denver basin. Thus, a well sited by a geologist was twice as likely to make a discovery as a well sited at random. The probability of discovery can be stated as

$$p(a) = \frac{Ca}{b} \, , \tag{1}$$

where $p(a)$ is the probability that the next well discovers a given field of area a; C is the efficiency of exploration; and b is the area where explorationists would be willing to site wells.

To simplify their analysis Arps and Roberts (1958) divided all oil fields into classes. The kth class contained all fields having between $(2^{k-1} \times 10^3)$ and $(2^k \times 10^3)$ bbl recoverable oil.

Fields within a given class were assumed to be of equal area. If there are n fields, each with area a, then the probability that the next wildcat will discover one of them is equal to nCa/b.

Because the probability of discovery of a field in a class is proportional to the number of fields in that class, the number of wells required to find half of the fields in a class is theoretically the same, no matter how many fields are in the class. For example, if there are 128 fields in the class then the number of wells required to find the first 64 fields is the same as the number of holes required to find the next 32 fields which is the same as the number required to find the next 16 fields, and so on. Thus the number of undiscovered fields declines as the negative exponential of the cumulative number of wildcat holes. This relationship can be summarized as follows

$$n_k(w) = n_k(\infty)\left[1 - \exp\left(-\frac{wCa}{b}\right)\right] \, , \tag{2}$$

where w is the cumulative number of exploratory wells; $n_k(w)$ is the number of discoveries in the class made by the w wells; and $n_k(\infty)$ is the ultimate number of fields to be discovered in the given class.

From their experience of exploration in the Denver basin, Arps and Roberts (1958) concluded that only $5 \cdot 7 \times 10^6$ acres of the basin (about 20% of the total area) was of interest to those searching for oil. Figure 2 shows predictions of the number of discoveries that would be made when 11 567 exploratory wells had been drilled, from the results of the first 2 673 wells. Also shown in figure 2 are the actual discoveries made by the end of 1974, when 11 567 exploratory wells had been drilled. The predictions are accurate enough to show that the Arps–Roberts model is a good description of the exploration process. From equation (2) we can see that a declining discovery rate is incorporated into the model in two ways. First, discoveries within a given class become less frequent as drilling proceeds, because the number of undiscovered fields declines exponentially. Second, fewer wells are required to find a given fraction of the fields in a class of large fields than in a class of small fields. These two effects mean that as exploration progresses, discoveries become less frequent and smaller.

A discovery-process model similar to the Arps–Roberts model was proposed by Drew et al (1980). In this model the measure of the extent of exploration was the area which had been exhausted, rather than the number of exploratory wells. The area exhausted is calculated from well locations and is based on the concept of the area of influence of an exploratory well (Singer and Drew, 1976). Because the area of influence of an exploratory well is defined with respect to targets of a given size and shape, the area exhausted is also defined with respect to these targets. Figure 3 shows a typical map on which calculations of the area exhausted are based. The crosses represent well locations, the ellipse at the right represents the size and shape of target fields being considered, and the contour lines represent different probabilities of discovery of such a field.

For example, if an oil field has the size and shape of the target ellipse in figure 3, and if the center of the ellipse is on the contour line marked say, 0.3, then for 30% of its possible orientations the target would be hit by one of the wells; for 70% of its possible orientations it would not be hit by any of the wells. The degree of exhaustion of points on the contour line is then defined to be 0.3. The degree of exhaustion at a point is thus defined relative to a particular size and shape of target. The area exhausted is then defined to be the average degree of exhaustion over the whole region, multiplied by the area of the region. It can be shown that an isolated well will exhaust an area equal to the assumed target area.

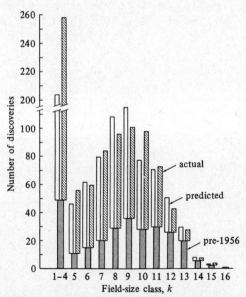

Figure 2. Data for exploration in 1949–1955 used to predict numbers of wells discovered in 1956–1974, for the Denver basin.

However, if a second well is sited near enough to the isolated well so that a single target could be hit by both wells, then the total area exhausted by by the two wells would be less than twice the area of the assumed target. Thus, as more wells are drilled, they begin to crowd each other, and the later wells exhaust less additional area than earlier wells.

In applying this measure of the extent of exploration to the history of discovery in a region, the oil fields are divided into size classes, such as those used by Arps and Roberts (1958), and defined earlier in this section. Then for each class, a representative target size and shape is chosen as the best approximation to the oil fields in that class. The representative target for each class of oil fields is then used to calculate the area exhausted for its class. The proposed model (Drew et al, 1980) establishes a relationship between the number of discoveries and the area exhausted on a class-by-class basis:

$$n_k(A_k) = n_k(b) \left(1 - \frac{A_k}{b}\right)^{C(k)}, \tag{3}$$

where $n_k(A_k)$ is the number of discoveries in class k as a function of A_k, the area exhausted with respect to class k; $n_k(b)$ is the number of class-k fields ultimately to be found in the total area of interest, b; and $C(k)$ is the efficiency of exploration for class k.

The assumption of a declining discovery rate appears in this model in three ways. First, if $C(k) > 1 \cdot 0$, then the richer parts of the region are explored early so that when, say, half the region has been exhausted, more than half of the targets within that class have been found. This is a consequence of equation (3), because the fraction of the fields in a given

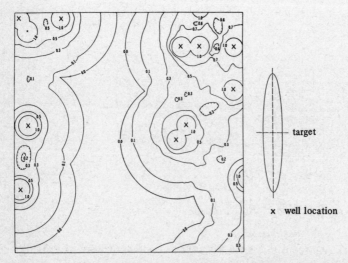

Figure 3. Typical physical exhaustion map for target shown (Singer and Drew, 1976). Contours refer to the degree of exhaustion.

class that are undiscovered is equal to $(1 - A_k/b)^{C(k)}$; this fraction is smaller than the fraction of the region that is unexplored, given by $(1 - A_k/b)$, whenever $C(k) > 1 \cdot 0$.

Second, from the way the area exhausted for a given class is calculated, it can be seen that later wells do not exhaust as much additional area as earlier wells, so that even if $C(k) = 1 \cdot 0$, the discovery rate within a given class (measured in oil yield per well) would decline. Third, from the way the total area exhausted is calculated, exploration is always further advanced for large targets than for small targets; thus, as drilling progresses, the average size of the fields discovered declines. In this model, just as in the Arps–Roberts model, discoveries are assumed to become less frequent and smaller as exploration progresses.

The effect of efficiency is shown in figure 4. This figure is a graph of $n_{12}(A_{12})$ versus A_{12}, that is, the cumulative number of discoveries versus the cumulative area exhausted for fields in the Denver basin containing

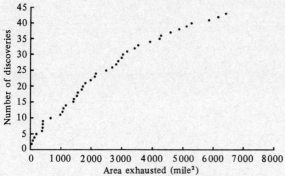

Figure 4. Denver basin: number of discoveries of oil fields containing $(2-4) \times 10^6$ bbl versus cumulative area exhausted, 1949-1974.

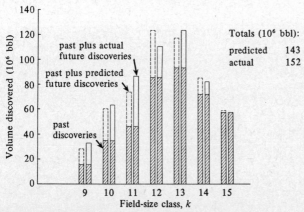

Figure 5. Predicted and actual discoveries of crude oil in the Denver basin from 1958 to 1974.

between 2048000 bbl and 4096000 bbl recoverable oil. Note that in the exploration of the first 3200 mile^2, 31 fields were discovered, but in the exploration of the next 3200 mile^2 only 11 fields were discovered. The two parameters b and $C(12)$ can be estimated directly from the drilling and discovery record, so that expert judgment need not be relied upon (Root and Schuenemeyer, 1980).

This method of analysis was applied to the portion of the Denver basin upon which Arps and Roberts (1958) had focused their study. The results are presented in figure 5, for the field-size classes 9–15. The number and timing of discoveries made by the first 4430 exploratory wells were used to forecast what would be discovered by the next 7137 exploratory wells. Note that the decline in the discovery rate which the model predicts did actually occur, and was more pronounced for the larger field-size classes.

Large-scale models

The fact that rates of discovery decline with increasing length of exploratory drilling has also been clearly demonstrated in an analysis of the returns to exploratory drilling in the western Canadian sedimentary basin (Ryan, 1973). In this analysis, Ryan plotted the rate of discovery of petroleum per new field wildcat versus cumulative new field wildcats for each of the seven major exploration plays which had unfolded in the basin before 1970 (figure 6).

The pattern of development within each of the seven plays is shown in figure 7. Comparison of figures 6 and 7 shows the importance of the development of new plays in maintaining a high discovery rate.

The first spike in figure 6 comes from the almost simultaneous initiation of three plays. The second small spike comes from a large discovery in an existing play. The third spike comes from the beginning of a new play, the fourth from the beginning of a new play, and the fifth from a large discovery in a known play. The last low peak comes from the development of two new small plays. In each play in figure 7 the same pattern is

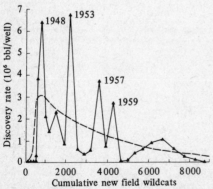

Figure 6. Discovery rate of recoverable crude oil in Alberta. (Source: Ryan, 1973.) The dashed curve is a theoretical smoothing calculated by Ryan.

present: high initial discovery rates followed by a rapid and then more gradual decline.

The speed of the decline of the discovery rate is greatly affected by the variation in the size of oil fields which occur in a given region. Once the large fields have been found, it takes a combination of many more small discoveries to find a similar amount of oil. In order to expose more fully the role of the field-size distribution in the declining rate of discovery, the field-size distribution for the Midland basin is examined. This basin, located in west Texas, covers approximately $35\,000$ mile^2. The distribution of the sizes of the 1957 oil and gas fields discovered in the numerous exploration plays which have taken place in this basin up to the end of 1974 is shown in figure 8. Note that this distribution is highly skewed. The largest single field contains $13 \cdot 4\%$ of the total petroleum discovered in the basin. The largest eighteen fields, those containing 10^8 bbl or more, make up $57 \cdot 8\%$ of the total. In fact, there is more petroleum in the two largest fields than in the 1890 fields which each contain 25×10^6 bbl or less. When the order

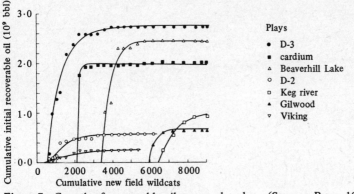

Figure 7. Growth of recoverable oil reserves by play. (Source: Ryan, 1973.)

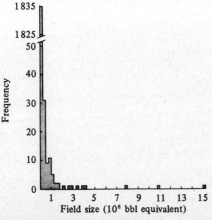

Figure 8. Field-size distribution for the Midland basin.

in which these deposits were discovered is examined, the eighteen largest fields are found to have been discovered relatively early in the exploration of this basin. All of these eighteen fields had been discovered by 1954, when only 5937 out of the 1974 cumulative total of 16014 exploratory wells had been drilled. Between 1955 and 1974, an additional 1352 fields were discovered from the drilling of an additional 10077 exploratory wells. The average size of these post-1954 discoveries was, however, small (see figure 9).

The Midland basin is not peculiar in having a few large fields and many little fields. This same skewed field-size distribution appears in the Denver basin, the North Sea, the Gulf of Mexico, western Canada, and the Alaskan North Slope. We know of no area which does not exhibit a highly skewed field-size distribution. The large fields are more easily discovered than the small ones; thus, even if the success rate for exploratory drilling remains constant, there will be a decline in the discovery rate as measured in barrels per well or in barrels per foot of exploratory drilling.

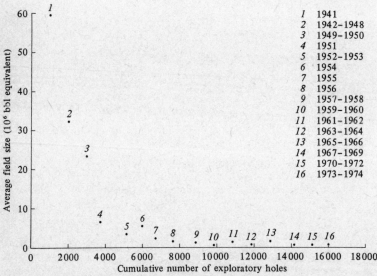

Figure 9. Midland basin: average size of fields discovered versus cumulative number of exploratory holes drilled.

Interrelationship between large-scale and small-scale models
The studies of single plays, and of basins with several plays, help explain qualitatively why the discovery-rate curve in figure 1 has its peculiar shape. As long as oil exploration can move into virgin territories, and when, simultaneously, the efficiency of exploration is improving, enough big fields will be found to keep the discovery rate high. Later, when almost all drilling is in basins which are already partially explored, the bulk of the

discoveries will be in the form of many small fields; the discovery rate will be low, and its decline much more gradual.

Another example of the decline phenomenon is shown in figure 10. The figure shows the discovery rate, measured in barrels of recoverable oil per exploratory well, for the non-Communist world outside the USA and Canada. This area has been drilled only sparsely compared with the lower forty-eight states of the USA; nonetheless the pattern of decline is already apparent.

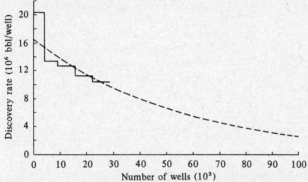

Figure 10. Crude-oil discovery rate versus cumulative exploratory drilling in the non-Communist world outside the USA and Canada; five-year averages 1951–1975. (Source: Root and Attanasi, 1980.)

Petroleum exploration can be studied on many different scales, from single plays up to a region covering more than half the world. The results of studies at different scales of observation reinforce each other. Analyses of single plays confirm the intuitive ideas that large fields are more easily discovered than small fields, and that the best discovery rate is at the beginning of the play. So in the exploration of a large area there is a strong incentive to look for new plays, even if there are several known productive formations which are not completely explored. When new plays are not being opened up fast enough, the discovery rate will decline. Small-scale studies give no indication as to when this decline can be expected to occur in the exploration of a large area, but they do show that the discovery rate can be kept high only by the continuous discovery of new productive formations. Small-scale studies suggest that there is a large random element in petroleum exploration, from which it would follow that plays having large areas are more easily discovered than small ones, as is the case with fields within a play.

The appearance of a decline in the discovery rate at every scale of observation increases our confidence in forecasting that this downward trend will persist.

References

Arps J J, Roberts T G, 1958 "Economics of drilling for Cretaceous oil and east flank of Denver-Julesburg basin" *Bulletin, American Association of Petroleum Geologists* **42** 2549-2566

Drew D J, Schuenemeyer J H, Root D H, 1980 "Resource appraisal and discovery rate forecasting in partially explored regions. Part A—an application to the Denver basin" professional paper number 1138, US Geological Survey, Reston, Va

Grossling B, 1976 *Window on Oil* (Times Newspapers Ltd, London)

Hubbert M K, 1967 "Degree of advancement of petroleum exploration in United States" *Bulletin, American Association of Petroleum Geologists* **51** 2207-2227

Root D H, Attanasi E D, 1980 "World petroleum resource estimates and production forecasts: implications for government policy" *Natural Resources Forum* **4**(2) 181-196

Root D H, Schuenemeyer J H, 1980 "Resource appraisal and discovery rate forecasting in partially explored regions: Part B—mathematical foundations" professional paper number 1138, US Geological Survey, Reston, Va

Ryan T J, 1973 "An analysis of crude-oil discovery rate in Alberta" *Bulletin of Canadian Petroleum Geology* **21** 219-235

Singer D A, Drew L J, 1976 "The area of influence of an exploratory hole" *Economic Geology* **71** 642-647

Alternative objectives for oil exploration by the United States government: the dilemma of the National Petroleum Reserve in Alaska

J C Davis, J W Harbaugh

Introduction

The history of the petroleum industry in the USA has been one of free enterprise, with profit-seeking corporations raising risk capital to finance exploration, production, and marketing. The profits from these activities, augmented in recent years by drilling-fund investment pools, have been used to finance industrial expansion. In general, the industry has not lacked capital to expand progressively as the demand for petroleum products has risen inexorably. Since the oil business began in 1859, it has operated in the USA without direct government support.

The attitude of the United States government toward the oil industry has been ambivalent; the industry's first decades coincided with a laissez-faire period, but an era of governmental constraints began with the 'trust-busting' actions which broke up the original Standard Oil Company in the 1920s. Since then, the industry has operated under an ever-increasing set of federal constraints which affect its every phase, including financing, transportation, safety practices, and marketing policies.

For many decades, the federal government has been deeply involved in the oil business as a mineral-rights owner, a relatively passive role equivalent to that of a private landowner who leases his land to an oil company and receives a royalty interest from the oil or gas that is produced. However, except for a brief flirtation with state ownership of oil refineries in some Populist states in the 1930s, there has never been widespread interest in government ownership or operation of any segment of the petroleum industry.

Serious discussion of a national, federally owned oil business did not begin until the recent series of international oil crises. Some members of Congress questioned the ability of private oil companies, particularly that of multinationals, to serve the needs of the country. It was suggested that a federal petroleum corporation should be established to supplement the spectrum of profit-seeking companies already in the industry. The concept of the Federal Oil and Gas Company (or 'FOGCO') was endorsed by some as being in the national self-interest, but bitterly denounced by others as unwise, uneconomic, and un-American.

Almost unknown to most Americans, a version of 'FOGCO' has been in existence since 1944. This federal enterprise has grown explosively since 1973, with expenditure for petroleum exploration totalling over a billion dollars. However, this 'FOGCO' produces no oil, and only a small amount of gas, which is sold to residents of Point Barrow, Alaska. These activities are conducted in the federal government's program of exploration in the National Petroleum Reserve—Alaska (NPR—A; formerly Naval Petroleum Reserve number 4), an area of 37000 mile^2 occupying about half of Alaska's Arctic slope. Here, the US government in effect has an exclusive exploration concession, and is operating in a manner similar to many nationally owned oil companies in other parts of the world.

Indeed, the pattern outside the USA is to establish national oil companies or to assume government ownership of private (usually foreign) corporations. Most of the West European countries have wholly or partially government-owned companies that compete in an otherwise free enterprise market system. Examples include ELF in France, AGIP in Italy, and BP in England. Within the third-world nations, the pattern is more that of national petroleum monopolies, such as PEMEX in Mexico, and the Brazilian giant PETROBRAS which is owned jointly by government and private interests. The OPEC nations are characterized by national companies which operate on an international scale; these are largely created from the former holdings of major private corporations: ARAMCO and the panoply of -VEN companies in Venezuela are examples.

The basic objective of a private corporation is to return the maximum amount of profit to its investors. This is true whether the company is engaged in the production of oil or the sale of hamburgers. However, national companies may have quite different objectives, and the profit motive may be weak or even nonexistent. For example, the primary objectives of many national companies in underdeveloped countries include the provision of full employment and a wide range of social services such as medical care. Many industrial nations use their government production companies to combat balance-of-payments deficits caused by massive imports of foreign oil. Considerations of profit are secondary to the achievement of a degree of freedom from dependence on foreign energy sources. Of course, in the planned economies of the socialist countries, the objectives of the petroleum sector are interwoven with the national objectives. The guiding principles of the socialist oil industry are difficult to examine in isolation from these greater considerations.

Within its exploration program for the National Petroleum Reserve—Alaska, the USA has established the nucleus of what could be a US national oil company, although it is not generally recognized as such. Congress created this entity without any realization of what it might become, without any clear-cut objectives or mandate, and in an atmosphere of confusion and contradiction. It is now in the midst of the largest exploration program ever undertaken in North America, conducting what is probably the most extensive seismic reconnaissance survey ever run on land by a single operator, and operating on an exploration budget of over a quarter of a billion dollars annually. This specific exploratory project on the Arctic slope of Alaska will presumably have a short life.

However, it provides an opportunity to study the consequences of alternative objectives that might be established for a government company chartered to explore for oil and gas within the USA.

The National Petroleum Reserve—Alaska

Naval Petroleum Reserve number 4 was set aside by Presidential order in 1923. It reserved the petroleum that might occur within a large block of land (figure 1) north of the Brooks Range in the Territory of Alaska, for the exclusive use of the US Navy. Unlike the other three Naval Petroleum Reserves established at approximately the same time (two in California and one in Wyoming), the Alaskan Reserve did not encompass any known occurrences of petroleum, but rather was based on the presence of numerous oil seeps that had been observed along the Arctic coast.

Because of the inhospitable environment, no effort was made to find oil in the Alaskan Naval Reserve until 1944. Spurred by wartime petroleum shortages, the Navy drilled a series of thirty-six wells, discovering the small Umiat oil field, the Gubik gas field, and a few other very small fields in the process. Approximately 3000 miles of seismic traverses were shot, mostly along the coast.

In 1973, the Navy was directed by Congress to explore thoroughly Naval Petroleum Reserve number 4 for oil and gas, and an appropriations bill of $1·2 billion was passed for this purpose. This sudden interest in the potential of the Reserve was kindled by the oil embargo placed on the US by the major OPEC nations, and by the discovery of the giant Prudhoe Bay field five years earlier. Prudhoe Bay lies within fifty miles to the east of the Reserve, and the possibility that other giant fields might be discovered on the Arctic slope was tremendously appealing both to the government and to the petroleum industry.

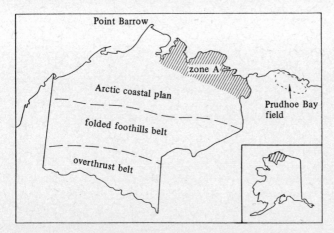

Figure 1. Major exploration plays within the National Petroleum Reserve—Alaska. Most drilling in the Reserve has been confined to zone A. (After Carter et al, 1977.)

In the initial year of exploration, the Navy acted as prime contractor for exploration activities, and followed a plan devised by Tetra Tech, Inc., a Houston-based geophysical consulting firm. In subsequent years, exploration has been conducted by Husky NPR—A, Inc., a specially established subsidiary of the Husky Oil Company. Tetra Tech has continued as the principal geophysical subcontractor, and exploration still proceeds substantially along the lines of their original plan.

The Navy operated under the assumption that its congressional mandate meant that it was to explore for, and to develop, commercial quantities of petroleum. It conducted its exploration program with much the same objectives of a large oil company, but was understaffed and hampered by inexperience. In addition, environmental restrictions prevented drilling except along the Arctic coast (in the so-called 'zone A'), and tantalizing giant structures detected by seismic reconnaissance remained untested. The seven wells drilled during the first three years of the exploration program were failures.

In 1977 Congress, no doubt dismayed by the lack of success under Naval operation, transferred control of the exploration program to the Department of the Interior. At the same time, Naval Reserve number 4 was redesignated National Petroleum Reserve—Alaska. The US Geological Survey (USGS) was assigned responsibility for the program, and a small group of geologists within the Geologic Division of the USGS was placed in charge of the exploration effort. Husky NPR—A continues as prime contractor, at least for the initial year of operation by the USGS, and the exploration program closely adheres to the original plan set out by Tetra Tech.

With the change of management from Navy to Department of the Interior, a major shift has occurred in the purpose of the exploratory effort. Indeed, there is disagreement within the Department of the Interior as to what the real intent of Congress is with regard to NPR—A, and how this intent should be translated into exploration objectives. A clear definition of these objectives is essential if the remaining NPR—A exploration program is to be conducted in a rational manner. More importantly, the resolution and definition of these objectives may have far-reaching consequences, not only with regard to the future of NPR—A, but also for development of other federal lands, and of the outer continental shelf.

The petroleum geology of NPR—A

Most of National Petroleum Reserve—Alaska lies within a structural basin, the Colville Trough, which is bounded in the south by thrust faults of the Brooks Range, and in the north by the Barrow Arch. The Colville Trough contains up to 30000 feet of sedimentary rocks, which range in age from Devonian to Holocene. A diagrammatic cross section from north to south, through Point Barrow, is shown in figure 2. Pre-Cretaceous sediments were derived from a northern source and include thick carbonates of Mississippian and Pennsylvanian age, and marine sandstones and shales of Permian to Jurassic age. The major reservoirs of the Prudhoe Bay super-giant oil field are developed in these units, where they have been brought into unconformable contact with Lower Cretaceous marine shales (Morgridge and Smith, 1972). Latest Jurassic, Cretaceous, and Tertiary sediments—mostly greywackes and shales—were derived from the Brooks Range to the south. The Cretaceous marine shales are organically rich and are potential petroleum source rocks. Prudhoe Bay petroleum is probably derived from these beds.

Additional fields geologically similar to Prudhoe Bay would be confined to the crest of the Barrow Arch, where pre-Cretaceous erosion might bring Mesozoic and Paleozoic reservoir rocks into contact with Cretaceous source beds. Unfortunately, the Barrow Arch passes into the Beaufort Sea north of NPR—A, so any reservoirs of this type probably lie beyond the boundaries of the Reserve.

Among the more promising plays that have been developed within NPR—A are those in the Arctic coastal plain, in the folded foothills of the Brooks Range, and in the southern overthrust part of the Brooks Range foothills. In the coastal plain, combination structural and stratigraphic traps may exist within Cretaceous sandstones which pinch out updip onto the flank of the Barrow Arch. These sandstones are analogous to the Cretaceous Mesa Verde group of the Rocky Mountains, but unfortunately tend to be poorly sorted with low porosity and permeability (Carter et al, 1977).

Enormous anticlinal structures have been detected seismically in the folded foothills region. If favorable reservoir rocks are present in the Lower and Middle Cretaceous interval, structurally controlled oil reservoirs of significant size may be present. Seismic anomalies also are present in deeper horizons, but the difficulties of deep drilling in the Arctic, plus the paucity of pre-Cretaceous source beds, make these prospects less attractive.

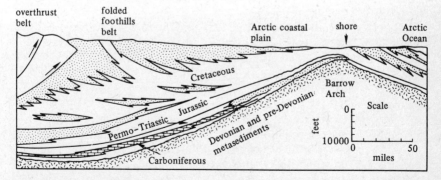

Figure 2. Diagrammatic cross section from the Beaufort Sea north of Point Barrow, south to the overthrust zone of the Brooks Range. (After Carter et al, 1977.)

The southern part of the Brooks Range foothills contains complexly faulted and folded structures, similar in style to those in the overthrust belt of Wyoming and Idaho. Gas reservoirs may be present in carbonates of Mississippian and Triassic age, or in Cretaceous sandstones. Structures beneath the overthrust zone have been detected, but their extreme depth, coupled with their distance from the coast, probably makes their testing economically unfeasible at present.

A variety of other plays have been delineated during the recent geophysical exploration of NPR—A. Certainly, an abundance of attractive structures and stratigraphic wedge-outs can be detected seismically within the Reserve. What is not yet known is whether adequate reservoir rocks are present in the subsurface, and whether petroleum has been generated from source beds.

Probabilistic nature of exploratory programs

All petroleum exploration programs have certain elements in common, a consequence of the geological nature of the setting in which petroleum is found, and of the procedures involved in searching for a hidden resource. Some of the general consequences of a drilling program can be deduced without regard for the express objectives of the program. These are concerned with the relations between exploratory drilling and the gain of information, petroleum, and financial return. The nature of geological variables also provides valuable insight into the form of a sampling (drilling) program to assess these variables.

Information gained from the drilling of wells

Figure 3 relates to information about a particular geological variable or variables that will be gained by drilling wells within a tract in NPR—A. (For our purpose, a *tract* may be defined as a geographical area within which the variable or variables being assessed are assumed to have approximately constant variances; the boundaries of the area may be established for convenience by economic, political, or other considerations.) The form of the information function depends upon the degree of spatial autocorrelation in the variables and upon the size of the tract. 'Autocorrelation' refers to the manner in which a variable behaves in a spatial sense; it is a measure of the statistical similarity between the values of a variable at a set of points, and the values at another set of points located a specified distance and direction away. Initially, before the first drilling activity, no information is available. The first exploratory well provides maximum information about the variables in question, and allows their means to be estimated. The second and subsequent wells provide additional information, so that the second and higher moments and autocovariances can be estimated; increasingly more precise estimates of these parameters can be made as additional wells are completed.

Initially, wildcat wells are located so far apart that they may be presumed to be independent, and each well contributes new information. However, the wells are drilled within a geographically confined space, so that the distances between well locations must become smaller as more wells are drilled in the tract. Eventually, additional wells will not be statistically independent, and variables measured in the wells will become increasingly autocorrelated with measurements in other wells. Hence, the contribution of wells drilled late in the exploration program to the total store of information will be smaller.

The same concept can be expressed in the form of the relative information content per well as a function of the number of wells drilled in a tract, as shown in figure 4. The first wildcat well contributes 100% new information, and the immediately succeeding wells are almost as valuable. Once spatial considerations constrain new wells to locations within the autocorrelated span of previous wells, however, the information content per well drops at a rate proportional to the degree of autocorrelation. The information function asymptotically approaches zero information content as the number of wells becomes infinite.

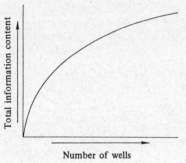

Figure 3. Information gained about geological variables as a function of the number of wildcat wells drilled into an exploratory tract.

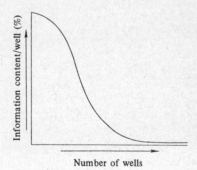

Figure 4. Relative information content of individual wells drilled successively in an exploratory tract.

Volume of petroleum discovered by the drilling of wells

It is axiomatic that wells must be drilled in order to discover or produce petroleum. However, the volume of oil contained within a geographical area is finite, and so the relationship between number of wells drilled and amount of oil found cannot be linear. The problem is analogous to sampling without replacement where the probability of a 'successful draw' is low, and follows a hypergeometric probability distribution. Expressed in terms of percentage of total volume of petroleum discovered as increasing numbers of wells are drilled, the relationship must have the form of an ogive curve (figure 5).

It is unlikely that any volume of oil will be discovered by the first drilling venture, because of the low probability of success attached to an individual wildcat well. Eventually, however, a discovery will be made. Because the probabilities of discovery are proportional to the areas of the reservoirs, larger fields will tend to be found first (Kaufman, 1974). If the well-known tendency for fields to be clustered is exploited, associated fields will be discovered in quick succession (Griffiths, 1962). This will cause the function shown in figure 5 to climb rapidly. Continued drilling, however, will discover smaller and smaller reservoirs, and the curve will become asymptotic to 100% at an infinite number of wells.

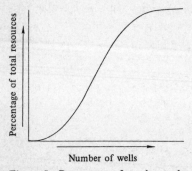

Figure 5. Percentage of total petroleum resources within an exploratory tract that will be discovered as successive wells are drilled.

Financial return from the drilling of wells
The implication of figure 5 is that continued drilling efforts will yield decreasing increments of oil. Wells drilled late in the development of the area, when most of the resource base has already been discovered, will have a very low (or negative) expected monetary value (EMV), because of the decreasing probability of discovering large volumes of oil. A plot of net economic return (measured in constant monetary units) versus number of wells drilled will have the approximate form shown in figure 6.

The economic-return curve will be expected to rise sharply after the initial commercial discovery, because these early wells have the highest probability of discovering the largest quantities of oil. With continued drilling, later wells will be expected to discover lesser quantities, and the economic-return curve will decline because of drilling and other exploration costs. Eventually, the value of the oil remaining undiscovered is less than the cost of exploration, and the economic return will become negative.

The shape of the curve is a function of the field-size distribution within the tract, costs per well, and the density of the well spacing. The economic-return curve has been extensively investigated by Drew (1967), who considered a variety of field-size distributions in real basins, and a succession of uniform sampling or drilling patterns. The relationship between number of wells and economic return is more complex for spatially irregular sampling, but holds to that described above in general.

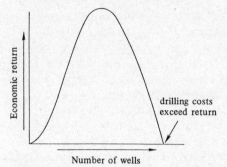

Figure 6. Economic return from drilling successive wells within an exploratory tract.

Geological variability through space
Geological properties, particularly those important as guides to petroleum exploration, are considered to be continuously variable in the spatial sense. This is why geologists prepare maps of various geological properties, and why these maps typically utilize contour lines which represent the form of a continuous surface. Examples of such geological properties or variables include the structural configuration of stratigraphic horizons, changes in the sand/shale ratio of sedimentary units, and variations in average porosity within reservoir rocks. Even though these properties usually are known only at points where samples have been taken, this information is extrapolated and interpolated to estimate values at all locations between the sample points. Features of the mapped surfaces, such as trends, are used as guides to petroleum prospects, and point estimates taken from these maps are used as input parameters for evaluation procedures such as the USGS Monte Carlo range-of-values program (Bernard and Akers, 1977).

  If a variable is spatially continuous, it must have a spatial autocorrelation function which is greater than zero for certain minimal distances. The autocorrelation at zero distance is one, and for some infinitesimal distance between points must be nearly one, because the points lie on an unbroken surface. The autocorrelation between sampling points decreases with increasing distance between points, so the general form of a spatial auto-correlation function for a geological variable is similar to that shown in figure 7, assuming that the variable is spatially stationary or free from trend. If a trend is present, it may be removed by subtracting a trend surface fitted by a least squares procedure, or by working with the differences between points (Box and Jenkins, 1970).

  Unfortunately very little is known in detail about the autocorrelation of geological properties. With the exception of research done on regionalized-variable theory by French geomathematicians (for example, Matheron, 1971), the importance of such information has not been appreciated by geologists.

There is also the very practical problem that most geological properties cannot be observed at will, because the rocks of interest may be deeply buried within the lithosphere.

In spite of the difficulty of obtaining spatial autocorrelations, they must be estimated in order to design rationally a geological sampling program. The number of samples which must be taken to achieve a specified level of precision is a function of the variance of the property being measured, assuming that all samples are independent. These samples should be taken without bias, by sampling either at random locations or according to a systematic sampling pattern. It has been demonstrated many times (for example, by Griffiths and Ondrick, 1968) that systematic sampling is more efficient (that is, converges more rapidly to the true parameters of the population) than random sampling. Therefore, an optimal sampling program would be one in which samples are collected uniformly across an area, but spaced at distances which insure that all samples are independent.

The most effective regular point pattern for uniform coverage of an area is a triangular grid in which every point is equidistant from six other points (a *face-centered hexagon*). The minimum distance between points should be twice the span of the autocorrelation function. This insures that all samples are independent, and that all intermediate points on the surface are related in some degree to the sampled points. Thus, if the autocorrelation function of the property being sampled is known, an optimal sampling grid can be specified.

This sampling pattern assumes that the geological property is isotropic, or free of grain or directional character. If this is not the case, the sampling pattern should be modified to account for the differences in autocorrelation for different directions. If one of the axes of the sampling pattern is set parallel to the direction of maximum autocorrelation, and

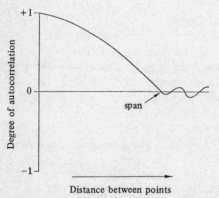

Figure 7. Autocorrelation function relating the degree of self-similarity of a stationary geological variable with distance between sampling points. At distances greater than the span, or distance to zero autocorrelation, all sample points are independent.

the spacing between sampling points adjusted to twice the span in this direction, then the lengths of the other axes of the hexagonal pattern should be shortened until they are proportional to twice the span of the autocorrelation in their corresponding directions. The result will be a 'flattened' or 'squashed' hexagonal network of sampling locations, whose axes are proportional to the degree of spatial variation in the property being measured.

Estimating the spans for most geological variables is a challenging task, because almost nothing is known about these statistical properties; within NPR—A, exploratory holes are too widely spaced for estimation purposes. However, the high-density seismic information currently available in NPR—A can provide information about the rates of change of certain variables. Since many geological properties are highly correlated, estimates based on seismically perceived variables may be used to estimate the auto-correlation functions of other variables. Examples of geological variables that can be measured on seismic cross sections include changes in structural elevation of reflecting horizons, thickness convergence within intervals bounded by reflecting layers, number of reflecting units within intervals, and changes in seismic properties along selected horizons. Properties such as seismic reflectance strength, phase angle, and amplitude are direct measurements of the physical characteristics of the rocks, although not in a form usually used by geologists. Of course, it is also possible that advanced processing techniques will extract more conventional lithologic properties from seismic profiles, and that these can be used to calculate autocorrelations.

It is probable that most of the properties that can be measured on the seismic profiles will have different autocorrelation functions. Some of these can be equated to the autocorrelation functions of more conventional geological properties (for example, a *bedding index* may be estimated from the number of reflectors within an interval). Resolving these into a single autocorrelation function in order to establish an optimum sampling spacing will require decisions as to the relative worth of these variables as indicators of petroleum. This assessment is expressed in the form of weights attached to the variables, which may then be used to calculate a multivariate auto-correlation function (Hannan, 1970).

Alternative objectives for federal exploration in NPR—A
The remainder of this essay is a theoretical analysis of objectives that might be set to guide exploration of the National Petroleum Reserve in Alaska. Four alternatives are defined, each representing an extreme in a continuum of possible objectives. While these objectives are specific to the NPR—A, they could be applied equally well to other large regions of of the USA or its territories which are to be explored with the aid of federal funds. These 'end-member' objectives are as follows.

Objective (a): to maximize knowledge of the petroleum potential of NPR—A. This is perceived by some as the intent of Congress when it mandated a Presidential study to 'assess' NPR—A.

Objective (b): to maximize the present worth, or current 'fair market value', of NPR—A. This is the objective currently used by the government in the leasing of federal lands.

Objective (c): to maximize the present worth of future petroleum production from NPR—A. This corresponds to the profit objective of a private corporation.

Objective (d): to maximize the production of petroleum in NPR—A. Under this objective, oil would be produced without regard to profit or loss.

These objectives are mutually exclusive alternatives; the pursuit of one precludes the pursuit of another with the same degree of dedication. Since all are desirable from the national viewpoint, an optimum choice must necessarily represent a compromise between two or more of the end members.

It should be emphasized that there is no 'right' or 'wrong' set of objectives in exploring for petroleum within NPR—A, as the government may legitimately establish goals which are different from those that would be set by a private company. It is imperative, however, that these goals should be sharply defined, so that exploration activities can be managed and directed to satisfy the chosen objectives in the most efficient manner.

In deducing the consequences of each of the end members, some basic assumptions about the nature of the NPR—A exploration program are necessary. Two of these assumptions are common to all end-member objectives.

Assumption 1 All federal funds appropriated for exploration within NPR—A will be expended. This assumption applies for a period extending into the early 1980s.

Assumption 2 All geological and geophysical information acquired by activities of the federal government in NPR—A will become public under the provisions of the Freedom of Information Act. This includes information that exists at present, as well as information to be acquired in the future.

The statistical considerations that are relevant to an exploration program in a sparsely drilled region, developed in the preceding section, must also be considered. Oil exploration is inherently probabilistic, and the application of formal statistical or probabilistic methods is an especially revealing way of deducing the consequences of certain exploration strategies.

The impact of alternative objectives on six aspects of the NPR—A exploration program are considered below. These areas include the geophysical exploration program, which to date has accounted for the bulk of the expended public funds; the drilling program, which will become increasingly important; the alternative land-use appraisal, which is mandated by Congress; and the program management. In addition, the flexibility allowed (or required) under different objectives is discussed specifically. Finally, the economic consequences of each alternative are considered.

Objective (a): to maximize knowledge of petroleum potential of NPR—A
Definition Maximizing *knowledge* about the petroleum potential of NPR—A involves obtaining unbiased estimates of geological properties which are most influential in the generation and entrapment of petroleum. The discovery of oil is a desirable but incidental benefit.
Geophysical effort All future seismic surveying effort within NPR—A should be devoted to obtaining essentially uniform coverage of the Reserve, bearing in mind local geological variability which may require modifications in the spacing of seismic lines. In general, 'detailing' of specific seismic prospects should not be done, with the possible exception that some closely spaced or replicate lines may be run to measure the degree of variability inherent in the seismic analysis and interpretation process. The general purpose of the seismic program should be to provide a uniform quality of seismic coverage over the entire NPR—A, so that no portion remains unappraised.
Drilling effort The specific objective of drilling within NPR—A should be to complement other sources of information (seismic, surface geology, subsurface information from outside the Reserve), so that the geology can be most effectively interpreted, and the oil potential forecast on the basis of the geological information. Thus, exploratory holes should be drilled primarily to obtain geological data, and their locations should be selected so as to maximize the amount of information obtained. Direct knowledge of the presence of oil is but one form of geological information, and although very important in this context should be treated in the same disciplined manner as the other forms of geological information, each weighted in an appropriate, consistent manner.

The most efficient pattern of exploratory holes is a face-centered hexagonal grid or 'squashed' hexagonal grid, with spacings between holes defined according to geological variability as expressed as a spatial auto-correlation function.
Alternative land-use appraisal Since the appraisal of oil potential in NPR—A will be uniform over its entire area, the forecast of those parts with low potential will be more reliable than under alternative objectives. Those areas of low forecasted potential would be the most logical to assign as wildlife refuges and nature reserves. On the other hand, the degree of uncertainty attached to areas of NPR—A which have the highest petroleum potential will be greater than under objectives (c) and (d).
Management requirements As most of the exploration planning can be done in advance of the exploration program, relatively little subsequent effort needs be devoted to major managerial decisions. Significant effort,

however, will be required for the scientific interpretation of the material recovered during drilling.

Program modifications The progress of the exploration program should not be influenced by the successful completion of a discovery well. Exploratory drilling should be regarded as a sampling program that is to be as impartial and unbiased as possible. Thus, even the discovery of a large petroleum reservoir should not cause the program to be modified. Stepout or development wells, if drilled, should be delayed until later, and would not be part of the exploration program as defined here.

Any modifications in the design of the drilling program should be limited to responses to changes in geological variability that are perceived as drilling progresses. Modifications may also be made to reflect changes in the relative weighting accorded to different geological factors that are useful in appraising the oil potential, and for which an improved understanding is obtained as drilling proceeds. The rate of the drilling program can be forecast with relative accuracy because the program is not strongly sequential. The outcome of one exploratory well should not have a significant influence on the decision to drill the next, and so on.

Economic consequences The appraised value of NPR—A at the conclusion of the exploration program can be defined as the sum of nonnegative expected monetary values (EMVs) attached to each of the areas surrounding locations that have been drilled. In addition a 'wild-card' factor must be assigned to the intervening areas which are beyond the immediate area of influence of individual exploratory wells. The wild-card factor must necessarily be appraised on the basis of regional exploration information.

The relative costs of drilling, seismic work, and technical services are highly predictable, because the exploration plan is subject to only minor modifications.

The likelihood of obtaining one or more large commercial successes (defined as those having a high EMV) is lower—perhaps much lower—than that for alternative objectives, as there will be no deliberate attempt to select areas of high oil potential. The overall probability of discovery of petroleum may be defined initially as the proportion of the area of NPR—A which is underlain by oil pools.

The frequency distribution of the sizes of any oil fields discovered will be biased toward large field size, but the size distribution will be less biased than under objectives (c) and (d).

Objective (b): to maximize the present worth of NPR—A
Definition Maximization of the present worth of NPR—A as a whole may be achieved through a leasing program which will require only limited exploration activity by the federal government. *Present worth* is defined as the sum of the bonuses for all tracts in NPR—A, plus the EMVs of future income derived from royalties.

Geophysical effort Seismic work by the federal government should be concentrated on those prospects which appear to have the greatest potential for petroleum at present. No additional seismic reconnaissance should be done except to delimit any favorable prospects now suspected south of the current area of seismic coverage. Seismic lines are to be run sequentially; if initial appraisal of a detailed line degrades a prospect, no additional lines should be shot over that prospect.

Drilling effort Drilling by the federal government should be conducted only on those tracts where drilling will result in a net increase in the EMV. Tracts should be drilled in a sequence that maximizes the potential increase in EMV. All subsequent prospects should be reevaluated after each drilling act, to determine whether the outcome has altered their EMVs. Note that a prospect with positive EMV under objective (c) may have a negative EMV under objective (b), because of the probability of losing the bonus price of adjacent tracts if a test should prove to be a failure. Compounding the risk is the possibility that a failure will depress the EMV of many other tracts as well. It is possible that the risk of depressing the fair market value of the entire NPR—A by continued failure of the drilling program is so great that all additional drilling by the federal government should be avoided.

Alternative land-use appraisal This objective is likely to provide the poorest appraisal of alternative land use, since its purpose is to develop the most *attractive* appraisal of petroleum potential possible consistent with known conditions. Sampling (namely drilling) is therefore likely to be highly biased and incomplete. If surplus exploration funds are diverted to non-conventional exploration activities, it is possible that suitable alternative uses may become apparent as a result of these activities.

Management requirements Because the exploration program and financial appraisal programs must be highly flexible, a significant proportion of the budget must be devoted to management and fiscal appraisal.

Program modifications Additional exploration work should be contingent upon an EMV analysis of NPR—A, in which the EMVs of all proposed exploration activities should be calculated. Exploratory actions should be conducted in order of decreasing EMV, excluding any actions that have negative EMVs. If all conventional exploration activities have negative EMVs, the exploration budget should be directed towards activities which will not contribute to a continued decline in net present value of the Reserve (laboratory analyses, studies of the surface geology in the Brooks Range, etc). All exploration activity of the federal government must be such that it will result in a net increase in the income to be subsequently derived from leases, including the sum of lease bonuses and future royalties. All future income (and expenses) will be discounted to yield present value. Thus, exploration activity that results in a decrease in net present value is to be avoided. The exploration program must be strongly sequential, so the rate of exploration may vary dramatically. Since all exploration activities are undertaken in order of decreasing EMV, and the results of one act may influence the EMVs of subsequent acts, the program should proceed one step at a time. In the event that all conventional exploration acts (drilling, reconnaissance or detailed seismic studies) have negative EMVs, the entire NPR—A should be offered for leasing. If all perceived prospects show negative EMVs and it is decided to lease the entire Reserve, all tracts should be offered simultaneously. If tracts are offered piecemeal, failure on the most promising tracts (which presumably would be leased first) would depress the bonus price of remaining tracts.

Economic consequences The net present value of NPR—A will be maximized with regard both to bonus payments for leases and to royalties on future production.

The nature of the expenditure (that is, the proportions devoted to well costs, seismic studies, personnel, field and laboratory studies, financial analyses, etc) is not highly predictable, and is subject to drastic changes. It is probable that the proportion devoted to seismic studies would be greater than under objectives (a) and (c). Sophisticated seismic processing to develop attractive stratigraphic prospects may absorb a high proportion of the exploration funds.

If the probability of success in drilling is relatively high, attempts should be made to discover petroleum in order to increase the net potential worth of NPR—A by demonstrating the presence of proven reserves. If the probability of success is relatively low, no attempt should be made to discover petroleum directly because of the depressing effect of probable failure on lease bonuses. Instead, efforts should be devoted to enhancing the attractiveness of prospects within the Reserve. Thus, objective (b) provides two alternative strategies for maximizing the net potential worth of NPR—A, whose selection depends upon the probability of drilling success.

Frequency distribution of field volumes will be highly biased toward large fields, even more so than under objective (c). The sample size of discovered fields is expected to be low.

Objective (c): maximize present worth of future petroleum produced from NPR—A

Definition This objective is to maximize the *financial gain* (or minimize the financial loss) from petroleum production, expressed as discounted net cash flows (present value analysis), assuming that the federal government operates the exploration program and provides the risk capital.

Geophysical effort Since the remaining areas to be included in the reconnaissance seismic net have low probability of financial success (because of high costs in remote areas, and generally unfavorable geological conditions), all reconnaissance in those parts of NPR—A that appear uneconomic should be avoided. All seismic effort should be devoted to detailed shooting over prospects in order of their apparent attractiveness.

The purpose is to refine probability estimates and to establish locations for wildcat wells.

Maximum effort should be made to refine and appraise seismic lines already shot, in order to achieve better interpretations, particularly of potential stratigraphic traps. This effort should include additional digital processing, and must include a major effort aimed at geological interpretation of seismic data from a lithological and stratigraphic viewpoint, as well as from a structural viewpoint.

Drilling effort Well locations should be defined by the perceived prospects, ranked in order of the EMVs, which, in turn, are based on the probability of drilling success and various cost factors. No effort should be made to achieve uniform, or even widespread, drilling coverage of NPR—A. In the event that no remaining prospects have positive EMVs, wildcat wells should be drilled, in order of increasing negative EMV, to the limit of the exploration budget.

Alternative land-use appraisal Since appraisal of oil potential is likely to be concentrated in the most favorable areas, little information will be gained about other potential uses of NPR—A in unexplored areas. However, the limits of potentially productive areas are likely to be sharply defined with relatively low uncertainty.

Management requirements Because the exploration program must be highly flexible, a significant proportion of the budget must be devoted to management.

Program modifications In-progress adjustments may be extensive, as the results of drilling a particular prospect may affect probabilities associated with similar prospects. So far as it is feasible, considering the logistical constraints encountered in NPR—A, drilling schedules should be modified to reflect any changes in the relative rank of EMVs of prospects. Changes in probability (hence in EMVs) resulting from additional seismic work should influence the drilling program. In the event that all available prospects have negative EMVs, the bulk of the exploration budget should be shifted momentarily to detailed seismic and additional seismic and geological analyses in an attempt to find new prospects with positive EMVs.

The drilling program should be strongly sequential, so the rate of exploration may vary dramatically. Since prospects are drilled in order of decreasing EMV, and the results from one well will influence the probabilities attached to all subsequent wells, an ideal program should involve drilling wells one at a time. This allows time for alternative prospects to be substituted into the drilling program, for the shooting of detailed seismic lines to define prospects, and for the drilling of development wells if a discovery is made.

Economic consequences The net present worth of NPR—A will be maximized with regard to future petroleum production.

As with objective (b), the nature of the expenditure is highly unpredictable and subject to drastic changes with time. Development wells (wells drilled in a field for production purposes) will be drilled as long as their EMVs exceed those of possible wildcat wells. Thus, it is possible that a major discovery would result in the redirection of much of the effort in NPR—A toward development of specific oil fields.

Probabilities for commercial success are the highest possible of the four alternative strategies. They are equal to the conditional probabilities developed relating the presence of oil and gas to the geological and geophysical exploration guides. The frequency distribution of field volumes will be highly biased towards large fields.

Objective (d): to maximize the amount of oil produced from NPR—A
Definition This objective seeks to maximize the *oil produced* within a fixed time period in NPR—A. To account for the uncertainty attached to individual prospects, the expected production volume (EPV) of the Reserve as a whole will be maximized.
Geophysical effort Seismic exploration should be concentrated on delineation of those prospects which have the highest potential resources, without regard for financial factors.
Drilling effort Funds should be spent on drilling of prospects in order of *expected production volume* (EPV) of oil returned, regardless of the EMVs of the prospects. The term 'expected production volume' is analogous to expected monetary value, and is obtained by multiplying the probability values attached to different production outcomes by the production in barrels, and summing. In order to achieve an exploratory drilling rate greater than that possible by government action alone, the NPR—A should be divided into tracts to be leased on a drilling performance basis. Royalties should be based on a sliding scale which rewards increased total production by an amount which exceeds the investment necessary to achieve this increased production. Government funds should be devoted to exploring for new prospects and to drilling those prospects not leased, in order of their perceived EPVs. The balance between seismic study expenditure and drilling expenditure will be determined by the number of prospects with relatively high EPVs which remain unleased. If few or no prospects have high EPVs, seismic work should be done in an attempt to delineate new prospects or to increase the EPV of known prospects. If many prospects with high EPVs remain unleased, remaining funds should be devoted to drilling.
Alternative land-use appraisal Oil exploration will not be constrained to the most favorable areas, so alternative land-use investigations will be more widespread than under objectives (b) or (c). However, a more uniform coverage will be obtained under objective (a).
Management requirements Because establishment of performance criteria and monitoring must be performed for a potentially large number of tracts, management requirements will be substantial.

Program modifications A high degree of flexibility in the exploration program is essential. The relative effort devoted to seismic studies versus drilling activities should reflect the degree of industrial participation in exploration, and the EPVs of the unleased tracts. The drilling program may be highly cyclic, with periods of government-sponsored seismic and geological exploration separated by lease sales and industry drilling activity. *Economic consequences* The undiscounted value of oil to be produced will be maximized. The nature of the expenditure (proportion devoted to drilling, seismic studies, personnel, etc) would be highly variable. Engineering analyses and performance monitoring costs will be significant.

The financial definition of 'success' is not applicable under this objective. The program may produce a relatively large number of fields that would not be considered commercially viable under other objectives. The frequency distribution of field sizes will be biased toward large fields, but less biased than under either objective (b) or objective (c).

Conclusions

The objective of the United States government in conducting an exploration campaign in NPR—A has yet to be clearly stated. The problem is that the optimum benefits to be derived from the NPR—A exploration program, or for that matter from any large federally funded exploration program, are difficult to define. By contrast, the objective of a private firm is simple in that it seeks to maximize its profits.

The current debate over an energy policy for the United States has been woefully shortsighted because so little effort has been devoted to defining national objectives with respect to the nation's energy resources. Once these objectives are clearly and consistently established, it will be much more feasible to forge policies that affect the nation's energy future. At present, for example, the federal leasing policies for offshore areas are designed to maximize immediate income in the form of lease bonuses through competitive sealed bids. Such a policy does not necessarily maximize gross income, and is probably strongly detrimental to the maximum development of the nation's offshore oil resources. As an alternative, it might be wiser to award offshore leases on an exploration performance basis.

If the federal government is to engage in the act of exploration itself, and to expend public funds for this purpose as it is currently doing in Alaska, it is imperative that its exploration objectives coincide with national energy policy. In turn, this policy must reflect an informed national consensus, based upon a clear understanding of the economic, political, and ethical consequences of various alternatives. Until this is achieved, federal exploration activities, both in Alaska and elsewhere, will continue to be ill-defined and unlikely to contribute to the nation's well-being.

References

Bernard W J, Akers H Jr, 1977 "Monte Carlo range-of-values program description" internal report, Alaska Office, Geological Survey Conservation Division, Anchorage, Alas.

Box G E P, Jenkins G M, 1970 *Time Series Analysis—Forecasting and Control* (Holden-Day, San Francisco, Calif.)

Carter R D, Mull C G, Bird K J, Powers R B, 1977 "The petroleum geology and hydrocarbon potential of Naval Petroleum Reserve No. 4, North Slope, Alaska" open-file report 77-475, Menlo Park Office, US Geological Survey, Menlo Park, Calif.

Drew L J, 1967 "Grid-drilling exploration and its application to the search for petroleum" *Economic Geology* **62** 698–710

Griffiths J C, 1962 " Frequency distribution of some natural resource materials" in *Proceedings, 23rd Technical Conference on Petroleum Production* circular number 63, College of Earth and Mineral Sciences Experimental Station, Pennsylvania State University, University Park, 16802 Pa, USA, pp 174–198

Griffiths J C, Ondrick C W, 1968 "Sampling a geological population" computer contribution number 30 (Kansas Geological Survey, Lawrence, Kan.)

Hannan E J, 1970 *Multiple Time Series* (John Wiley, New York)

Kaufman G M, 1974 "Statistical methods for predicting the number and size distribution of undiscovered hydrocarbon deposits" in *Proceedings, AAPG Research Symposium on Methods of Estimating Volume of Undiscovered Oil and Gas Resources, Stanford University, 21–23 August 1974* (American Association of Petroleum Geologists, Tulsa, Okla) pp 247–310

Matheron G, 1971 "The theory of regionalized variables and its applications" *Cahiers du Centre de Morphologie Mathématique de Fontainebleau, France* volume 5 (Centre de Morphologie Mathématiques, Fontainebleau, France)

Morgridge D L, Smith W B Jr, 1972 "Geology and discovery of Prudhoe Bay field, eastern Arctic slope, Alaska" in *AAPG Memoir 16: Stratigraphic Oil and Gas Fields—Classification, Exploration Methods, and Case Histories* (American Association of Petroleum Geologists, Tulsa, Okla) pp 489–501

Recommended reading

Drew L J, 1975 "Linkage effects between deposit discovery and postdiscovery exploration drilling" *Journal of Research of the United States Geological Survey* **3**(2) 169–180

Grayson C J Jr, 1960 "Decisions under uncertainty, drilling decision by oil and gas operators" Division of Research, Graduate School of Business Administration, Harvard University, Cambridge, Mass

Harbaugh J W, Doveton J H, Davis J C, 1977 *Probability Methods in Oil Exploration* (John Wiley, New York)

Singer D A, 1975 "Relative efficiencies of square and triangular grids in the search for elliptically shaped resource targets" *Journal of Research of the United States Geological Survey* **3**(2) 163–168

Singer D A, Drew L J, 1976 "The area of influence of an exploratory hole" *Economic Geology* **71** 642–647

Singer D A, Wickman F E, 1969 "Probability tables for locating targets with square, rectangular, and hexagonal point-nets" publication 1-68, College of Earth and Mineral Sciences Experimental Station, Pennsylvania State University, University Park, Pa

Part 4

Relations

Productivity

Intensity of Supervision

0

-MTR-

Analysis of asymmetric relationships in geological data

R A Reyment, C F Banfield

Introduction

The relationship between two geological specimens i and j can be quantified in many ways. Such quantities, a_{ij} say, are calculated from measurements made on the specimens, and these can be many and varied, depending on what aspects of the specimens are of interest. If the relationship between the two specimens is symmetric then the quantities a_{ij} and a_{ji} will be equal. Geologists are familiar with a variety of multivariate statistical methods, including principal components analysis and multidimensional scaling, which can be used to analyse square symmetric matrices holding such values. Because distance is usually metric in nature and symmetric by definition, many of these statistical techniques aim at representing the values a_{ij}, or some simple transformation of them, as distances on a plot in which the distance d_{ij} ($= d_{ji}$) represents the relationship of specimen i to specimen j.

There are, however, many problems in geology where the relationship between two specimens is nonsymmetric, and $a_{ij} \neq a_{ji}$. Asymmetric data are commonly generated in the quantitative interpretation of spatial relationships of all kinds of geological maps: for example, counts on the number of times formation i is the nearest unlike neighbour of formation j, or the number of times sediment i is surrounded by sediment j. Here, the sites have a known geographical relationship, and the significance therefore lies in the asymmetry of the observations. Another example is provided by thin sections of rocks which also form a kind of map in which relationships between neighbouring mineral species may be of interest, as in *Gefüge*-analysis. Other examples can be found in problems of spatial paleoecology, microfacies analysis, and the distribution of glacial drift. All of these have the common factor that they can be represented as some kind of map.

A square nonsymmetric matrix, **A**, whose rows and columns are classified by the same specimens (see, for example, table 1), cannot be treated by the same methods as used in the standard multivariate analysis of square symmetric matrices. Attempts at analysing asymmetric data have been made by using nonmetric multidimensional scaling for analysing the symmetric matrix $\frac{1}{2}(A + A^T)$; this approach was exploited by Shepard (1963) in his study of perception of Morse-code signals, but the method does not consider the asymmetry in the data. The asymmetry may sometimes be caused by so-called 'noise', but in geological applications it may have an intrinsic meaning.

Gower (1977) has given attention to methods of analysing the asymmetry in square nonsymmetric matrices whereby the differences between a_{ij} and a_{ji} can be represented graphically. Among the solutions examined by him, we have chosen that of the canonical analysis of asymmetry as being the most applicable to geological-map interpretation in its broadest sense. This technique should not be confused with spatial modelling, to which it bears no relation. We present here an application of Gower's method to a sedimentary environmental pattern, which is perhaps elementary in nature, but helps to illustrate how the technique might be employed in more complex situations.

Description of the method

Given an $n \times n$ nonsymmetric matrix, A, whose rows and columns classify the same n specimens, then M $[= \frac{1}{2}(A + A^T)]$ is a symmetric matrix, N $[= \frac{1}{2}(A - A^T)]$ is a skew-symmetric matrix, and $A = M + N$. Gower (1977) shows that matrices M and N can be analysed separately, and their results considered together to form an overall representation of the asymmetry values in A. However, if M is of little importance or is unknown, the analysis of N alone can still be considered.

The canonical form of N can be obtained by using the singular-value decomposition of N [see, for example, Golub and Reinsch (1970), or Wilkinson and Reinsch (1971)]:

$$N = USV^T ,$$

where S is a diagonal matrix containing the singular values s_i ($i = 1, 2, ..., n$), U is an orthogonal matrix, and V is the product UJ, where J is the elementary block-diagonal skew-symmetric matrix made up of 2×2 diagonal blocks $\begin{pmatrix} 0 & 1 \\ -1 & 0 \end{pmatrix}$. If n is odd, then the final singular value, s_n, will equal zero, and the final diagonal element of J will equal one. Because N is skew-symmetric, the singular values will be in pairs, and can be arranged in descending order of magnitude: $(s_1 = s_2) > (s_3 = s_4) > (s_5 = s_6) ...$.

Gower (1977) shows that the pair of columns of U corresponding to each pair of equal singular values, when scaled by the square root of the corresponding singular value, holds the coordinates of the n specimens in a two-dimensional space—a plane—that approximates the skew-symmetry. The proportion of the total skew-symmetry represented by this plane is given by the size of the corresponding singular value. Hence, as the first singular value, s_1 ($= s_2$), is the largest, the first two columns of U give the coordinates of the n specimens in the plane accounting for the largest proportion of skew-symmetry. If s_1 is large compared to the other s_i ($i > 2$), the plot of the first plane will give a good approximation to the values in N. If s_1 is not comparatively large, then other planes are also required to give a good approximation. Because of the nonmetric nature of skew-symmetry, the values of N are represented in the planes by areas of triangles made with the origin. Hence, the area of triangle $(i, j, 0)$ defined by points i, j, and the origin is proportional to the ijth value in the matrix N, and

$$\text{area}(i, j, 0) = -\text{area}(j, i, 0) \ .$$

The canonical analysis of asymmetry can, in some respects, be regarded as an analogue of the familiar method of principal components analysis. In principal components analysis, distance is explained by differences in transformed values, relative to principal axes (one axis for each latent root), whereas the skew-symmetry here is explained by areas of triangles relative to principal planes (one plane for each pair of singular values).

The plot or plots of the n specimens formed from analysis of the symmetric matrix M may also be considered with the plane or planes obtained by the method described above, as together they give a complete graphical representation of the asymmetric values in A. The plots for M can be obtained by many methods, including multidimensional scaling (see, for example, Kruskal, 1964; Gower, 1966). This will provide a set of coordinates for the n specimens that hopefully represent the symmetry values of M with good approximation in two or three dimensions. If two dimensions are sufficient, then these coordinates can be plotted to produce a graphical representation in which the distances between the specimens represent the values of M. If two dimensions are insufficient, a third dimension, or even more, will be necessary.

The data

The data used for illustrating the application of Gower's method to a typical geological problem based on a map were taken from McCammon (1972). The material consists of observations on the nearest-neighbour relationships between the following seven environmental categories in the

Table 1. Matrix [a] of the number of nearest unlike neighbours for depositional environments in the Mississippi Delta. (Source: McCammon, 1972, page 424.)

Category	(1) natural levee	(2) point bar	(3) swamp	(4) marsh	(5) beach	(6) lacustrine	(7) bay sound
(1)	–	117	286	148	0	2	0
(2)	38	–	5	2	0	0	1
(3)	301	10	–	175	1	138	12
(4)	538	3	168	–	29	320	281
(5)	0	0	0	9	–	0	8
(6)	2	0	168	292	0	–	20
(7)	0	1	147	617	161	25	–

[a] The asymmetric entries reflect the widely differing spatial relationships between the environments, hence order in the geological pattern. Symmetric entries would be indicative of geological disorder.

Mississippi Delta region of southeastern Louisiana: (1) natural levee; (2) point bar; (3) swamp; (4) marsh; (5) beach; (6) lacustrine; (7) bay sound. Further information on these variables can be obtained from McCammon (1972).

The observations on the spatial relationships between these depositional environments were extracted from a map of the delta (McCammon, 1972, page 423). Each environment is represented by a distinct variety of sediment. The matrix of nearest unlike neighbours is shown in table 1.

Asymmetry analysis of the data

The 7×7 asymmetric matrix listed in table 1 was first transformed to a matrix A of proportions (table 2) by dividing each value by its row total. Note that this transformation would not usually be necessary, but McCammon's matrix had no diagonal elements so the transformation was used to ensure constant known diagonal elements for the later analysis of matrix M. From the matrix A, the skew-symmetric matrix N, $N = \frac{1}{2}(A - A^T)$, shown in table 3 was computed.

The pairs of singular values from the singular-value decomposition of N, and the corresponding coordinate vectors, are given in table 4. It will be seen from the relative sizes of the pairs of singular values that representation of the skew-symmetry needs at least four dimensions—that is, two planes

Table 2. Asymmetric matrix of table 1 transformed to proportions.

Category	(1) natural levee	(2) point bar	(3) swamp	(4) marsh	(5) beach	(6) lacustrine	(7) bay sound
(1)	1·0000	0·2116	0·5172	0·2676	0·0000	0·0036	0·0000
(2)	0·8261	1·0000	0·1087	0·0435	0·0000	0·0000	0·0217
(3)	0·4725	0·0157	1·0000	0·2747	0·0016	0·2166	0·0188
(4)	0·4018	0·0022	0·1255	1·0000	0·0217	0·2390	0·2099
(5)	0·0000	0·0000	0·0000	0·5294	1·0000	0·0000	0·4706
(6)	0·0041	0·0000	0·3485	0·6058	0·0000	1·0000	0·0415
(7)	0·0000	0·0011	0·1546	0·6488	0·1693	0·0263	1·0000

Table 3. Skew-symmetric matrix formed from matrix of table 2.

Category	(1) natural levee	(2) point bar	(3) swamp	(4) marsh	(5) beach	(6) lacustrine	(7) bay sound
(1)	0·0000	−0·3073	0·0223	−0·0671	0·0000	−0·0003	0·0000
(2)	0·3073	0·0000	0·0465	0·0206	0·0000	0·0000	0·0103
(3)	−0·0223	−0·0465	0·0000	0·0746	0·0008	−0·0660	−0·0679
(4)	0·0671	−0·0206	−0·0746	0·0000	−0·2539	−0·1834	−0·2195
(5)	0·0000	0·0000	−0·0008	0·2539	0·0000	0·0000	0·1506
(6)	0·0003	0·0000	0·0660	0·1834	0·0000	0·0000	0·0076
(7)	0·0000	−0·0103	0·0679	0·2195	−0·1506	−0·0076	0·0000

corresponding to singular values $0 \cdot 4241$ and $0 \cdot 3078$—to represent it with good approximation. The distributional properties of the singular values are as yet unknown, but an ad hoc approach might be to require values to sum to more than 80% of the total for an acceptable approximation. Here, a plot of the first two dimensions (that is, the first plane) alone would not show all the skew-symmetry in the material—only $50 \cdot 7\%$, in fact—and the second plane, corresponding to the second pair of singular values, has to be considered as well to improve the overall fit of N to $87 \cdot 5\%$ (see figure 1).

The plot of the first plane, corresponding to the first pair of singular values, is shown in figure 2. The largest skew-symmetric value of N is n_{12},

Table 4. Singular values and coordinates in the corresponding planes, obtained from the skew-symmetric matrix of table 3.

Singular values and corresponding planes [a]							
(1)	(2)	(3)	(4)	(5)	(6)	(7)	
$0 \cdot 4241$	$0 \cdot 4241$	$0 \cdot 3078$	$0 \cdot 3078$	$0 \cdot 1050$	$0 \cdot 1050$	$0 \cdot 0000$	
first plane		second plane		third plane			
(1)	$-0 \cdot 0891$	$-0 \cdot 1489$	$0 \cdot 4995$	$0 \cdot 1779$	$-0 \cdot 0276$	$-0 \cdot 0116$
(2)	$0 \cdot 1314$	$-0 \cdot 0237$	$0 \cdot 1993$	$-0 \cdot 5010$	$0 \cdot 0205$	$0 \cdot 0120$
(3)	$0 \cdot 1205$	$-0 \cdot 0351$	$0 \cdot 0429$	$0 \cdot 1057$	$0 \cdot 2455$	$0 \cdot 0391$
(4)	$0 \cdot 3788$	$-0 \cdot 4673$	$-0 \cdot 0821$	$0 \cdot 0376$	$0 \cdot 0053$	$-0 \cdot 0859$
(5)	$0 \cdot 2354$	$0 \cdot 3610$	$0 \cdot 0771$	$0 \cdot 0467$	$0 \cdot 0194$	$-0 \cdot 2361$
(6)	$0 \cdot 2054$	$0 \cdot 1893$	$0 \cdot 0476$	$0 \cdot 0382$	$0 \cdot 1161$	$0 \cdot 1510$
(7)	$0 \cdot 3785$	$0 \cdot 1248$	$0 \cdot 0432$	$0 \cdot 0947$	$-0 \cdot 1722$	$0 \cdot 1315$

[a] Categories (1)-(7) are as described in previous tables.

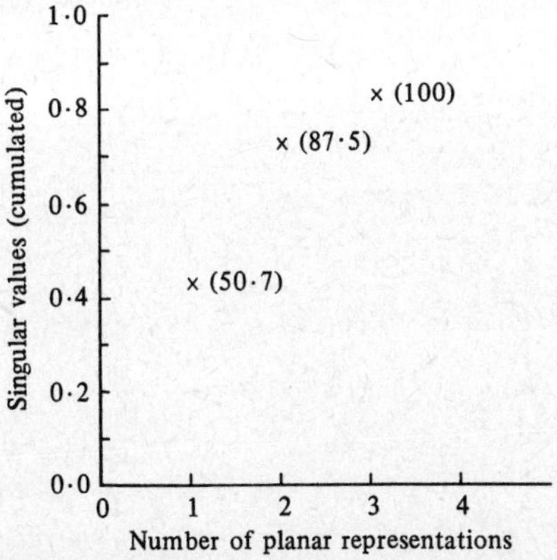

Figure 1. Plot of cumulative singular values (one from each pair) against the number of corresponding planar representations. The percentage of the total of the singular values is shown in brackets for the inclusion of each plane.

which equals 0·3073. The corresponding triangle formed by joining
environments (1) and (2) to the origin, and here denoted (1, 2, 0), has a
relatively small area and, in fact, there are several triangles that greatly
exceed it in area. The triangle with the greatest area is (4, 5, 0),
corresponding to the second-largest skew-symmetric value (0·2539). This
shows that the first planar representation alone is not sufficient to give a
good approximation to the values of N. The second plane (figure 3),
however, discloses that (1, 2, 0) is the triangle with the largest area, and
that all other triangles have relatively small areas. So this plane, as
expected from the singular values, accounts for the area not explained by
the first planar representation.

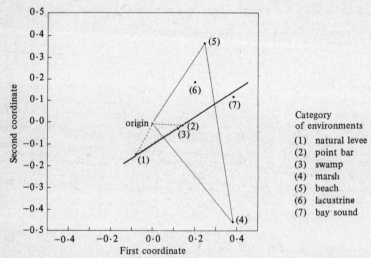

Figure 2. Plot of the coordinates for the seven environments in the plane corresponding
to the first pair of singular values.

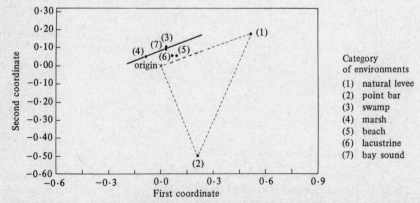

Figure 3. Plot of the coordinates for the seven environments in the plane corresponding
to the second pair of singular values.

The plot of the first two dimensions (figure 2) shows that all points have a large asymmetry with point 4 [that is, environment (4): marsh], because all triangles including point 4 have large areas. There seems to be a collinear relationship between points 1, 3, 2, and 7, in that order (that is: natural levee, swamp, point bar, bay sound). Gower (1977) shows that collinear points will have equal skew-symmetry values with a point on a line through the origin parallel to the collinearity, because these points will form triangles having the same base and with equal height hence equal area. Unfortunately the line through the origin and parallel to the collinearity passes through no other environment, so the theory cannot be exemplified in this planar representation (it can be seen, however, in figure 3). Points 2 and 3 (point bar, swamp) lie close together in this plane, and thus appear to have similar skew-symmetry values with the other environments because triangles formed by other environments with the origin and these two points will have very similar areas. However, in the second plane this misrepresentation is corrected, as points 2 and 3 are not close.

In figure 3, the supplementary information of the second plane is dominated by the triangle (1, 2, 0). Environments (1) and (2) share the largest asymmetry, suggesting a possible genetic relationship. Inspection of McCammon (1972, figure 1) indicates this to be identifiable with the immediate depositional environment of the Mississippi River. The points for environments (3), (4), and (7) lie on a line parallel to that joining the origin to environment (1), indicating that in this planar representation, and this one only, these environments show equal skew-symmetry with the natural levee environment. The close proximity of swamp with bay sound, and beach with lacustrine shows that these pairs have similar skew-symmetries with the other environments.

Table 5. Principal coordinates obtained for the symmetric matrix derived from the data of McCammon (1972).

	Latent roots [a]						
	(1) 1·5372	(2) 1·1909	(3) 0·8086	(4) 0·7107	(5) 0·5897	(6) 0·1629	(7) 0·0000
(1)	0·6275	−0·1240	−0·2031	−0·1038	−0·3053	−0·2434	−
(2)	0·5829	−0·4592	0·4623	−0·0332	0·1939	0·1452	−
(3)	0·3256	0·3957	−0·5505	0·1937	0·2273	0·1518	−
(4)	−0·3191	0·1951	0·0526	−0·3383	−0·4770	0·1843	−
(5)	−0·5247	−0·4387	−0·0402	0·5819	−0·1464	−0·0166	−
(6)	−0·1451	0·7078	0·4614	0·1355	0·1325	−0·1300	−
(7)	−0·5472	−0·2767	−0·1825	−0·4357	0·3751	−0·0913	−

[a] Categories (1)-(7) are as described in previous tables.

In order to reify further the relationships outlined by the graphical representations of the skew-symmetry, we analysed the symmetrical part, M, of the asymmetric matrix A by principal coordinate analysis, which is a form of metric multidimensional scaling due to Gower (1966). The results of the calculations are given in table 5. The plot of the first pair of coordinates, illustrated in figure 4, represents 54·5% of the total symmetry between the depositional environments. Because these coordinates are plotted in a Euclidean space, we can give a metric interpretation to the interpoint distances, which represent the similarities between the environments. As might be expected, bay sound and beach are comparatively close together, showing that the degree of organization of their spatial relationships is similar; natural levee and point bar are also similar. Lacustrine is seen to be very different from beach and point bar, and most similar to swamp and marsh.

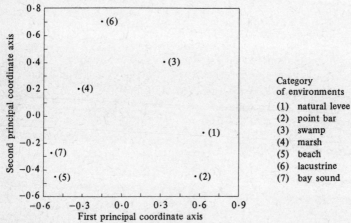

Figure 4. Plot of the principal coordinates for the seven environments, accounting for 54·5% of the total symmetry.

Concluding remarks
Despite the rather elementary nature of the geological problem considered here, it is evident that the method of Gower (1977) offers a means of analysing and graphically displaying data in such a way as to aid geological interpretation. Plots of specimens can reveal pattern and structure which is difficult to discern in tables of figures alone, and here we have obtained plots both for the skew-symmetric and the symmetric parts of the assymmetric relationship. The main area of geological application seems to be within the sphere of map interpretation, broadly viewed, although further avenues will probably be opened up. Research is obviously needed into the distributional properties of the singular values, so that significance tests can be used in deciding the number of planar representations for good approximation of the skew-symmetry values.

Acknowledgements. We wish to thank John Gower for bringing the topic of this paper to our attention and for patiently answering many questions concerning the interpretation of plots in nonmetric space. The calculations were made at Rothamsted by means of a macro written in the GENSTAT language (Nelder et al, 1977) by Banfield. We also wish to thank Richard B McCammon (Reston, USA) for supplying information on the data used here.

References

Golub G H, Reinsch C H, 1970 "Singular value decomposition and least squares solutions" *Numerische Mathematik* **14** 403-420

Gower J C, 1966 "Some distance properties of latent roots and vector methods in multivariate analysis" *Biometrika* **53** 325-338

Gower J C, 1977 "The analysis of asymmetry and orthogonality" in *Recent Developments in Statistics* Eds J Barra et al (North-Holland, Amsterdam) pp 109-123

Kruskal J B, 1964 "Multidimensional scaling by optimizing goodness of fit to a nonmetric hypothesis" *Psychometrika* **29** 1-27

McCammon R B, 1972 "Map pattern reconstruction from sample data: Mississippi Delta region of south eastern Louisiana" *Journal of Sedimentary Petrology* **42** 422-424

Nelder J A, and others, 1977 *GENSTAT Manual* Rothamsted Experimental Station, Harpenden, Herts, England

Shepard R N, 1963 "Analysis of proximities as a technique for the study of information processing in man" *Human Factors* **5** 19-34

Wilkinson J H, Reinsch C H, 1971 *Handbook for Automatic Computation, Volume 2: Linear Algebra* (Springer, New York)

Comparison functions and geological structure maps

D F Merriam, J E Robinson

Introduction

Determination of similarities between contoured maps may be based on several factors including size, shape, or orientation of any or all sets of component features. Before attempting to compare maps, the analyzer must define his objectives, and determine what is being compared. Also one must consider the differences between a single-valued comparison function that gives an overall comparison, and a spatial comparison or function which produces a map on areal goodness-of-fit that permits a two-dimensional similarity interpretation.

Many methods of map comparison have been proposed. The simplest method is to overlay the two maps and visually compare them. Quantitative methods have usually been equally simplistic. One manner is to take corresponding data points from each map and to plot them on a scatter diagram to determine the degree of accordance (Mirchink and Bukhartsev, 1959). Other more involved techniques include computing numerical approximations of the surfaces and using their representations for the comparison. These techniques include: the fitting of low-degree polynomial trend surfaces and use of the coefficients (Merriam and Sneath, 1966); use of the computed matrices of the grids for direct comparison (Miller, 1964); use of the least-squares summation equations (Rao, 1971); comparison of positive and negative areas on the residual maps (Merriam and Lippert, 1964); and matching values at grid positions on spatial filtered maps (Robinson and Merriam, 1972). Fourier maps and power-spectra functions (Esler and Preston, 1967) have also been used as numerical descriptors of surfaces. All of these techniques have drawbacks and limitations, and we believe that the procedure outlined here is an improvement and circumvents some of the problems with the other approaches.

Comparison of structural maps

Structural contour maps are generally compared casually and on visual appearance. If all contours on the map are identical then the problem is trivial; however, where the maps contain conflicting trends, only some of which may occur in both maps, the problem is more difficult. Not only must the type of similarity be defined, but also the type of the components being compared must be declared.

Structure surfaces are usually composed of a variety of features, ranging from the large-scale regional trends to local anomalies that are barely within the resolution of map control. Overall configuration of the map is influenced most strongly by the large features. Where regional dips are steep, local features may be distorted or completely masked. The smaller the feature, the more difficult it is to define in any sequence of maps.

Where the comparison is based on selected attributes, such as features within a spatial scale range, then visual appearance is not satisfactory and some form of quantification is necessary. Because comparison of maps is also dominated by those features that are large and of high amplitude, the selected features to be compared must first be extracted and displayed as new maps.

Selective display of specific features can be accomplished best by spatial-filtering techniques. In this method (Robinson, 1968; Robinson et al, 1968), the selected range of features are considered to be composed of a set of frequency components. A spatial filter is designed that will pass only the desired component frequencies. This filter is then convolved with the map, which produces a new map containing only the desired range of features. The filtering process passes the desired features on the basis of their frequency-component content, and rejects all other features containing frequencies outside this range. The resultant new map contains only the desired features, usually displayed as positive and negative departures from a mean zero elevation. If the filter is correctly designed, the position of the features is retained, and they are not distorted on the resultant map. Radially symmetric filters allow desired features to pass regardless of directional properties. Other filters include discrimination on the basis both of size and of trend. The design, use, and limitations of filtering geological data are discussed by Robinson and Merriam (1972).

Data sets
The data used here are gridded subsea elevations for three horizons in the western part of Kansas:
(1) top Lansing Group (Pennsylvanian);
(2) top Arbuckle Group (Cambrian–Ordovician);
(3) top Precambrian rocks.
The data were taken on a six-mile grid from published structure contour maps (Merriam et al, 1958; Merriam and Smith, 1961; Cole, 1962). Information on the data sets is given in table 1, and the contour maps as generated by the SURFACE II program (Sampson, 1975) are shown in figure 1.

Table 1. Data sets studied here.

Horizon	Grid interval (miles)	Number of E–W grid points	Number of N–S grid points	Total data points	Contour interval [a] (feet)
Top Lansing	6	57	29	1653	50
Top Arbuckle	6	57	29	1653	100
Top Precambrian	6	57	29	1653	100

[a] Refers to digitized maps.

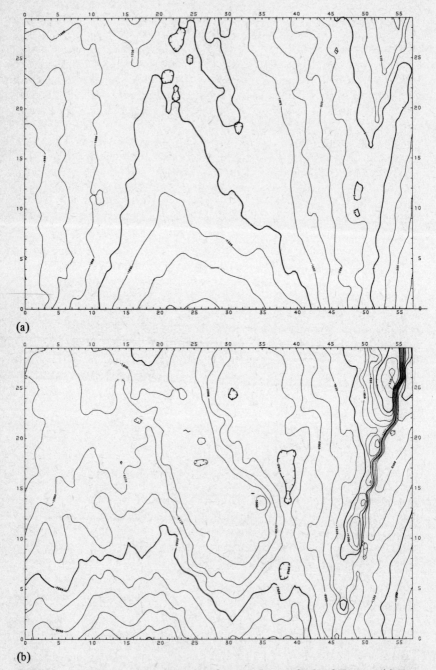

(a)

(b)

Figure 1. Contour maps of original data for western two-thirds of Kansas. (a) Top Lansing (Pennsylvanian); (b) top Arbuckle (Cambrian–Ordovician); (c) top Precambrian. Contour interval: 250 feet.

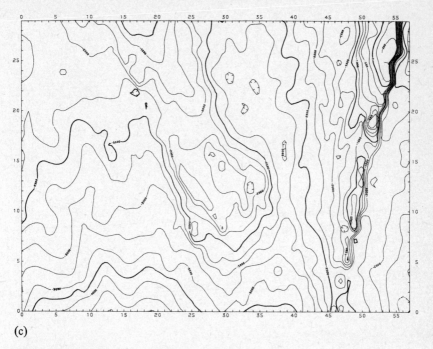

(c)

Figure 1 (continued)

Technique

The original structure maps were filtered with a 7 × 7 band-pass filter which will pass features with widths of 1·5 to 6 nondirectional units. This filter also removes the average elevation and any regional trends as well as any high-frequency noise in the original data. The filter is given numerically in table 2; the filtered maps are shown in figure 2.

The filtered maps were then rescaled for ease of computation. The same scaling factor was used on all three maps to insure that relative amplitude information was obtained. In this example, the factor was determined by normalizing the highest amplitude map to a maximum range of −1 to +1.

Table 2. 7 × 7 filter, pass widths 1·5 to 6 units, nondirectional.

−0·012	−0·014	−0·031	−0·040	−0·031	−0·014	−0·012
−0·014	−0·042	−0·031	−0·018	−0·031	−0·042	−0·014
−0·031	−0·031	0·053	0·137	0·053	−0·031	−0·031
−0·040	−0·018	0·137	0·255	0·137	−0·018	−0·040
−0·031	−0·031	0·053	0·137	0·053	−0·031	−0·031
−0·014	−0·042	−0·031	−0·018	−0·031	−0·042	−0·014
−0·012	−0·014	−0·031	−0·040	−0·031	−0·014	−0·012

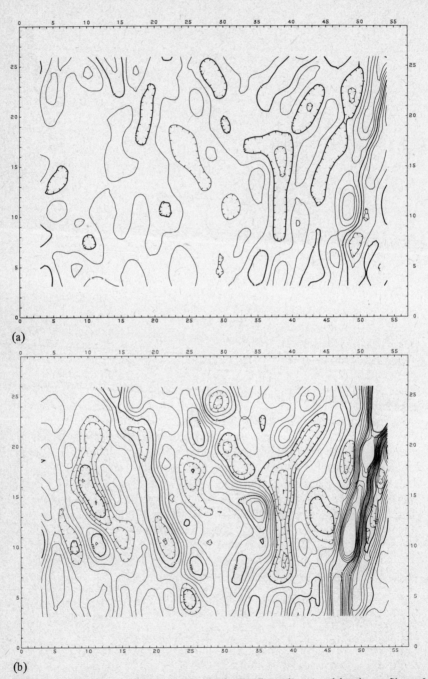

(a)

(b)

Figure 2. Filtered maps obtained by using the 7×7 nondirectional band-pass filter of the map horizons and area of figure 1: (a) Lansing, (b) Arbuckle, and (c) Precambrian. Contour interval: 50 feet.

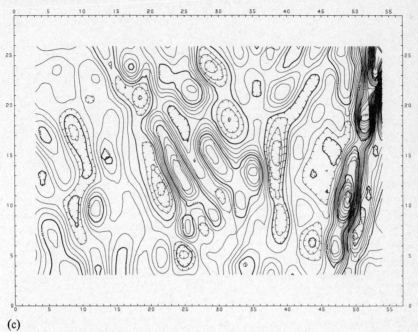

(c)

Figure 2 (continued)

These prepared data sets were then multiplied element by element [they could also be added or subtracted, see Davis (1973)], and a similarity map computed. The similarity-function map has elements, X_{ij}, given by

$$
X_{ij} = \begin{cases} +|A_{ij}B_{ij}|^{\frac{1}{2}}, & A_{ij}B_{ij} > 0 \; ; \\ -|A_{ij}B_{ij}|^{\frac{1}{2}}, & A_{ij}B_{ij} < 0 \; ; \end{cases} \tag{1}
$$

where A_{ij} and B_{ij} are the elements of the maps to be compared. With n rows and m columns, there will be nm elements of the first map $[A_{ij}]$, the second map $[B_{ij}]$, and the similarity-function map $[X_{ij}]$. Simple cross multiplication produces a new map that is no longer a linear function of the input maps. Similar features of high amplitude are given undue precedence over equally similar features of lower amplitude. Taking the square root of the product values restores a form of linearity, and produces a better perspective of spatial similarity between the component features.

The resultant maps of the cross-multiplication operation given by equation (1) have the following characteristics. A flat area on one map compared with a high or low on the other map would give a near-zero value of the similarity function X_{ij}. Two corresponding highs or lows would result in a positive value; coincidence of a high and a low would result in a negative value. These relations are shown graphically in table 3, along with results for two maps compared by cross addition and subtraction.

At the time that the maps are cross multiplied, a correlation coefficient, r, is computed:

$$r = \sum_{i,j} X_{ij} |X_{ij}| \Big/ \left(\sum_{i,j} A_{ij}^2 \sum_{i,j} B_{ij}^2 \right)^{\frac{1}{2}} . \qquad (2)$$

Because the filtering operation removes the mean, and because X_{ij} is the signed square root of the cross multiplication, the calculated coefficient r in equation (2) is the Pearson product-moment correlation coefficient (Pearson, 1896). The coefficient is dependent on the linear relationship between the two maps. Thus the comparison of two identical maps would give $r = +1$, whereas two absolutely opposite maps would give $r = -1$. Where the maps contain only small features on horizons separated by a considerable rock thickness and a number of geological events, low values of the correlation coefficient are expected and small differences in the degree of correlation may be geologically, if not statistically, significant. Correlations based on the large regional features, such as basins, that are less prone to refract or die out with tectonic and depositional change, will give high values of r, and may require a correspondingly large change in the degree of correlation to indicate significant variation. Thus statistical significance may differ markedly from geological significance. However, digitized large-scale maps have a high number of degrees of freedom, because of the extremely large size of the original data set. Confidences in the t test for the three maps are better than 99·99%.

Table 3. Resultant map features from comparison of two maps by multiplication, addition, or subtraction.

Comparison function	Resultant map feature		
	positive	zero	negative
Multiplication	⌢ ⌣	⌣ ⌐ =	◡⌒
Addition	⌢ ⌒	◡ ⌐ =	⌣ ⌣
Subtraction	⌒ ⌣	⌢ ⌣ =	⌣ ⌐

Key: ⌢ ≡ high; ⌣ ≡ low; — ≡ flat.

Results

The configuration of the original data reveals the major structures of Kansas (Merriam, 1963). The asymmetrical northeast-trending Nemaha Anticline is prominent (figure 1). Just to the west of this is the Salina Basin (to the north) and (to the south) the northern end of the Sedgwick Basin which in turn are separated from the Hugoton Embayment (in the far west) by the Central-Kansas-Uplift/Cambridge-Arch/Pratt-Anticline complex. The features are more pronounced on top of the Precambrian, and become less so in the younger beds.

The major structures within the pass-range of the filter are well displayed on the filtered maps (figure 2). Similar sized structures of low amplitude that are masked on the original maps are also easily seen. Large structures and regional trends are completely absent. Of course, other filters can be designed to display features of any desired trend. However, for similarity maps and functions, only those filters that remove the average elevation should be applied. The filtered maps are smaller than the original ones because information is lost from the edge of the maps in a border of width equal to one-half the filter width. The structural grain is emphasized on these maps, which show the strong northeast, northwest, and north–south trends. The north–south trend is not so obvious on the original data maps; in fact, Robinson and Merriam (1972) determined that the north–south and east–west set of trends may be more important in the structural development of Kansas than was previously supposed. This is partly because the northeast and northwest set is more prominent and masks the other. The width of these filtered structures is from 9 miles to 36 miles, but they may be of any length.

The cross-multiplied maps (figure 3) are also reduced in each direction by a length equal to one-half of the width of the filter. Remember in interpreting the maps that the positive areas are produced by the coincidence of high *or* lows, the near-zero areas are the result of a flat area on one map and a high or low on the other, and the negative areas are produced by the correspondence of a high *and* a low (table 3).

The largest positive values are along the Nemaha Anticline as would be expected. Parallel to the Nemaha Anticline are a series of northeast-trending structures. Farther west, the coincident structures seem to have a more north–south orientation. Some of the negative areas in the western part of the map may be the result of a lack of data on the deeper horizons; thus the contoured maps of the original data are not as accurate as would be needed for detailed studies.

Visually, the three cross-multiplied maps all show approximately similar patterns. This similarity is verified quantitatively by the correlation coefficients (table 4). The Precambrian and Arbuckle surfaces are the most similar, followed by the Precambrian and Lansing; least alike are the Arbuckle and Lansing surfaces. These relationships are what would be expected from other evidence and from visual inspection.

Table 4. Comparison of the configurations of the different stratigraphic horizons, using the correlation coefficient *r*.

	Lansing	Arbuckle	Precambrian
Lansing	–		
Arbuckle	0·823	–	
Precambrian	0·794	0·901	–

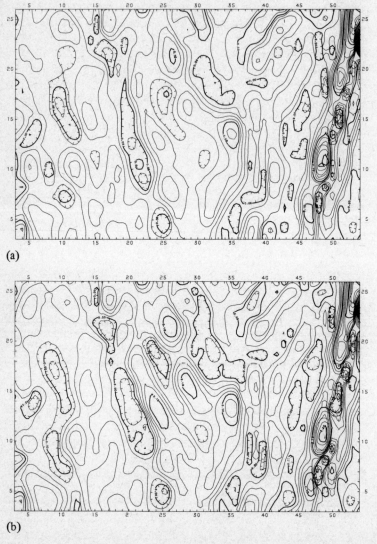

(a)

(b)

Figure 3. Cross-multiplied comparison maps made by use of equation (1) of same area in figures 1 and 2. Surfaces compared: (a) Lansing and Arbuckle; (b) Lansing and Precambrian; (c) Arbuckle and Precambrian. Contour interval (normalized units): 0·02.

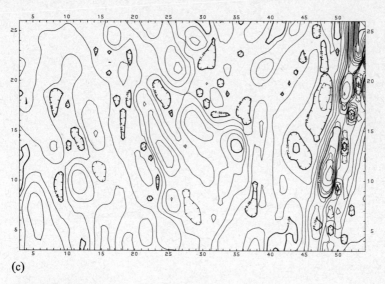

(c)

Figure 3 (continued)

Acknowledgements. We would like to thank William Burroughs for assistance in the programming and processing of the data. Janice Potak typed the manuscript. J C Brower and J C Davis kindly read an early version of the manuscript.

References
Cole V B, 1962 "Configuration of top of Precambrian basement rocks in Kansas" Kansas Geological Survey, Oil and Gas Investigation number 26, map
Davis J C, 1973 *Statistical Analysis of Geological Data* (John Wiley, New York)
Esler J E, Preston F W, 1967 "FORTRAN IV program for the GE 625 to compute the power spectrum of geological surfaces" Kansas Geological Survey, Computer Contribution 16
Merriam D F, 1963 "The geologic history of Kansas" *Kansas Geological Survey Bulletin* (162)
Merriam D F, Lippert R H, 1964 "Pattern recognition studies of geologic structure using trend-surface analysis" *Colorado School of Mines Quarterly* **59** (4) 237-245
Merriam D F, Smith P, 1961 "Preliminary regional structural contour map on top of Arbuckle rocks (Cambrian-Ordovician) in Kansas" Kansas Geological Survey, Oil and Gas Investigation number 25, map
Merriam D F, Sneath P H A, 1966 "Quantitative comparison of contour maps" *Journal of Geophysical Research* **71** (4) 1105-1115
Merriam D F, Winchell R L, Atkinson W R, 1958 "Preliminary regional structural contour map on top of the Lansing Group (Pennsylvanian) in Kansas" Kansas Geological Survey, Oil and Gas Investigation number 19, map
Miller R L, 1964 "Comparison-analysis of trend maps" in "Computers in the mineral industries, part 2" *Stanford University Publications, Geological Sciences* **9** (2) 669-685
Mirchink M F, Bukhartsev V P, 1959 "The possibility of a statistical study of structural correlations" *Dohlady Akademii Nauk SSSR* (English translation) **126** 495-497
Pearson K, 1896 "Regression, heredity and panmixia" *Philosophical Transactions of the Royal Society of London, Series A: Mathematical and Physical Sciences* **187** 253

Rao S V L N, 1971 "Correlations between regression surfaces based on direct comparison of matrices" *Modern Geology* 2 (3) 173–177

Robinson J E, 1968 *Analysis by Spatial Filtering of Some Intermediate Scale Structures in Southern Alberta* doctoral dissertation, Department of Geology, University of Alberta, Edmonton, Alberta, Canada

Robinson J E, Charlesworth H A K, Kanasewich E R, 1968 "Spatial filtering of structural contour maps" *23rd International Geological Congress (Czechoslovakia), Proceedings* Section 13, 163–173

Robinson J E, Merriam D F, 1972 "Enhancement of patterns in geologic data by spatial filtering" *Journal of Geology* 80 (3) 333–345

Sampson R J, 1975 "SURFACE II graphics system" in *Series on Spatial Analyses* number 1 (Kansas Geological Survey, Lawrence, Kan)

Natural systems

R G Craig

It is not uncommon in geological investigations to encounter the hypothesis that a particular set of variables comprise a *system* (see, for example, Chorley, 1962). This is a convenient assumption, in that it lends a coherent structure to all further investigations of the variables in question. Unfortunately, demonstration that such a system exists often is not achieved, or employs ad hoc procedures with little to recommend themselves as generally useful tools.

Such methods are less than satisfactory because they contain an inherent bias towards the recognition of systems where none exist. Indeed, the bias is such that, unless extraordinary measures are employed, the researcher is virtually certain of accepting the existence of some system within the variables. Thus, in fact, the number of natural systems which exist may be far fewer than the number that have been reported in the literature.

Identifying systems
'Systems' are defined as a set of elements, together with the relations between those elements (Mesarovic and Takahara, 1975). Typically one is concerned only with non-null relations. Indeed, the types of relations considered are quite often restricted to linear functions, although nonlinear relations are not totally ignored. Regardless of the type of relation involved, it is usual to include only those elements which are related to at least one other element. In fact, more restrictive conditions are applied.

If an element, X, is considered to be related to another element, Y, whenever both are related to any third element, Z, then the relation is called *transitive*. It is typical to find natural systems described whose relations are assumed to be transitive. When such is the case, the term *system* is usually restricted to apply only to sets of elements, together with their relations, such that each element is related to every other. An example might clarify this notion.

According to Ashby (1971), it is convenient to express the system as a set of points (corresponding to the elements), with lines (corresponding to the relations) connecting the points. Such a representation is used in figure 1. Each of the three diagrams could represent a system, depending upon the liberality of the relations allowed. Typically, null relations are ignored, so that figure 1(b) is not considered a system. If one wishes to avoid disjoint 'subsystems' as in figure 1(a), it is necessary that all points be related to all others as described above. Under such a definition, only figure 1(c) qualifies as a system. This is the concept most commonly applied, and I will restrict my attention here to systems so defined; although the problem to be explained will occur regardless of the definition used.

The 'system hunter' has at his/her disposal a number of methods which allow the determination of relations between elements. Three strategies that are applied are: correlation analysis (for example, as in the least-squares approach), path analysis (Heise, 1975), and cluster analysis. With such powerful tools available, it might seem that the task is practically trivial. One need only make observations on the elements (variables) of interest and determine whether or not a relation exists. From these elements, the maximally connected subset would then define the system.

Of course, we recognize that (at least in a Newtonian world) "everything is related to everything else", so that our exercise is always incomplete. However, in many cases one can legitimately ignore very weak relations and restrict attention to elements transitively related to all other elements at some minimal level. Innumerable instances exist to show that such a heuristic is often useful. The number of systems recognized in this manner increases at a regular pace. At the same time, there is a certain degree of dissatisfaction with these 'systems analyses' (Berlinski, 1976). Thus it is important to consider the possibility that spurious identifications might arise when one is following such procedures.

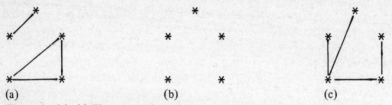

Figure 1. (a)-(c) Three possible systems.

Misidentifying systems

The conditions described above seem to be, without argument, *necessary* for systemhood; but are they *sufficient*? Unfortunately, a simple counterexample shows that nonsystems can satisfy this condition.

Consider an arbitrary set of observable variables, and compute their relations. Remove from that set any points not related to at least one of the others. What remains satisfies the above conditions, yet is almost certainly not a system. For example, two problems can occur:
(a) the set of variables lacks essential elements;
(b) the set of variables contains extraneous items which may be related to other elements but are not essential elements of the system under consideration.
It would be useful to have available tools capable of detecting such discrepancies. Such is the object of this essay.

Systems may contain points which are connected to more than one additional element; typically systems will have a number of such points.

The number and exact nature of the relations which occur are precisely
determined by the physical constraints of the situation. Although one
may, on occasion, mistakenly include some element as a member of the
system, and, on occasion, fail to recognize an essential element, one must
expect, in the long run, to have defined the system correctly. If this is
not true, any efforts at 'systems thinking' are futile. Thus, we do not
intend to be 'fooled' by random coincidences which might temporarily
indicate systemic relations between arbitrary elements. And, more
importantly, we would prefer to have some means of distinguishing
accumulations of elements from true systems.

A situation prone to such confusion occurs when sets of variables are
collected with the idea of testing for relations. Such a situation is not
uncommon when the researcher is involved in exploratory research and is
willing to admit the possibility that a system exists but cannot specify its
exact nature.

In such a case, a prudent policy would be to assume that the elements
under consideration do not constitute a system. Under this hypothesis,
H_0, one can make very precise statements about the kinds of relations
which occur, because they should arise totally at random. That is, the
overall set of relations should follow no particular pattern, although very
real relations may still exist among individual components.

In its simplest application, such an hypothesis will tell us precisely the
number of relations which will be identified, for there are a finite number
of relations which can be defined on any (finite) set of elements. For
example, with five points, the possible combinations of five relations are
shown in figure 2. The precise rule which allows computation of this
number is known as Polya's enumeration theorem (Harary, 1972). It is
based upon the view of systems as graphs, which follows directly from the
representations of Ashby (1971). The computations are difficult, being

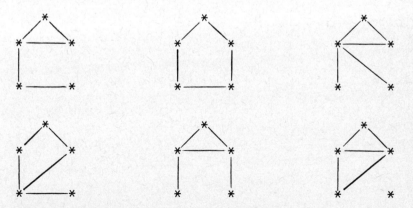

Figure 2. Possible combinations of five relations which can be defined on five
elements of a 'system'.

based upon the underlying algebraic structure of graphs, which is the group. However, these computations have been performed for many of the graphs that systems investigators are likely to encounter. Results for five elements are given in table 1.

Such information is not, in itself, sufficient for the analysis, as the method will only be required in the case where each of these points has at least one connection. Thus, we are interested in the number of connections which can be expected given that all points are connected. The possible configurations for a set of five connected points are shown in figure 3. Thus there are twenty-one ways in which a configuration of elements could appear to be a system, when five elements are involved. Conversely, there are eleven ways in which the configuration will appear not to be a system. Thus the probability of concluding that one is observing a system, when in fact the elements may or may not comprise a system (with

Table 1. Number of systems that can be defined on five elements for each given number of total relations observed.

Number of relations	Number of (generalized) systems	Number of connected systems
0	1	0
1	1	0
2	2	0
3	4	0
4	6	3
5	6	5
6	6	5
7	4	4
8	2	2
9	1	1
10	1	1
Total	34	21

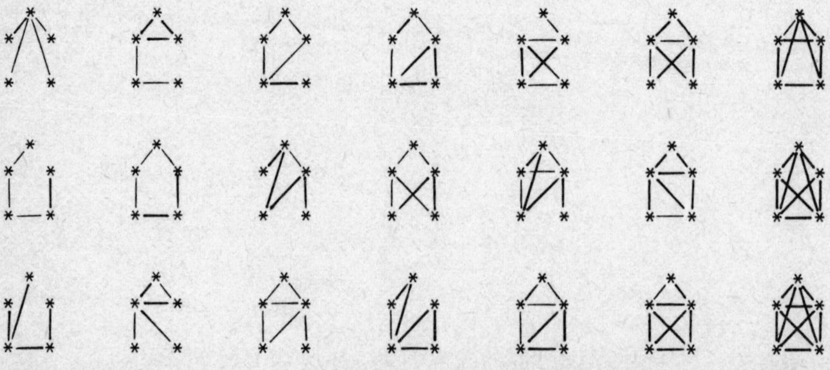

Figure 3. Possible connected configurations which can be defined on five elements.

individual probabilities of 0·5), is 0·62. One will therefore be biased towards the assumption that a system exists (hypothesis H_α). This should be adequate warning for any 'system hunter' that recognition requires careful rules of procedure.

Whereas Polya's enumeration theorem predicts the distribution of connectedness to be expected when examining elements not known to comprise a system, a different distribution is to be expected if, in fact, the elements *are* a system. In particular, the observed pattern will be a member of the set of connected graphs on n elements. The method of enumerating these graphs has been derived by Harary (1955). Their numbers can be expressed in terms of the total number of graphs on p points. Solutions are available for small values of p [Sloane (1973, page 75, sequence number 649); for $p < 10$, see also Stein and Stein (1967)], and are listed in table 2.

With this in mind, one can formulate the relative probability of each hypothesis (H_0 and H_α) given any observation, and from this determine the decision to be made for a given error probability. For example, if a set of five elements is known to comprise a system then q, the number of relations in that system, must be a member of the set $\{4, 5, 6, ..., 10\}$. The distribution of systems for various values of q are listed in table 3.

Table 3 shows that the expected value of q given that one is examining a system, $E(q|s)$, is approximately six. If seven or more relations are observed, one is reasonably confident that the set does comprise a system. Conversely if five or fewer relations are observed it is unlikely that one is dealing with a system [$p(s) < 0·38$]. If six relations are observed, the matter is no less uncertain than before the analysis began. A decision in this case should not be based on the mere observation that these relations are inferred.

Unfortunately this is not the most serious dilemma facing the system hunter. If we are willing to admit at the beginning of our investigation

Table 2. Number of graphs and number of connected graphs (systems) on p points ($p < 10$).

p	Number of graphs	Number of connected graphs	Probability of observing a connected graph
1	1	1	1·00
2	2	1	0·50
3	4	2	0·50
4	11	6	0·55
5	34	21	0·62
6	156	112	0·72
7	1044	853	0·82
8	12346	11117	0·90
9	2274668	261080	0·95
10	12005168	11716571	0·98

that not all of the elements under consideration will necessarily be members of the system that we are 'uncovering', then it follows that, from an assemblage of n elements, the system that we ultimately identify may only contain m elements, where $m < n$. For example, in the above case five elements were considered, each having an assumed (*a priori*) probability of 0·50 of actually being a component of that system. Thus we could easily accept a final identification of the system as one with four or perhaps three elements. A system with two or fewer elements may not be sufficiently interesting to warrent attention.

In this case the combinatorial aspects of the problem create great hazards for the researcher. For example, there are six systems of four elements, and four systems of three elements, which can occur in any set of five elements. Thus, of the thirty-four possible distinct patterns of relations which occur, a total of thirty-one contain configurations which

Table 3. Probabilities, $P(q|s)$, of observing a system for a given number of relations, q, between five elements.

q	Number of graphs	Number of systems	$P(q|s)$	$qP(q|s)$
4	6	3	$\frac{3}{21}$	$\frac{12}{21}$
5	6	5	$\frac{5}{21}$	$\frac{25}{21}$
6	6	5	$\frac{5}{21}$	$\frac{30}{21}$
7	4	4	$\frac{4}{21}$	$\frac{28}{21}$
8	2	2	$\frac{2}{21}$	$\frac{16}{21}$
9	1	1	$\frac{1}{21}$	$\frac{9}{21}$
10	1	1	$\frac{1}{21}$	$\frac{10}{21}$
Total	26	21		$E(q|s) = \frac{130}{21} = 6\frac{4}{21}$

Table 4. Probability of observing a system of connections that involve at least one-half of the elements, when connections are formed at random.

Number of elements, n	Number of systems	Number of connected systems with $>n/2$ elements	Probability of observing a system including at least $n/2$ elements
1	1	1	1·0000
2	2	2	1·0000
3	4	3	0·7500
4	11	9	0·8182
5	34	29	0·8529
6	156	141	0·9039
7	1044	992	0·9502
8	12346	12109	0·9808
9	274668	273183	0·9946
10	12005168	11989754	0·9987

would be judged as systems. Thus, in such an exploratory program, in which it is assumed *a priori* that a system occurs with probability $0 \cdot 50$, one will in fact conclude with probability $0 \cdot 91$ that a system exists. Even if attention is restricted to systems which contain at least one-half of the original elements, the bias is still overwhelming (table 4).

Thus an approach which attempts to identify systems by the existence of a set of correlations between observable elements is inherently biased in favor of incorrectly 'discovering' such a 'system'. An objective evaluation of proposed systems cannot be achieved solely on the basis of observed relations.

An example

Melton (1958) analyzed fifteen variables which he considered to represent the significant geomorphic, surficial, and climatic elements of the drainage system. On the basis of an analysis of the correlation structure of these variables, he determined that forty-two significant relations exist. One of these was discarded on the basis of rational arguments, leaving forty-one relations. These he arranged hierarchically into several subsets so that the total system structure could be defined. In this, he was guided by the results of a Q-mode cluster analysis. Rearranging and simplifying the system in conformance to reasonable geomorphic expectations, he suggested (Melton, 1958, page 457, figure 16) that the essential elements of the drainage system could be described by using five of the original variables (see figure 4). Six additional hypothetical variables were added to complete the system and to clarify the relations. It is of interest to consider the probability of discovering a system comparable to that of figure 16 in Melton (1958), within a set of variables which in fact do not contain a true system.

It is reasonable to confine our attention to the original variables defined by Melton (figure 5), as the existence of the other elements and their relations to the remaining variables are hypothetical. Excision of the hypothetical variables leaves a residue of three disconnected subsystems [figure 5(a)], two of which are single elements. Even if the five elements are considered to be related through the hypothetical elements, as in figure 5(b), the system remains exceedingly simple, compared with the great diversity of systems which can be defined on a set of fifteen elements. Indeed, there are 31 426 485 969 804 308 768 distinct graphs which can be defined on fifteen points! The probability that a particular correlation matrix will contain a relational structure which can be confused with a true system is greater than $0 \cdot 99$. Given that systems with fewer than fifteen elements are, in this case, acceptable models of the stream system, the researcher is virtually certain to identify some system. Regardless of whether the system identified is otherwise reasonable, the structural conclusions based upon analysis of the correlation matrix cannot be taken as support for any conclusions concerning the existence of a true system.

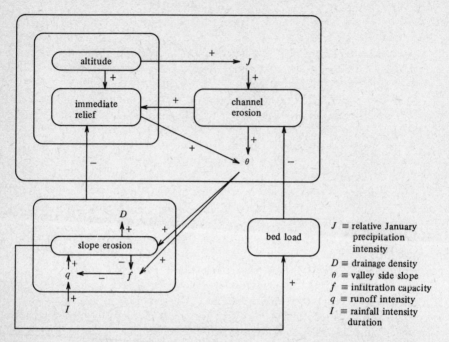

Figure 4. Correlation structure of morphometric properties of stream basins. (After Melton, 1958, page 457.) Note that I was not one of the original variables actually measured by Melton.

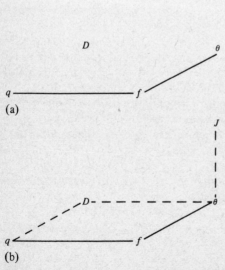

Figure 5. (a) Relations on the original variables measured by Melton (1958, page 457); (b) relations on original variables when connected by hypothetical relations.

Recommendations

A distinction should be made between the search for systems on the one hand, and the specification of the structure of a known system on the other. Because correlational studies (and all analogous procedures) will almost always suggest the existence of systems, such suggestions are useless. However, if one knows that a certain set of elements does comprise a system, then correlation-type studies might be useful to suggest the precise set of relations in that system. The system hunter must fashion other weapons to allow selective capture of the right prey. Three methods can be suggested to aid this kind of search. First, one should specify clearly, before any correlation-type analysis is performed, the minimum number of elements (out of the original set) that will be deemed a satisfactory system. This number must not be too small, or a 'system' will almost certainly be found. Another way of stating this principle is to stipulate that one must be quite sure of the system elements. It would appear that a minimum probability of membership of $0 \cdot 75$ is necessary before a variable could be added to the set to be studied. A much higher probability, say $0 \cdot 95$, should preferably be used. This consideration is of value only when the number of elements to be considered is reasonably small (less than nine, say); it makes little difference otherwise. One must also state *a priori* any known or mandatory relations which must be included in the outcome of the correlation study. This will significantly improve the odds that decisions will reflect real conditions.

A second useful method which can be used to weed out inappropriately chosen systems is to examine the history of the elements. If the correlational properties can be calculated for more than one time period, the recognition of spurious systems is much less likely. Experiments with the system should be considered, in order to test the predictions of the correlational model. A powerful method would be to perform intervention analysis of the results of the experiments.

Finally, one should attempt to quantify the information content and information flow of the system, as natural systems which are homeostatic should tend to preserve the information content: "Information is neither created nor destroyed".

References

Ashby W R, 1971 *An Introduction to Cybernetics* (Methuen, Andover, Hants)
Berlinski D, 1976 *On Systems Analysis* (MIT Press, Cambridge, Mass)
Chorley R J, 1962 "Geomorphology and general systems theory" Professional Paper 500-B, US Geological Survey, Reston, Va
Harary F, 1955 "The number of linear, directed, rooted, and connected graphs" *Transactions of the American Mathematical Society* 78 445–463
Harary F, 1972 *Graph Theory* (Addison-Wesley, Reading, Mass)
Heise D R, 1975 *Causal Analysis* (John Wiley, New York)
Melton M A, 1958 "Correlation structure of morphometric properties of drainage systems and their controlling agents" *Journal of Geology* 66 442–460

Mesarovic M D, Takahara Y, 1975 *General Systems Theory: Mathematical Foundations* (Academic Press, New York)

Sloane N J A, 1973 *A Handbook of Integer Sequences* (Academic Press, New York)

Stein M L, Stein P R, 1967 "Enumeration of linear graphs and connected linear graphs up to $p = 18$ points" Report LA-3775, Los Alamos Scientific Laboratory of the University of California, Los Alamos, N Mex.

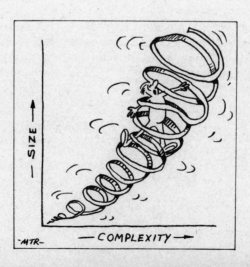

Order and randomness in geological processes

F E Wickman

The definition of mathematical geology apparently is a matter of opinion. To most geologists it seems to be the analysis of large amounts of geological data by statistical methods. Thus, its main purpose is to extract information or summarize data, with the focus of interest being on ordering and classifying these data. To others, mathematical geology corresponds to mathematical physics—that is, the use of mathematics in the study of geological models. It is this second definition of mathematical geology which I will develop in this essay.

Although the term 'geological process' will have the same sense here as that commonly used in textbooks of physical geology, it is the particular model under consideration which determines the beginning and end of a process, as well as the size and the time scale on which the process is operating. From the point of view of physics and chemistry, geological processes are extremely complex phenomena. It is therefore difficult to construct simple and realistic models. For this reason it is practical to divide the process into portions which are simpler to describe by using physical and chemical laws. The number of divisions needed may be great. Each portion will be called a *box* or a *state*, and will consist of a mechanical and/or chemical process. This description of a geological process has similarities to descriptions of industrial processes. Let us therefore compare them.

In industry, efficiency, uniformity of the products, and high productivity are essential. Each step of a process is carefully planned and controlled so that its physical and chemical conditions are optimized to give maximum performance. A factory must also have a steady supply of raw materials. Errors in design and construction of a plant are corrected, failures are repaired, and parts are replaced. Furthermore, storage space must exist for the raw materials, as well as for the finished products prior to their sale and removal. Finally, the company operating the factory must comply with environmental and occupational safety regulations.

In nature, organisms perform operations which are, in many respects, analogous to those performed in the factory. Despite the fact that for geological processes other rules seem to govern, it is instructive to pursue the analogy a bit further.

A chemical engineer distinguishes between flow and batch processes. A similar distinction can be made in geology: geological processes can be characterized as a sequence of superimposed continuous and step processes. The transport by a river of dissolved salts, although varying with time, is an example of a flow process, whereas a rain shower (characterized by its erosional and transportational properties) is an example of a batch process.

Clearly the size and frequency of batches often vary in an unpredictable manner. In nature, the physical and chemical conditions of a state are not varied systematically until they give the maximum possible yield. Instead they are determined, in the main, by local (or almost local, on a global scale) factors. The result of a state of a process may thus vary from excellent to extremely poor.

In a factory, the storage space is limited; unless finished products are removed, production must eventually cease. In nature, similar situations exist—that is, a process or reaction may be inhibited or stopped by its products. Chemical precipitation in rock fissures is an example. In other cases an excellent transport system may exist, such as a river network in a region with a humid climate. Such a river system serves to remove weathered material rapidly; this permits weathering to continue. Sometimes the termination of a process will be the beginning of another process. For example, frequently a failure, mechanical or otherwise, occurs in a geological process. The failure itself may trigger a repair process—fissure filling, for example. In general, however, a failure stops the process.

We have already modeled a geological process as a collection of states whose number and kinds are not fixed. The process is instead a branching process with a number of choices at each branching point. It may well happen that a step in a natural process totally or partially annihilates the results of the previous states of the process. The existence of such states, which may occur in a geological process, may be regarded as natural equivalents of design errors.

This sketchy evaluation of some features of nature's geological factories may serve as a basis of a general model of geological processes. Each state is characterized by specific physical and/or chemical processes. The model of a state is a set of differential or other equations from physics and chemistry. The number of free parameters may be large, on account of the complexity of the process.

The process of a state may be called *deterministic* for the following reason. If we study those natural examples of the state which exist today, we can determine (assuming it is technically possible) the free parameters for each individual case. Since these are not natural constants, we cannot predict their numerical values, but have to determine them. The parameters are unlikely to be completely independent of one another, so that there are statistical rules for their respective values. Studies of present processes are, of course, fundamental for our understanding of their mechanisms. However, geology is mainly concerned with what has happened and, to some extent, with what will happen. We have no means of determining the exact parameters of such past or future processes. Processes taking place in the interior of the earth are not even accessible to detailed study. Lacking the ability to determine the parameters, one may model a state as a *stochastic* process, where the free parameters become random variables.

Each free parameter will thus have a frequency distribution. The study of these distributions, and the possible relationships between the distributions of different free parameters, is clearly a major problem for geologists in the future.

Another complication is that, with geological phenomena, there may be nothing left of the state. For example, either the process may destroy evidence of its own existence as it proceeds, or subsequent processes may overprint earlier ones. The important problem is therefore not the parameters themselves, but how they relate the input and output of a state. We can therefore regard the state as a 'black box'. The input into a state is a variable, and is described by a frequency distribution. The output is, in the same way, described by another frequency distribution. The distributions are of great interest for geology in general and for catastrophes in particular. Discussions, such as those by Wolman and Miller (1960) on the importance of size and frequency for geological processes are of interest in this context.

The scale on which natural processes and states operate varies tremendously. We may think of the amount of sediments stored in a small delta at the end of a rill formed by a short rain shower, and of the huge amount of sediments stored in the delta of the Mississippi River. Still greater differences are possible if we include processes having cosmic causes. We have already mentioned that the model may be viewed as a branching process. It is the branching process that we compare with the real world and therefore call a geological process. The number of choices at a node is usually greater than two, with a probability of occurrence assigned to each path. These probabilities are functions of factors such as time, location, environment, composition of the material being manipulated in the process, and the previous history of the process. We have previously compared the partial processes to black boxes; in the same way we may compare the total process to a random walk among black boxes.

However, the walk is not completely free, but is guided by an important principle for geological processes: they have to proceed in an orderly way, as in a queue; the degree of queue discipline is dependent on the scale of the process. It is not possible at present to formulate the principle in a simple sentence, but some aspects of this concept will be discussed below.

A geological process, in the sense that has been used here, consists of the collection of geological events of a definite character and of a limited extension and duration. The activity may include not only mechanical and chemical processes, but also the transport, generation, and consumption of heat. The temperature rise in a rock close to an igneous intrusion may serve as an example. The activity appears to be limited to discontinuities, moving boundary layers, or sometimes inhomogeneities. The geological process does not act throughout the entire volume simultaneously; instead, the activity is concentrated in some parts of the material and proceeds in space and time in a regular, irregular, or erratic fashion.

However, this statement must be qualified by taking into account the volume of material considered. If we consider a volume of only a few cubic centimetres, we certainly have activities for which the total volume is involved. On the other hand, if we consider processes on the scale of the delta of the Mississippi River, it is clear that the activity at any moment only refers to a small portion of the total delta. However, it is not clear how this coupling between size and active portion can be defined more precisely. Some examples will now be given which illustrate the queueing principle, and which refer to 'well-sized' processes.

Erosion of a hill works at the surface layer and not randomly inside the hill. Even if caves and dolines are forming within the hill, the process is still working at a surface layer.

During regional metamorphism, as the temperature rises the existing mineral assemblage may become unstable, and chemical reactions will yield a new stable assemblage of minerals. The rise in temperature is a slow process which can be described as the movement of isotherms into the volume under consideration. The reaction, in principle at least, starts in a particular volume element when the proper isotherm passes through it.

In general, geological processes thus proceed in an orderly fashion, somewhat like serving customers in a queue. Two extreme cases occur; in both the process as such performs the ordering.

(1) The spot of activity is more or less fixed, and the material is moving to this spot. The ordering depends on the limited capacity of the active spot (server) and the transportation (arrival) time.

(2) The activity (server) is moving in relation to the material (customers).

The service will be a time sequence which depends on directions, distance, and so on. Even if this principle of queueing appears trivial, it is the validity of this principle that is responsible for the fact that we have a geological record covering almost 4×10^9 years. The reason is simply that the formation of covering layers preserves the covered ones (that is, upsets the queueing order for some geological processes).

Let us now go back to the geological processes described by the models indicated. One aspect of any model is as follows. A geological process may be defined as having a beginning and an end. However, sometimes it is not easy to define a natural and unique beginning or end. For example, a volcanic eruption may start with an explosion, but is that the beginning of the process? There is certainly a long process before the explosive activity starts. In a way, the only processes which stop are those where every trace of activity is annihilated. In nature, we actually observe processes in all stages at the same time. This problem has to be taken care of in a model.

We have also something we may call *null processes*. These are the periods when, from the point of view adopted, nothing happens to the geological entity considered (for example, a rock protected from erosion by a cover of sediments).

One of the great advantages of constructing models of geological processes is that frequency distributions other than the standard ones will appear. The fashion of using only a few frequency distributions limits the researcher. A single example will suffice to demonstrate this.

Many processes consist in principle of dividing a certain amount of material into portions. Two kinds of such processes are illustrated by crushing a nut and pouring coffee from a coffee pot. In the first process the whole piece is involved, and once the process has started it is brought to the end. In the second, the process goes on for some time and can be stopped at any moment. Both processes have many geological equivalents.

In the first kind of process there is no ordering principle at work, whereas in the second process ordering clearly exists. What are the distribution functions related to these two processes?

In the first process, it is natural as a first approximation to assume that the parental mass is divided into two portions and that the fraction of the parental mass received by each portion is distributed uniformly in the interval [0, 1]. The density function is thus one.

In the second process, the distribution will depend on the duration of the process and its intensity. The simplest assumption regarding its duration, X, appears to be that the probability of its ending within the time interval Δt is constant; that is, we have an exponential distribution. The intensity, Y, is assumed to be constant during the duration X, but its magnitude is governed by an exponential distribution (linear growth during an exponentially distributed time interval). The mass produced, Z, is then given by

$$Z = XY .$$

We assume here that the parameters of the two exponential distributions are so large that we can neglect the fact that the parental mass is finite.

If we write the density functions for X and Y as $\lambda \exp(-\lambda x)$ and $\mu \exp(-\mu y)$, the density function $Z(t)$ is given by

$$Z(t) = \lambda\mu \int_0^\infty \frac{1}{x} \exp\left(-\lambda x - \frac{\mu t}{x}\right) dx ,$$

or

$$2\lambda\mu K_0(2\lambda^{\frac{1}{2}}\mu^{\frac{1}{2}}t^{\frac{1}{2}}) ,$$

where K_0 is a modified Bessel function. In these two related processes, two different mechanisms give quite different density functions.

We have written down the density of only a single event. In nature, we usually observe the result of an unknown number of events. The density function will then be different from that of the single event. For example, in the first case the density was simply one, but assuming that the number of splittings is random and subject to a Poisson distribution with parameter α, the density function f(t) of that process, given by

$$f(t) = \exp(-\alpha)\left\{1+\left(\frac{\alpha}{\tau}\right)^{\frac{1}{2}} I_1[2(\alpha\tau)^{\frac{1}{2}}]\right\}, \qquad \text{where} \quad \tau = \ln(t^{-1}),$$

is a rather complicated function of the modified Bessel function I_1.

This example shows the variety of distributions and densities which will surface when models are constructed. We are still far from the day when we can discuss geological processes in quantitative terms. Too many factors are unknown. In the past, the lack of an absolute chronology of the Earth's history made any attempt impossible. This obstacle is now rapidly disappearing. Hopefully within the next decade we will have some feeling for the history of the Precambrian era, and for the rates of different geological processes during various epochs. This will give a better foundation for modelling. It appears to me that some generalized queueing models will be useful in the construction of geological models.

Reference

Wolman M G, Miller J P, 1960 "Magnitude and frequency of forces in geomorphic processes" *Journal of Geology* **68** 54–74

ONE MORE RUN LIKE THAT ONE, HIGGINS, AND I'M GOING TO GET YOU.

A structural language of relations

P Gould

"The glutted must be taught to regard a third as a gift of God, a perfect fourth as an experience, and a fifth as the supreme bliss. Reckless gorging undermines the health. We thus see how necessary it is to preserve contact with the simple original."

Carl Nielson, 1925, Levende Musik

"Today, it is becoming increasingly obvious that a different kind of mathematics from that of the classical calculus is needed ... [It] will include set theory, discrete mathematics ... abstract algebra ... group theory and topology."

J C Griffiths, 1966 "Future Trends in Geomathematics"

I hope it is not necessary to apologize for being a geographical Judas goat hiding between the sheepskin covers of a volume in the geosciences. To do so would acknowledge the relevance of the archaic intellectual fiefdoms that have divided human knowledge since the 18th century, regions separated by schizophrenic fissures in academic throught that even imply a partition between a set of elements labelled the *sciences* and another called the *humanities* (Gould, 1977; 1978a). We have tolerated such partitional thinking too long: imaginative and creative inquiries in all areas of human endeavor are characterized by an explication of relations and a deep concern for structural questions. It is worth noting immediately the way in which the word *structure* appears in every area of intellectual inquiry: we speak of the structure of a personality, a molecule, a ballet, a bridge, a soil, a sonnet, a quartet, a university, trade relations, a time series, a region, an experiment, a town, a novel ..., the list is literally endless, and geologists have traditionally evinced great interest in the subject. The problem is that when we try to define, pin down, and operationalize the intuitively valid, but usually vague, notion of structure (what a mathematician would formally term an *explication*—Schreider, 1975), we often experience great difficulty. Too often the result is a 'well-you-know' appeal to intuition, accompanied by hand waving and growing feelings of exasperation and intellectual defensiveness. We have all experienced these feelings, because we have all experienced the frustration of trying to express, of literally to *press out* (Stokes, 1973), a concept that is difficult to put into words.

And here, perhaps, is a clue to our difficulties: the notion of structure may be impossible to explicate in words, in the language of our everyday discourse. Fortunately, we are not confined to words, for we have a rich

range of other languages to express our intellectual inquiries, our aesthetic
sensibilities, and perhaps even our ethical concern. However, rather than
trying to explicate the notion of structure in such languages as music or
dance, this essay is concerned with a language of structure written in
algebraic topology, with some translations into the language of graphics.
I suspect that none of us are very good at algebraic topology yet, and
such terms as mapping, face operator, graded pattern, homotopy, and
filtration seem strange and discomforting. For this reason, plus my own
inadequacies and the obvious constraints of space, I shall confine myself to
an overview and introduction, feeling free to mention existing applications
in fields as diverse as medicine, urban aesthetics, regional planning,
taxonomy, and the game of chess. Since structural questions are so
fundamental, I am sure you will have no trouble in constructing at least a
few injective mappings from the specific examples to your own problems
in the geosciences.

Learning about relations
I take it for granted that all explanation involves relating things to things:
indeed, the idea of relationships is contained in the notion of explanation,
and I challenge you to think of any form of explanation which does not
involve some process of linking something to something else. Explanations
in any field involve connecting things and finding coherent and replicable
patterns and structures. This being so, it is with an awesome sense of
human perverseness, not to say sheer bloody-mindedness, that we discover
that virtually no one is ever taught anything about relations. Or if they
are, it is because they are pushed through an archaic mathematical
curriculum that replicates the history of the discipline (Simon, 1969), and
exposes them to a distressingly limited and constrained form of relation—
namely, the function (Hammer, 1969, page 136).

One wonders why this should be so: perhaps it is partly a matter of the
extraordinary progress, and therefore the terrifying intellectual seductiveness,
of the physical sciences which makes us follow in their footsteps like sheep,
borrowing their older and highly constrained mathematical approaches,
instead of thinking through our own requirements. Of course, in following
their 19th century footsteps we are only half a century out of date,
because fifty years ago physicists were exploring new algebraic structures,
and today one can think of many physics, depending upon which algebra
one chooses to map one's observations (Atkin, 1965). Some algebraic
structures are more robust, and capable of providing us with deeper
insights, than others. In the same way, there are, of course, many geologies
(is this the first time the plural form has ever appeared in print?), although
most remain unexplored.

Or perhaps it is because our educational system represents a mapping of
the few insights, and many failures, of one generation onto another (Laing,
1969). Like sorting out and dampening the perturbations of Victorian

attitudes to sex, it may take four or five generations to filter the factual middens from our curricula, and reorganize our knowledge into small conceptual nuggets genuinely worth teaching. I suggest we shall be helped in this task by moving to more general forms of mathematics which contain, and provide some perspective upon, the more specific forms that are traditionally taught.

A function (with which we are all familiar) is a mapping is a relation (figure 1): if we start with the latter, more general, expression we shall lose nothing; if a function exists it will appear as a highly constrained and special form of relation (Open University, 1971). But the reverse is not true, and often by approaching a set of empirical data in the 'functional' state of mind that convention and tradition require, we force the world into a highly restrictive mould, and may destroy important pattern, structure, and therefore insight, in the process. What, then, is a relation? An infinite number of them may be defined, but we will confine ourselves to those called *binary* relations. The restriction is hardly damaging at this point, given the fundamental nature and importance of binary relations in mathematical languages (Hammer, 1969, pages 142–143).

First, some basic terminology: suppose we have a set of well-defined elements, say X, and we consider all the pairs of the form $\langle x_i x_j \rangle$. These pairs are *ordered*, which means that we distinguish between $\langle x_i x_j \rangle$ and $\langle x_j x_i \rangle$. If we represent the set of ordered pairs as the Cartesian product $X \otimes X$, then a binary relation, say λ, may be considered as a subset of all the ordered pairs, and we write $\lambda \subseteq X \otimes X$. So a binary relation is a pair $\langle \lambda, X \rangle$, where λ is the set of pairs defining the relation on the *support X*. For example, a geographer or economist might be interested in a set of countries C, and the relation between them defined in terms of trade exceeding a certain value, say θ, each year, so that $\tau \subseteq C \otimes C$, if annual trade $\tau_{ij} \geqslant \theta$. More often, however, we are concerned with relations between different sets, say X and Y, and we define such a relation as $\lambda \subseteq X \otimes Y$. We can make this a little more intuitive and less abstract by thinking of each relation λ as an incidence matrix Λ in which

$$\Lambda = [\lambda_{ij}] , \quad \text{where} \quad \lambda_{ij} = \begin{cases} 1 & \text{if } \langle x_i y_j \rangle \in \lambda , \\ 0 & \text{otherwise.} \end{cases}$$

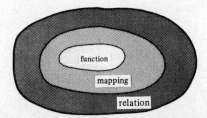

Figure 1. Functions and mappings embedded within more general relations.

For example:

λ	y_1	y_2	y_3	y_4	y_5	y_6
x_1	0	0	1	1	1	1
x_2	0	1	1	1	1	0
x_3	0	1	1	1	1	0
x_4	0	0	0	0	0	0
x_5	1	1	0	1	0	0

Notice something: $\lambda \subseteq X \otimes Y$ does not contain an ordered pair with x_4, since x_4 is not λ-related to any element of the set Y (figure 2). For example, if λ contained $\langle x_4 y_2 \rangle$ (mentally put in a one in the incidence matrix Λ), we would have a *mapping*, in this case a many-to-many mapping.

If each element of X were λ-related to a *unique* element of Y, we would then have a *function*, say f. We have three possibilities (figure 3): (a) if all the elements of Y are used we have a *surjective* mapping, but the

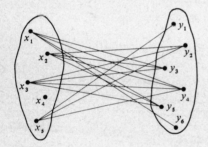

Figure 2. The relation between X and Y.

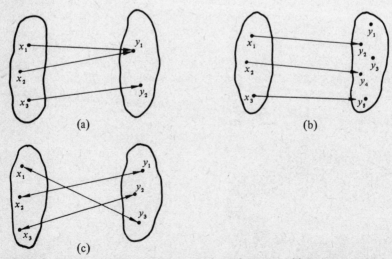

Figure 3. The three possible mappings which are functions. (a) Surjective; (b) injective; (c) bijective.

inverse f^{-1} does not exist because y_1 is not uniquely defined; (b) all the elements of Y are not used and we have an *injective* mapping, but f^{-1} does not exist because y_1 and y_3 are not defined; and (c) there is a one-to-one correspondence or *bijective* mapping so f^{-1} exists. We see now why cryptographers are interested in bijective mappings whose inverses are difficult to find (Diffie and Hellman, 1976).

Do you see now why a function is a mapping is a relation, but not the reverse? And is it not strange that we usually focus our attention on the most highly constrained type of relation, a bijective mapping, and then add the even more stringent requirement that f and f^{-1} are linear? For example, the conventional data matrix of N observations on two variables X and Y is nothing more than a list of ordered pairs $\langle x_i\, y_j \rangle$ from the Cartesian product $X \otimes Y$ (figure 4). Perhaps such a data matrix (list) might even be represented by Λ, or graphically as in figure 4.

Faced with such rich information about the relations between elements of two sets, we slap on one of an infinite number of possible linear bijective mappings called f (otherwise known as a regression line), assume f^{-1} exists and slap that on for good measure, and then look up the cosine of the angle between them (called a correlation coefficient). Since we have probably destroyed much of the information in the data set by such a violent assault, we should not be too surprised if the 'fit' is less than satisfactory, and our contribution to human knowledge only that structure which finally oozes through the linear filter that we perversely insist upon placing between us and our observations about the 'real world'.

At the foundation of the language of structure (Atkin, 1972a; 1974a; 1977a) is the fundamental concern for relations between finite sets of elements. The property of finiteness should not disturb us: *all* our data are finite, and even when a data set appears to us as a continuous trace (the recording of a heartbeat or an earthquake), we break it up into finite discrete pieces for digital computation. Despite the fact that we still teach the calculus to deal with continuous problems, all physics appears to be capable of being recast into finite form, often with highly fruitful and

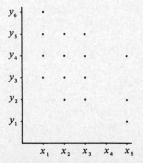

Figure 4. The relation λ as a subset of the Cartesian product $X \otimes Y$.

deep insights into the physical processes themselves (Greenspan, 1977; Wheeler and Patton, 1977). Indeed, the traditional calculus which students are still made to slog through like a *rite de passage*, may appear shortly (in a hundred years?) as nothing more than a mildly interesting, three-hundred year old dead end. The continuum does not exist in reality; it depends on perfectly legitimate mathematical definition, but its appearance is a result of a human perceptual problem. It is worth noting that the analysis of such a problem of the human brain led to the creation of tolerance spaces in which weak relations were mandatory (Schreider, 1975; Zeeman, 1965).

A concern for relations between finite sets of elements places an apparently simple and obvious, but in reality often excruciatingly difficult, task upon us: we have to think about, and define with great clarity, the elements of the sets; and we have to have sufficient insight, imagination, and creativity to define the relations between the sets. In brief, we have to know what we are talking about. This is a crucial point: the language of structure is not another technique like the trend-surface, factor, spectral, fill-in-the-blank analyses that have successively rolled by during the past two decades, but a general topological language that can express and absorb such specialized techniques by equivalence relations. It represents not a paradigmatic shift in the totally erroneous sense of Kuhn (1962), but an enlargement of viewpoint absorbing older approaches as specialized and constrained cases (Stegmüller, 1976). If such a claim seems extravagant, one has only to observe the extraordinary power and ability of the topological viewpoint to order, structure, and relate large areas of mathematics itself, and see how the use of such higher order languages is shaping the physical sciences and certain areas of philosophy today (Hammer, 1969; Kocklemans, 1972).

The structure of a discipline
Before considering an actual example that may be of direct interest to geoscientists, let us try to get some intuitive grasp of the algebraic topological language of structure with a hypothetical example. We cannot explore all the rich mathematical development here, but simply get a 'feel' for some of the terms and concepts. Suppose we have a well-defined set P of people {A, B, ..., G}, who might form a traditional discipline or perhaps a department in a university. Suppose, further, that we define a set I of professional interests {1, 2, ..., 16}, noting in passing that these are all at the same *hierarchical level*. This means that we should not have mathematics in our set as well as matrix algebra and differential equations, since the latter two elements are contained in the former. Mathematics itself is a *cover*, at the $(N+1)$-level, for matrix algebra and differential equations at perhaps the N-level or even lower. We shall not explore this most important idea further (Laborit, 1977), but simply note that (1) the language of structure requires us to sort our sets very carefully

into proper hierarchical levels, and (2) many of our conceptual muddles and problems of empirical data analysis seem to come from mixing concepts at different levels. We must not confuse elements of sets with sets of elements, or we will face deep logical contradictions of the sort that Russell characterized as the Barber Paradox (Atkin, 1977b).

We can represent our two sets P and I as the rows and columns of an incidence matrix Λ, and see how each element (person) of P is λ-related to the elements (intellectual interests) of I.

Intellectual interests, set I

λ	1	2	3	4	5	6	7	8	9	10	11	12	13	14	15	16
A	1	0	0	0	0	1	0	0	1	0	0	0	0	0	0	1
B	0	0	0	1	0	0	0	0	1	1	0	0	0	0	0	0
C	0	1	0	0	0	0	0	0	0	0	0	1	0	1	0	0
D	0	0	0	0	0	0	1	0	0	0	0	1	0	1	1	0
E	0	0	1	0	1	0	0	1	0	0	1	0	1	0	0	0
F	0	1	0	0	0	1	0	0	0	0	0	0	0	0	0	0
G	0	0	0	1	0	0	0	0	0	0	0	0	0	1	0	0

(row label: People, set P)

It is easy to see from the matrix itself how the relation λ is a subset of the ordered pairs $\langle p_j i_k \rangle$ of the Cartesian product, or $\lambda \subseteq P \otimes I$. Now each person is defined for the specific purpose at hand, namely the analysis of a discipline, as a set of *vertices* (points) in the set of interests. Thus we can represent each person as an n-dimensional geometric figure called a *simplex* (figure 5), defined by $n+1$ vertices.

For example, A is defined as the three-dimensional, or 3-simplex, σ_A^3, where $\sigma_A^3 = \{1, 6, 9, 16\}$ and appears as a tetrahedron; F is the 1-simplex σ_F^1, and $\sigma_F^1 = \{2, 6\}$; E is the 4-simplex, represented by $\sigma_E^4 = \{3, 5, 8, 11, 13\}$, lying, or rather working, in a four-dimensional space purportedly represented by the simplex with dashed lines to the fourth dimension (since the latter is difficult to visualize without considerable transcendental meditation), and so on for all the other people. The set of simplices taken together form the *simplicial complex*, $K_P(I; \lambda)$, but notice

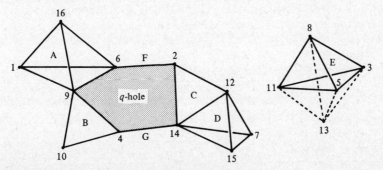

Figure 5. The geometric representation of $K_P(I, \lambda)$, the structure of a discipline.

that we can just as easily consider the *conjugate complex*, $K_I(P; \lambda^{-1})$, in which the intellectual interests are represented as simplices defined on the set of people vertices, with elements $p_j \in P$.

Although Λ defines our relation between P and I, we are not limited to such binary matrices: Λ might be the more familiar data matrix with rational or integer elements, z_{jk}, which represent measures of intensity. We can, however, choose a *slicing parameter*, θ, and define our binary relation as $p_j \lambda i_k$ iff $z_{jk} \geqslant \theta$; this means that all values of z_{jk} greater than or equal to θ can be thought of as defining a binary relation by being converted to ones, whereas all the rest are set at zero. We also have the option of choosing a set of slicing parameters, $\{\theta_j\}$, perhaps different for each row, or even a set $\{\theta_{jk}\}$, one for each ordered pair of Λ. Naturally the *structure* of our discipline, and the structure of intellectual interests (notice that we now have an operational definition of structure), will change according to the set of slicing parameters we choose. But intuitively this is as it should be: if we include even the most casual interest of the people concerned, we would expect to get a highly connected structure; if we choose a higher slicing parameter, so that only intense interests remain to define our relation, we would expect to get a much less connected, more highly fragmented, discipline.

In our example (figure 5), the subset of simplices $\{\sigma_A^3, \sigma_B^2, \sigma_C^2, \sigma_D^3, \sigma_F^1, \sigma_G^1\}$ forms a connected structure, whereas $\{\sigma_E^4\}$ is completely disconnected. If we imagine our discipline moving along Time's Arrow, we can see how it might change: perhaps person σ_E^4 is defined by interests that are 'ahead of their time', but as $K_P(I; \lambda)$ moves it has the possibility of (1) picking up new simplices which share a subset of the vertex set $\{3, 5, 8, 11, 13\}$ which defines σ_E^4; (2) developing new vertices in the set I which may or may not form connective intellectual tissue between the elements of P; (3) forming new connections representing growing interests by the simplices; or (4) losing vertices by the retirement of a simplex. We can see that if σ_E^4 retires while disconnected, the interests defining him or her will be lost— perhaps until they are 'rediscovered' by yet-unborn simplices far in the future. The history of science is replete with such examples. Notice, too, that a set of interests defining our individual may be disconnected at one time, but may become part of the complex at a later date, as the once-eccentric ideas are accepted and incorporated literally into the structure of the discipline.

We can define the *eccentricity*, e, of each person or simplex in a way that conforms closely to our everyday use of the word. We call the dimensionality of a simplex top-q or $\hat{q}$, and the dimensionality at which it begins to connect with other simplices bottom-q or $\check{q}$. Then one measure of the eccentricity is:

$$e = \frac{(\hat{q}+1) - (\check{q}+1)}{(\hat{q}+1)},$$

giving a scale of eccentricity from zero to one. A simplex deeply embedded in a complex does not stick out like a sore thumb and would have a low, probably zero, eccentricity. Another simplex, with very conservative ideas long given up by his or her colleagues, or with very advanced ideas not shared by the majority, would have a very high eccentricity. The appearance and disappearance of a multidimensional and highly eccentric simplex such as Newton or Einstein has severe structural consequences, since the geometry of the intellectual space is radically altered. This is because the complex $K_P(I; \lambda)$, and its conjugate $K_I(P; \lambda^{-1})$, literally define the geometry of the space, the multidimensional *backcloth*, which must be strong enough to bear the traffic of the discipline. For example, one sort of traffic on this structural backcloth might be professional ideas and discourse. If an idea is very complex, and requires a combination of professional competence or interest in $\{3, 5, 8, 11, 13\}$ for its expression, then σ_E^4 can talk only to himself and no traffic in these ideas flows from him to others in the discipline. In fact, very little intellectual traffic can flow on this particular structure. Since most of the simplices share only one vertex, only very simple ideas involving single shared interests can flow, the exception being σ_C^2 and σ_D^3 who share $\{12, 14\}$. Of course, this is just a simple, pedagogic example, and no real discipline or department would have such a weak structure. At the same time, it illustrates how a backcloth can sometimes produce considerable *obstruction* to traffic, so that perhaps we might think in terms of *what the geometry allows*.

The idea of obstruction is clearly brought out in a structural, or q-analysis[1]. Very simply, in the initial stages of a q-analysis of structure, we can think of ourselves putting on successive pairs of spectacles with lenses that filter out objects with certain dimensional characteristics but not others. In $K_P(I; \lambda)$, for example, our lenses for $q = 4$ allow us to see only the simplex σ_E^4, and to all intents and purposes the other simplices of lower dimensionality are invisible (table 1). At $q = 3$ we change to

Table 1. q-Analysis of $K_P(I; \lambda)$.

q	$K_P(I; \lambda)$
4	$\{E\}$
3	$\{A\}, \{D\}, \{E\}$
2	$\{A\}, \{B\}, \{C\}, \{D\}, \{E\}$
1	$\{A\}, \{B\}, \{C, D\}, \{E\}, \{F\}, \{G\}$
0	$\{A, B, C, D, F, G\}, \{E\}$

Obstruction vector: $\{\overset{4}{0}\ 2\ 4\ 5\ \overset{0}{1}\}$

Eccentricities: ecc(A) = 0·750; ecc(B) = 0·666; ecc(C) = 0·333; ecc(D) = 0·500
ecc(E) = 1·000; ecc(F) = 0·500; ecc(G) = 0·555

[1] An analysis of structure happened to be called a q-analysis by Atkin, but it has nothing to do with the traditional Q-mode analysis in factor analysis.

lower-dimensional lenses which let us see σ_A^3, σ_D^3, and of course σ_E^4, but these simplices are all disconnected since they do not share four vertices that define their three-dimensional existences. At $q = 2$, the triangles σ_B^2 and σ_C^2 enter, but the structural *components* are still disconnected, and only at $q = 1$ does the component $\{C, D\}$ appear, showing that these simplices are connected at this q-level to form a small q-chain. Finally, at $q = 0$, a 0-level q-chain appears with $\{A, B, C, D, F, G\}$, but $\{E\}$ is still off by itself.

A global measure of structure is provided by the structure vector Q, $Q = \{\overset{4}{1}\ 3\ 5\ 6\ \overset{0}{2}\}$, where the superscripts denote the q-value, with an associated obstruction vector $\hat{Q}$ given by $Q - U = \hat{Q} = \{\overset{4}{0}\ 2\ 4\ 5\ \overset{0}{1}\}$. We can interpret the latter as a literal measure of the obstruction to traffic moving on a backcloth defined at a particular q-level. For example, at $q = 1$, the corresponding element in $\hat{Q}$ is 5, indicating the high degree of fragmentation or structural disconnection, making even one-dimensional traffic (professional discourse requiring two interests) impossible.

In the conjugate $K_I(P; \lambda^{-1})$, the simplices representing the intellectual interests of the discipline are defined by the people themselves (figure 6). From this perspective the structure of the discipline is very simple, since most of the interests are represented by only one or two people. The most popular is the simplex σ_{14}^2, an interest shared by C, D, and G, which appears as a triangle in the complex. The least popular are the 0-simplices, five of which are represented by the highly eccentric E. If we put on our dimensional spectacles for a q-analysis (table 2), only σ_{14}^2 appears in the structure at $q = 2$, and at $q = 1$ the discipline is still highly fragmented, with σ_{14}^2 and σ_{12}^1 being joined by the common vertices of C and D.

In fact, σ_{12}^1 appears as a face of σ_{14}^2, so we would expect it to have zero eccentricity. Any traffic on this structure is highly constrained by the geometry: for example, if we consider professional papers on a particular interest coauthored by two people, only C and F can work together on interest 2, and so on. Our obstruction vector indicates the high degree of constraint to greater collaboration at this q-level. At $q = 0$, the intellectual interests join to form a q-chain, but those represented only by E form one component, so that even at this level there is still some obstruction.

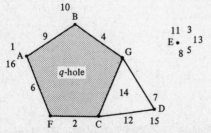

Figure 6. Geometric representation of $K_I(P; \lambda^{-1})$, the conjugate structure of a discipline.

Both in $K_P(I; \lambda)$ and in $K_I(P; \lambda^{-1})$ the people and interest simplices are connected in such a way that distinct holes appear in the backcloths. These have been termed *q-holes* and, paradoxically, they seem to be analogous to solid objects in our more familiar, but limited, Euclidean space E^3. If one thinks about it for a moment, a solid object, such as a tree in E^3, is actually a hole in the space, and unless you are in a *q*-hole destroyer, such as a tank, you cannot pass through it, but must move around it. In the same way, the spaces describing the geometry of our discipline have holes in them which can be conceived as solid objects. They are, in fact, traffic generators, since traffic (perhaps ideas and collaborative papers) has to move around these structures. Only a highly eclectic person in $K_P(I; \lambda)$, or an $(N+1)$ cover discipline (such as systems analysis or algebraic topology?) in $K_I(P; \lambda^{-1})$, can fill the *q*-hole and stop the essentially wasteful and dysfunctional traffic going around and around these objects in the intellectual space.

As a stage aside, how often do students complain of going around in circles, of repeating materials in courses, of doing everything over again from only a slightly different angle? The structures of our curricula are full of *q*-holes, and too seldom do we appear to provide our students (or ourselves?) with $(N+1)$-level *q*-hole fillers. How much of university education is the noise of traffic going around disconnected *q*-holes in the structural backcloth of archaic and outmoded curricula?

If complexes $K_P(I; \lambda)$ and $K_I(P; \lambda^{-1})$ are structural backcloths, traffic can be considered as a pattern or mapping, π, which associates each simplex with an object, often one of the rational or integer numbers Z. Formally[2]

$$\pi: \{\sigma_i^p; \ 0 \leqslant p \leqslant N, \text{ all } i\} \to Z \ ,$$

and since the simplices are *graded* by their *q*-values, so the pattern is also graded, or

$$\pi = \pi^0 \oplus \pi^1 \oplus \pi^2 \oplus \dots \oplus \pi^t \oplus \dots \oplus \pi^N \ .$$

Table 2. *q*-Analysis of $K_I(P; \lambda^{-1})$.

q	$K_I(P; \lambda^{-1})$
2	$\{14\}$
1	$\{2\}, \{4\}, \{6\}, \{9\}, \{12, 14\}$
0	$\{1, 2, 4, 6, 7, 9, 10, 12, 14, 15, 16\}, \{3, 5, 8, 11, 13\}$

Obstruction vector: $\{\overset{2}{0} \ \overset{}{4} \ \overset{0}{1}\}$

Eccentricities: $\mathrm{ecc}(1, 3, 5, 7, 8, 10, 11, 12, 13, 15, 16) = 0$; $\mathrm{ecc}(2, 4, 6, 9) = 0 \cdot 500$;
$\mathrm{ecc}(14) = 0 \cdot 333$

<hr>

[2] In algebraic topology, there are perfectly good reasons why the dimensional notation p is subscripted, and the identifier i is superscripted. However, in keeping with conventional scientific usage, I have reversed this specialized mathematical notation.

Thus, π^t might be a pattern of integers on the t-dimensional simplices, so it constitutes a set function defined on the $t+1$ subsets of the vertex set. Any change in pattern can be considered as a result of a t-force, and since the pattern is graded, so the force must also be graded. We might call this the Newtonian view of things, in which a change in pattern values due to t-forces is analogous to a force of gravitation for which movement (change) measures the gravity—or gravity 'causes' the movement (the tautology in conventional physics is nicely highlighted).

For example, consider traffic on $K_P(I; \lambda)$ resulting from a mapping of the simplices onto the integers, which might represent numbers of students working with people, the number of letters of inquiry received, the number of job offers, the professional articles published, and so on. If

$$\pi: \{A, B, C, D, E, F, G\} \rightarrow \{5, 2, 1, 0, 2, 0, 0\},$$

we have a *pattern polynomial* at some moment of time of

$$5\langle X_1 X_6 X_9 X_{16}\rangle \oplus 2\langle X_4 X_9 X_{10}\rangle \oplus 1\langle X_2 X_{12} X_{14}\rangle \oplus 0\langle X_7 X_{12} X_{14} X_{15}\rangle$$
$$\oplus 2\langle X_3 X_5 X_8 X_{11} X_{13}\rangle \oplus 0\langle X_2 X_6\rangle \oplus 0\langle X_4 X_{14}\rangle.$$

Suppose at a later time, the pattern has experienced an incremental change in π, say $\Delta\pi$, so that the pattern polynomial is now

$$4\langle X_1 X_6 X_9 X_{16}\rangle \oplus 2\langle X_4 X_9 X_{10}\rangle \oplus 0\langle X_2 X_{12} X_{14}\rangle \oplus 0\langle X_7 X_{12} X_{14} X_{15}\rangle$$
$$\oplus 6\langle X_3 X_5 X_8 X_{11} X_{13}\rangle \oplus 0\langle X_2 X_6\rangle \oplus 1\langle X_4 X_{14}\rangle.$$

These changes can only be due to t-forces acting on the backcloth; in other words:
a repulsive 3-force on $\langle X_1 X_6 X_9 X_{16}\rangle$ of $\Delta\pi^3 = -1$,
a repulsive 2-force on $\langle X_2 X_{12} X_{14}\rangle$ of $\Delta\pi^2 = -1$,
an attractive 4-force on $\langle X_3 X_5 X_8 X_{11} X_{13}\rangle$ of $\Delta\pi^4 = +4$,
an attractive 1-force on $\langle X_4 X_{14}\rangle$ of $\Delta\pi^1 = +1$.
If necessary, we could conceive of changes being constrained by the time available to answer inquiries, upper limits being placed on the number of students admitted, and so on; perhaps

$$\sum_{p=0}^{\hat{q}} \Delta\pi^p \leqslant 0.$$

Of course, the integer 4 now associated with σ_A^3 need not be associated with all the vertices; the letters of inquiry, the papers published, the students working with person A, may involve only a subset of the vertices, so that these may also form a graded pattern, say

$$\pi_A = 2\langle X_1 X_6 X_{16}\rangle \oplus 1\langle X_9\rangle \oplus 1\langle X_1 X_6 X_9\rangle.$$

Equally, suppose the pattern on σ_B^2 is also graded to

$$\pi_B = 1\langle X_9 X_{10}\rangle \oplus 1\langle X_4\rangle,$$

and no faculty member can handle more than three inquiries/papers/ students/etc. Then σ_A^3 will look to see who can help, and notes his connections through vertex $\langle 9 \rangle$ to σ_B^2. Thus we are now considering the *star* formed by σ_A^3 and σ_B^2, and if σ_B^2 accepts the additional load the result is a change in pattern owing to a graded *t*-force acting on the star. If, on the contrary, the graded pattern on σ_B^2 is

$$\pi_B = 2\langle X_9 X_{10} \rangle \oplus 1\langle X_4 \rangle ,$$

then a change in pattern $\pi_A + \Delta\pi_A$ can only produce a change $\pi_B + \Delta\pi_B$ if further zero-dimensional traffic is produced by σ_B^2 off-loading onto σ_G^1. If σ_B^2 and σ_G^1 did *not* share $\langle 4 \rangle$, the structure would not permit a change in pattern. In this sense, we have an Einsteinian view of change being that permitted by the *geometry*. If the geometry will not permit a solution, the only possibilities are (1) to increase the work loads, (2) to help people to keep learning by increasing their vertices, and so expand their professional dimensions, or (3) to encourage a new simplex with the appropriate vertices to enter the structure. Unlike the apparent geometrical constancy of the spaces of conventional physics, the geometry of the multidimensional spaces of other disciplines may not remain the same. We may well be able to consider a change in *structure*, as well as a more straightforward change in graded pattern.

For example, suppose person D starts thinking about intellectual interest $\langle 11 \rangle$, and adopts it to enlarge his structure to the simplex $\langle 7, 11, 12, 14, 15 \rangle$ (which is the *Lefschetz Prism* based on $\langle 7, 12, 14, 15 \rangle$ and $\langle 11 \rangle$). In this case, he is now 0-connected to person E, and they can properly talk about $\langle 11 \rangle$ together. Perhaps as a result of this zero-dimensional traffic, E becomes interested in $\langle 7 \rangle$, possibly because D wants to talk about $\langle 7, 11 \rangle$ ideas. Although E has $\langle 11 \rangle$, his $\langle 7 \rangle$ is too weak for the one-dimensional traffic to flow. Thus, he studies $\langle 7 \rangle$ until he assimilates it into his structure to form the Lefschetz Prism $\langle 3, 5, 7, 8, 11, 13 \rangle$. In this way, the desire for one-dimensional discourse has encouraged D and E to become 1-connected, because without such a connection they both know that communication at this more complex level is impossible. In this case, we have the shared face $\langle 7, 11 \rangle$, perhaps the start of further structural changes integrating E into the backcloth—or *vice versa* depending upon one's point of view.

After each structural change, consisting either of greater connectivity as more interests are shared, or less connectivity as old interests are dropped, we could consider each person as a *strain-pair*. This would provide us with an indication not only of traffic changes and the associated attractive and repulsive *t*-forces, but also of the changes in the dimensionality of each simplex or person. Thus the history of an entire discipline can be considered as a dynamic backcloth upon which both pattern and geometric changes have produced the particular intellectual configuration or geometry of the present day. That geometry constrains the possibility of traffic until the

t-forces, and the patterns themselves, result in *structural* changes that allow higher dimensional complexes and also the freer flowing traffic that they allow in turn.

Applications of the language of structure

Because the example I have chosen to introduce basic terms and concepts may appear somewhat fanciful, although perhaps not totally absurd to many with university experience, it is worth noting how q-analyses of structure have made considerable contributions to an extraordinarily wide spectrum of problems. The wide diversity of applications should not surprise us too much if we remember that the very notion of structure appears in practically every area of human inquiry and discourse. Structural questions, in which various relations are deeply embedded, appear throughout the geosciences, and perhaps these examples from widely different fields will encourage some to look at their own discipline and subject matter with new eyes.

After several years of exploring cohomology theory in physics, Atkin published two remarkable papers raising similar topological questions of structure in the human sciences, disciplines characterized by much softer forms of data (Atkin, 1971; 1972a). One of the first applications came from an attempt to help urban planners 'harden up' planning policy by getting them to define the elements of the sets with which they were dealing, and so raising the possibility of describing structural matters in an effective, unambiguous, and essentially manipulable algebraic topological language (Atkin, 1974b; 1975a). For example, a small, highly-prized Tudor village in Suffolk (Atkin, 1972b), and a street of early Victorian row houses in Southend-on-Sea, generated backcloths from the relations between the sets of houses and the sets of appropriate architectural features. Traffic on such a backcloth might be the numbers of tourists seeking out the picturesque, the amounts of money they spend, or even a person's eye 'traveling' over such a structure and lingering for various time intervals. In these particular cases, architectural homogeneity led to a highly connected simplicial complex, and one can easily imagine what tearing down a Tudor cottage and substituting a modern hamburger stand would mean in terms of visual eccentricity, alterations in the geometry, and the wrenching changes in the pattern polynomials brought about by the strong t-forces involved. Similarly, historical analyses of whole towns have been undertaken, such as 16th century Saffron Walden (Atkin, 1973a), and more modern Southend-on-Sea between 1910–1972 (Atkin, 1973b). The urban areas were divided into appropriate lozenges, or cover sets, to examine both the structural change and the magnitude of the t-forces that brought about the alterations in the physical geometry within which people live their daily lives.

An examination of the structure of a university was also conducted, partly because it had gone through 'bad times' in the early 1970s and partly because it appeared that a lack of connectivity both in the physical and in the human structures might be exacerbating the problem of students versus faculty versus administration (Atkin, 1974c; 1977a). Whether questions involved the possible conversion and change in the use of a building, or the pressing demands and takeover actions of students, a distinct lack of hierarchical filtration in the appropriate structures was apparent. Committee structures produced large q-holes in the backcloth and, since these were the equivalent of solid objects in our more familiar Euclidean space, they caused traffic (in the form of reports and memoranda) but prevented concrete decisions and action. Many faculty members felt a sense of alienation, feeling intuitively that involvement in such a structure was a waste of time. As a result, pseudo-committees developed outside the system to fill the q-holes, but someone like the Vice-Chancellor, a simplex of high dimensionality, was not necessarily acting in a malevolent way by taking apparently arbitrary and high-handed decisions. On the contrary, the *structure*, the connective geometry of the university, virtually required that he act in such a way.

In medicine, structural analyses have aided in the treatment of brain-damaged elderly people (Gedye, 1974) and the diagnosis of difficult maladies such as Behcet's disease (Chamberlain, 1976), and work is continuing on the application of q-analysis to the structure of relations between people, and to the graphic expression of such relations in the remarkable maps of the psychiatrist Mulhall, which are nothing less than Ashby's state diagrams highlighting the connective tissue (relations) between members of a family under various emotional states (Mulhall, no date). Positive and negative feedback loops have obvious interpretations, but more remarkable are some recent analyses from individual patients and their responses to sets of 'open-ended' questions in clinical interviews. In one case, the simplicial complex describing the response structure of a man experiencing quite dysfunctional difficulties showed how ostensibly his feelings toward homosexuality in general were open and literally 'on the surface'. However, in the conjugate complex the same responses were deeply buried, with the obvious clinical interpretations.

It is in the game of chess that some of the most impressive insights have been achieved by properly describing for the first time the relevant structures. If, at one level, chess simply appears to be 'a game', we must recall that this old and terrible field of combat has been the subject of intense research and analysis for the past quarter of a century in the area of artificial intelligence. Such research may literally prove to be vital in the future, since the human brain cannot deal with the extreme combinatorial complexity characteristic of modern life, and requires all the prosthetic help it can get.

The explosive combinatorial features of chess have directed most research to simplistic tree-searching methods, in which bigger and faster computers, combined with some intriguing, but essentially simple, algorithms, search down a few more branches. Such tree searching or *tactical* play is characteristic of patzers, amateurs like you and me, and does not resemble at all the thinking of a grandmaster. A master player thinks in terms of *positional* play, which is essentially structural, being concerned with *relations between sets*. With one of England's grandmasters, Atkin has succeeded in providing a structural description of a chess game at the positional level by forming the simplicial complexes between the pieces and the squares. In a sense he has literally provided an algebraic description of the microgeography of a chessboard and its aggressive inhabitants (Atkin and Witten, 1975; Atkin et al, 1976). A q-analysis at the $(N+1)$-positional level isolates a few squares, which can then be examined at the N-tactical level, each possible move being evaluated in terms of its overall structural contribution at the higher, positional level. Classical games have been analyzed in which the structure vector indicates a weakening of one player's position up to ten moves before a master chess analyst can point to a definite weakening of positional play (Atkin, 1978).

In statistics, a restatement of problems and conditions in a topological language appears to point out weaknesses arising from earlier over-simplifications. We have already seen how regression represents replacing a relation λ with a highly constrained mapping, so that a correlation coefficient is really a measure of the degree to which the simplicial complexes of the data can be replaced by two collections of disconnected zero-dimensional simplices. If we return to our example (figure 4), the cosine between such linear mappings is only $r = 0.52$, because we have discarded much of the information in the data set. In contrast, we can define a *structure coefficient*, h, the cosine between the two structure vectors, $Q_{KX} = \{2\ 2\ 1\ 1\}$, and $Q_{KY} = \{1\ 1\ 1\ 1\}$, so that $h = 0.95$, indicating a much higher degree of structural dependence than the simpler, traditional form. It should also be noted that h may well be nonzero when r takes the value zero, as in the case of a perfect linear relationship, $Y = a$. In such a case, the traditional r would be zero, although h assumes the value of $1/n$, corresponding intuitively to the idea that one complex contains only $1/n$ as much structure at its conjugate (Atkin, 1974a, page 185). Furthermore, if it is argued that traditional regression makes explicit assumptions (seldom met) about error terms, then it may be more fruitful to recast these approaches directly in terms of the *structure-distorting* effects of noise on a simplicial backcloth. Thus, noise, or error, consists of unwanted patterns on a complex, resulting in a distortion of structure. Finally, recent work on the structure of sample spaces (Atkin, 1977b) has literally opened up questions never before posed in traditional statistical theory. By considering the sample space as the power set $E = P(X)$ (where $X = \{X_1, X_2, ..., X_n\}$—the n elementary events), the relation $\lambda \subseteq E \otimes X$ contains all the possible events as subsets of X. This means that $K_E(X; \lambda)$ is an $(n-1)$-dimensional simplex with all its faces, so that the probabilities distributed by the binomial theorem correspond to a graded pattern on the structure. In any theory relying upon probability distribution functions, including the Poisson and normal distributions derived from the binomial, such a structure must play a fundamental role. If we interpret probability as likelihood, an observation of an event represents a change in pattern $\pi' = \pi + \Delta\pi$, implying (1) that our sense of likelihood is constantly subject to t-forces, and (2) the geometry must be strong enough to carry such traffic, which will be true only if the obstruction vector of $K_E(X; \lambda)$ is zero. Empirical studies point "... strongly to the breakdown of this criterion" (Atkin, 1977b, page 130). In the conjugate $K_X(E; \lambda^{-1})$, traffic is a pattern of probabilities that the vertices $\{X_i\}$ will define event $\{E_j\}$—the usual appeal to Bayes's Theorem. But $K_X(E; \lambda^{-1})$ will rarely be described by a zero obstruction vector, and an ungraded pattern of probability is conceptually poverty-stricken.

Nothing said above should be construed as the decline and fall of the Fisherian Paradigm à la Kuhn, but rather a Stegmüllerian absorption into an enlarged paradigm by equivalence relations. From this more general and richer viewpoint, the earlier and unperceived inadequacies are simply shown to be oversimplified and highly constrained examples. At the same time, this enlarged viewpoint should remind us that science made excellent deterministic progress for three hundred years before overenthusiastic proponents implied that statistical methodology was the embodiment of the scientific method itself.

Nor is the use of the language of structure confined to areas traditionally labelled as scientific. Since a concern for binary relations is at the heart of the language, and since these are essentially *qualitative*, the language of structure holds considerable promise for examining areas of human thought where difficulty has been experienced in expressing structural matters. For example, an analysis of television programming operationalizes the difficult-to-define notion of programming structure (Gould and Johnson, 1978), and by finding proper cover sets even analyses of modern paintings have been undertaken (Atkin, 1974a). A q-hole has been found in a Shakespearean sonnet consisting of key words which appear to cause traffic on the aesthetic backcloth (Atkin, 1978), and my wife and I have discovered similar objects in the space of some of the two-part and three-part inventions of J S Bach. Such objects would imply a lack of connection between certain notes at specified q-levels, so that sequences of notes (melody?) must follow certain transitional rules or constraints imposed by the geometry of the fugue.

It is, however, in the area of traffic design and regional planning (Johnson, 1975; 1976; 1977), that some of the most provocative insights have been obtained, mainly by demonstrating that most planning directives are so ambiguous that no one can put them into operation, foresee their consequences and counterintuitive effects (Forrester, 1971; Chadwick, 1977), or evaluate the 'results'. Often only cursory attempts are made to define the elements of the sets, or the connecting relations, and it often appears that the fundamental job of describing complexity is so inadequately undertaken that it seems that no one knows what they are talking about (Atkin, 1977b). Only one example must serve, but it will lead directly to a similar problem in the geosciences.

Partitional thinking
In an analysis of regional unemployment in East Anglia (Atkin, 1975b), it was discovered that the Ministry of Labour, responsible for the bijective mapping of the set of unemployed people to the set of available jobs, had classified the set of approximately 12000 job descriptions into a set of 223 conventional, and by traditional definition, nonoverlapping subsets labelled Occupation Unit Groups. However, it is obvious that many different sorts of jobs share a large number of similar, more fundamental skills, and one would hope that an effective job classification scheme would provide a great deal of flexibility in matching the skills offered by unemployed people to those required by potential employers. With a flexible classification-reference system, unemployed people could move easily into jobs with similar ranges of skills, and, perhaps with a little retraining and work experience, have the human dignity of holding a job once again.

Unfortunately, two hundred years of Linnaean partitional thinking had done its work; instead of the terms of the occupation groups being *cover sets* at the $(N+1)$-level, they represented those rare cases of *partitions* that are so seldom appropriate. Someone had sat down with roughly 12000 jobs at the N-level and 'classified' them by throwing them into 223 boxes at the $(N+1)$-level. The result was all satisfactory and tidy because it conformed to the *unquestioned* notion of a 'good' classification, namely a partition. Since everybody had been brought up to believe that a taxonomic scheme was 'good' if it represented a partition, no one questioned the appropriateness of the archaic partitional thinking itself. It is difficult to get out of a trap if you do not know you are in it. The result was that an unemployed bricklayer went to the local employment office, an official looked in the box labelled "bricklayers", found none, and said, "sorry, no job for you today". What was not realized was that partitioning the set of jobs had done great violence to a highly inter-connected *structure*. The Ministry of Labour had undertaken a totally inappropriate, many-to-one surjective mapping. constraining the assignment to a *function*, when what was required was *at least* a many-to-many mapping, with a possibility of having job descriptions at the N-level not assigned at all to a descriptor at the $(N+1)$-level. In other words, allowance should have been made for the possibility of a nonpartitional *relation* between the two sets. In fact, the simplicial complex $K_{J(N+1)}(J_{(N)}; \lambda)$ was a highly connected structure of 223 simplices defined on the 12000 vertex set of job descriptions. The classical classificatory thinking that had been applied had torn this structure apart, making it practically impossible for traffic (searching for jobs with roughly similar skills), to move on the backcloth. High values in the obstruction vector would have indicated the dysfunctional effect of such thinking, and the very notion of *forcing* a partition on a set intuitively contains within it the idea of t-forces distorting a backcloth describing the multidimensional employment space.

Such partitional thinking is very common: children draw distinctions between their toys and place them in different boxes; scientists group animals and plants, and then argue for years about the appropriateness of their categories. It may even be a fundamental act of the human brain to draw a distinction (Spencer-Brown, 1962), but such thinking may also produce nonsense: we know fundamentally that the world does not come to us in the little, nonoverlapping boxes of the Reductionist. Even geographers know that lines on maps, separating the wheat and corn belts, are not partitions but overlapping zones where sets of areal simplices are connected over vertex sets of agricultural land uses. We have agricultural cover sets on the Earth's surface, not partitions. Recent research in television confirms that exploring and interrogating the structure of television programming is far more meaningful for policymaking than impressing the traditional partition upon a set of highly connected programmes (Gould and Johnson, 1978; 1980; Gould, 1978b).

Classification schemes are an impress of the human brain on a set of data, they have no 'reality', and when they consist of partitional acts they destroy the richness of ambiguity. We know we do violence to the data: we often feel the t-forces and acknowledge the stress of assigning an object or observation to one box or another, when we know it should be partly in both, or in neither. Nevertheless, on we plunge, devising discriminate functions, or numerical taxonomic algorithms, to do the dirty work for us, as if these pathetically constrained functions and FORTRAN sausage machines can absolve us from the responsibility of making a partitional decision. But we must remember that Pontius Pilate was a bureaucrat— hardly a model for a scientist. No one questions the intellectual hand-me-down of performing the act of partition in the first place.

Yes, I know: it reduces variety (Ashby, 1956) and complexity, but at least in certain areas we have today prosthetic aid in the form of a computer to help us handle creatively the additional complexity that relational approaches imply, rather than to use the computer simply to carry out the same operations as before (Beer, 1974; 1975). All the work in the area of information retrieval, all the sincere, but nonoperational gropings toward fuzzy sets, represent a renewed recognition of the richness of ambiguity as the 'glue' that holds our data structures together. If there really are partitions at one q-level, we shall find them; but to rip apart such structures unthinkingly with information-destroying partitional thinking should hardly be encouraged. Otherwise the next generation will think just like us—which means that little progress will have been made.

Do foraminifera assemblages exist—at least in the Persian Gulf?
The data sets we shall now examine were called to my attention by a
pedagogic example of a cluster analysis of nonquantitative data (Bonham-
Carter, 1967). The data matrix recorded the amounts of 20 foraminifera
(columns) over 70 grab samples (rows) from the bottom of the Persian
Gulf. The data sets were extracted from a monograph by Houbolt (1957),
and considerably reduced in two respects. First, only 70 out of the
possible 137 observations made by Houbolt were chosen (presumably
because only a small pedagogic example was required); and, second, the
'semiquantitative' (actually ordinal) scale was collapsed from six to four to
fit a convenient binary coding scheme. The association between each pair
of samples was measured by Jacquard's coefficient, which consists of a
simple ratio of positive matches (the same foraminifera found in both
samples) to the sum of the positive and mismatches, ignoring negative
matches. Thus the association could range from $0 \cdot 0$ to $1 \cdot 0$. The cluster
algorithm made several 'passes' over the matrix of coefficients, each time
grouping the most closely associated grab samples, and substituting a
weighted mean value to represent the 'clusters' thus formed. The
conventional taxonomic tree was produced, and the branches sliced at
distances, or association values, ranging between $0 \cdot 34 - 0 \cdot 41$ to produce
seven 'clusters', A, ..., G.

A number of things are worth calling attention to in passing: (1) although
the cluster analysis was purportedly for nonquantitative (qualitative?)
data, the calculation of a coefficient of association immediately substituted
a number upon which various arithmetic operations are conducted, such as

Table 3. q-Analysis of the reduced (70 × 20) foraminifera data set forming $K_S(F; \lambda)$
sliced at $\theta \geqslant 1$.

q	$K_S(F; \lambda)$
13	{RC400(A)}
12	{RC361(D)}, {RC400(A)}
11	{RC263(A)}, {H357(D)}, {CH359(D)}, {RC361(D)}, {C362(A), RC400(A)}
10	{RC262(A)}, {RC263(A)}, {H357(D)}, {CH359(D)}, {CH360(D)}, {RC361(D)}, {C362(A), RC400(A)}, {C399(A)}, {C540(A)}
9	{RC262(A)}, {RC263(A), C362(A), C399(A), RC400(A)}, {RC329(D), H357(D), CH359(D), RC361(D)}, {CH338(D)}, {CH360(D)}, {RC368(A)}, {RC372(A)}, {C540(A)}
8	{RC262(A), RC263(A), RC329(D), CH338(D), H357(D), CH359(D), CH360(D), RC361(D), C362(A), RC367(D), RC372(A), TM398(A), C399(A), RC400(A), C540(A)}, {C318(A)}, {RE348(B)}, {RC368(A)}, {TM397(A)}, {RC433(A)}
7	{12A, 1B, 9D}, {H326(D)}
6	{13A, 2B, 13D}, {C322(D)}
5	{15A, 5B, 1C, 17D, 4E}, {RM114(C)}
4-0	all connected

the calculation of a 'weighted mean'; (2) the algorithm is a deterministic partitional machine—any set of numbers can be substituted, including those simply chosen at random, and a taxonomic tree produced, and then cut across some branches to produce a partition; and (3) comparisons "... are not easily made between qualitative variables" (Bonham-Carter, 1967, page 2), although with the language of structure there is no problem in considering the conjugate binary relation, λ^{-1}.

The data sets of 70 samples and 20 foraminifera were coded 1 (single to rare, in Houbolt's original terminology), 2 (6–100 specimens), and 3 (100 specimens to more than one quarter of the rock sample), to conform with the semiquantitative scale, and an initial slicing parameter $\theta \geqslant 1$ chosen to define the relations $K_S(F; \lambda)$ and $K_F(S; \lambda^{-1})$. In the complex $K_S(F; \lambda)$, we are examining the structure of connections between the 70 grab samples, each one of which we can think of as a multidimensional simplex defined by vertices in the foraminifera set. In the q-analysis of the structure (table 3), we can note immediately that some very high dimensional samples are in the complex. For example, $\sigma^{13}_{RC400(A)}$ is a thirteen-dimensional simplex defined on 14 out of the possible 20 foraminifera vertices[3].

At q-levels $\geqslant 10$, only a few sample simplices appear, and all are disconnected with the exception of {C362(A), RC400(A)}. At $q = 9$, the start of what we might term A and D components appear, containing a few of the samples designated by the cluster algorithm, but the hope for a partition is removed at $q = 8$. At this still-high dimensional level, 15 of the sample simplices defined by 9 foraminifera are all highly connected to form a single component, while 5 others appear as their own disconnected components. Yet it is precisely this highly connected structure of 15 simplices, defined by a truly *qualitative* relation, that is torn apart by the pseudometric operations of the deterministic partitional algorithm. Can greater conceptual violence be conceived? At lower q-levels, $q \leqslant 7$, virtually everything is connected into a coherent structure, which means that the algorithm, and the partitional thinking it represents, is literally shredding a complex in which most of the simplices share the presence of many similar foraminifera. If we want to understand how things (samples) are related, why do we tear apart the very connective tissue that would allow us to explore the relationships involved?

Even when we slice at $\theta \geqslant 2$, so that our relation $\lambda \subseteq S \otimes F$ is now defined by the 'common' (6–100 specimens) or even greater presence of foraminifera, only very weak evidence for a single partition appears at a high q-level (table 4). Of course, our structure is less connected, since we

---

[3] An alphanumeric label in table 3 refers to one of Houbolt's original designations, while a capital letter in parentheses denotes the group into which the cluster algorithm placed the sample. To save space, the former are only recorded for $q \leqslant 7$, unless a sample appears in its own component. For example, at $q = 7$ there were 12 samples from group A, 1 from group B, and 9 from group D connected into one component.

have sliced away all the single or rare (2–5 specimens) relations, but even now there is only the hint of a D-component. In the two A-components, simplices both of the B and the D groups are already highly connected over 5 foraminifera. At $q = 3$, a large well-connected structure appears, and all the simplices are embedded in the complex at $q \leqslant 1$. As for the conjugate complex $K_F(S; \lambda^{-1})$, whether we slice at $\theta \geqslant 1$ or $\theta \geqslant 2$, all the foraminifera simplices display a highly connected structure at *all* q-levels, but I shall reserve discussion of these implications for the more detailed analysis below.

It might be argued that it is not really fair to take a pedagogic example devised simply to illustrate the use of an algorithm in numerical taxonomy. Accordingly, a detailed study was made of the data sets reported in Houbolt's original monograph, based upon the 137 grab samples for which full information was provided, by using an ordinal scale of six degrees of intensity[4]. Thus, we are considering a series of relations on the 137×20 incidence matrix defined by the set $\theta_i = \{1, 2, 3, 4\}$ of slicing parameters.

The full analysis cannot be given here for reasons of space, although it may be possible to present it elsewhere, but the main points are easy to make: even with the enlarged data set it does not matter what slicing parameters we choose, the conclusions about the highly connected structures of the $K_S(F; \lambda)$ complexes are essentially the same. Whether we define the relation between the elements of our sets weakly, $\theta \geqslant 1$, or strongly, $\theta \geqslant 4$ (so few 5s and 6s were present that it was not worth slicing at $\theta \geqslant 4$), the clusters of the grab samples partitioned by the algorithm result from tearing highly connected structures apart. Even when slicing at $\theta \geqslant 1$, a mixed component of A and D samples are linked in one component at $q = 9$, and by $q = 6$ virtually total connection has occurred. A similar result occurs at $q = 6$ when $\theta \geqslant 2$.

Table 4. q-Analysis of the reduced (70×20) foraminifera data set forming $K_S(F; \lambda)$ sliced at $\theta \geqslant 2$.

q	$K_S(F; \lambda)$
7	{CH360(D)}
6	{H357(D), CH360(D)}, {RC361(D)}
5	{H357(D), CH359(D), CH360(D)}, {RC361(D), RC367(D)}, {RC400(A)}, {RC433(A)}
4	{TM151(D), TM256(D), RC327(D), H339(D), H356(D), H357(D), CH359(D), CH360(D), RC361(D), RC367(D)}, {RC262(A), RC263(A), RC265(B)}, {RC264(A), RC400(A), RC433(A)}, {RC329(D)}, {C430(A)}
3	{11A, 4B, 12D}, {RM114(C)}, {C540(A)}
2	{13A, 5B, 7C, 17D, 1E}, {P84(C)}, {RM101(C)}
1, 0	all connected

[4] All data were compiled from Houbolt (1957), enclosures I, V, Va, Vb, together with the accompanying legend sheet.

I think the major point has been made: what is of equal interest is the way in which the structural analyses at $\theta \geqslant 3$ (common) and $\theta \geqslant 4$ (abundant) highlight structural facts that are simply obliterated by partitioning the set of samples. At $\theta \geqslant 3$, the components $\{257\}$, $\{357(D), 359(D), 360(D)\}$, and $\{361(D)\}$ appear at $q = 6$, indicating that there are five simplices with very strong relations defined by seven, and in two cases eight, foraminefera. Four of these were taken within 2–3 km of each other on the Shah Allum Shoal and were defined by relatively large quantities of *Miliolidae, Textularia, Rotalia, Cibicides, Amphistegina, Heterostegina, Heterostegina qatarensis*, and *Operculina*. Similarly, at $q = 5$, six out of the nine samples falling into a 'D component' were all taken on a single traverse in water deeper than 30 fathoms south of the Shah Allum Shoal, and at $q = 4$ one component appears to include bottom samples that were found near the Halul, Cable Bank, and Shah Allum salt domes—samples sharing strong (recall $\theta \geqslant 3$) relations which the cluster algorithm partitions into A and D groups. Perhaps these areas represent specialized environments leading to considerable proliferation of forms.

A second component at $q = 5$ contains samples taken on the same traverse between 30–40 fathoms, samples which are partitioned by the algorithm to A and B groups, even though they share large quantities of five foraminifera. At $q \leqslant 3$, almost total connection appears, although the cluster 'machine' assigned to different groups samples taken at (1) adjacent locations, (2) in the same depth of water, and (3) strongly connected by four foraminifera. Additional anomalies could be cited: for example at $q = 2$, the adjacent simplices $\sigma^3_{540(A)}$ and $\sigma^2_{539(E)}$ taken in 10–11 fathoms are partitioned into A and E groups. One is reminded strongly of Boyce's evaluation of numerical taxonomy for hominoid classification (Boyce, 1964). Dependent upon which cluster algorithm was employed, female chimpanzees, orangutans, and gorillas were clustered with male and female human beings before joining the male partners of their own genera. Of course, one must be broadminded these days, but clearly partitional thinking can produce strange bedfellows.

When $\theta \geqslant 4$, we slice out 51 samples from our structure, since our relation is now defined only by the presence of at least 'abundant' specimens. Three small components, $\{256(D), 260, 437\}$, $\{357(D), 359(D)\}$, and $\{540(A)\}$, appear at $q = 3$, but it is worth noting that 256, and 357 and 359, which the algorithm places in the same D cluster, were found 140 km apart, the first in 14 fathoms, and the latter two near the Shah Allum Shoal at 18 and 28 fathoms respectively. At $q = 2$, six components appear, but they already contain mixtures of A and D groups, and at $q \leqslant 1$ total connection occurs. It is worth noting that the algorithm's group C, containing five samples taken at 1–5 fathoms in Doha Bay, all join at once, and their zero eccentricities indicate they are all deeply embedded in the complex.

The so-called dendogram groups found by the cluster algorithm "correspond quite closely to the original assemblages of Houbolt" (Bonham-Carter, 1967, page 7), and by plotting the samples against the depth of water in fathoms, it is implied that the assemblages are a *function* of depth (figure 7)[5].

From the former's plot (Bonham-Carter, 1967, page 12), it is easy to see that there is a total overlap between groups A and D, E and F, and C and G, and considerable overlap between them all, suggesting perhaps that we have a strongly connected structure—which is exactly what the q-analysis implied. We have, in fact, two questions here: (1) what is the validity of the assemblages formed by a partition on the set of grab samples, and (2) is the variation of foraminifera with depth of water significant. Let us consider each in turn.

To a geographer rushing in like a fool where even highly specialized palaeontologists tread lightly, it appears that the idea of an assemblage should be reasonably straightforward. Presumably various types of foraminifera are found together to the exclusion of others; these can be readily identified by skilled and patient observers, and, given information about various amounts of foraminifera, different observers should be able to reproduce each other's results. After all, this is what science is about. Unfortunately, no matter how hard I try, I cannot find the assemblages in the data.

Houbolt lists *Rotalia–Elphediella, Rotalia–Cibicides, Cibicides, Textularia–Miliolidae, Heterostegina*, and *Rotalia–Elphidium* assemblages; but consider

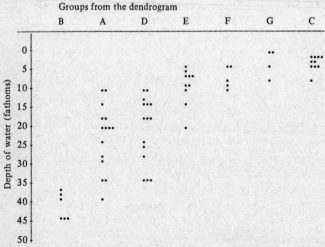

Figure 7. Distribution of dendogram groups with depth of water (from Bonham-Carter, 1967, page 12).

[5] This plot has been reproduced because it contained all 70 of the grab samples selected for the smaller, pedagogic analysis. The similar plot of Houbolt (1957, page 76) contained only 93 of the possible 137 samples, although it is not clear why 44 were left out.

$K_F(S; \lambda^{-1})$ defined by the set of slicing parameters, $\theta_i = \{1, 2, 3, 4\}$. If assemblages occur, if various types of foraminifera do truly occur together to the exclusion of others, we should find partitions of the set, F, of foraminifera. No such partitions occur (table 5). Instead we have a highly connected structure, no matter whether we consider the high q-level of 133, at which point the single 133-simplex *Miliolidae* appears, or all subsequent q-levels ≥ 5, at which point the seldom found foraminifera *Sorites* enters the analysis. At successively lower q-levels each foraminifera joins immediately, or appears as a separate component for just a few levels before it is absorbed into the complex. With the exception of eight unitary terms, the 133 element obstruction vector is zero, and all the eccentricities of the foraminifera are $\leq 0\cdot03$, with the exception of the ubiquitous *Miliolidae*, whose eccentricity is equal to $0\cdot133$.

The same conclusions hold true at all slicing parameters: we have a series of highly connected structures described by virtually zero obstruction vectors and zero eccentricities. With the exception of *Miliolidae*, none of

Table 5. q-Analysis of $K_F(S; \lambda^{-1})$ sliced at $\theta \geq 1$.

q	$K_F(S; \lambda^{-1})$
133–119	$\{Miliolidae\}$
118–117	$\{Miliolidae\}, \{Textularia\}$
116–78	$\{Miliolidae, Textularia\}$
77	$\{Miliolidae, Textularia\}, \{Cibicides\}$
76–71	$\{Miliolidae, Textularia, Cibicides\}$
70–68	$\{Miliolidae, Textularia, Cibicides, Rotalia\}$
67	$\{Miliolidae, Textularia, Cibicides, Rotalia\}, \{Nodophtalmidium\}$
66–56	$\{$all above$\}$
55	$\{$all above$\}, \{Operculina\}$
54–45	$\{$all above$\}$
44	$\{$all above$\}, \{Amphistegina\}$
43–41	$\{$all above$\}$
40–39	$\{$all above $+ Elphidium\}$
38	$\{$all above$\}, \{Peneroplidae\}$
37–36	$\{$all above$\}$
35–33	$\{$all above $+ Heterostegina\}$
32	$\{$all above$\}, \{Eponides\}$
31	$\{$all above $+ Gaudryina + Discorbis\}$
30–25	$\{$all above $+ Heterostegina\ qatarensis\}$
24	$\{$all above $+ Elphidiella\}$
23	$\{$all above $+ Nonion\}$
22–20	$\{$all above $+ Globigerina\}$
19–9	$\{$all above $+ Bolivina\}$
8–6	$\{$all above $+ Reussella\}$
5–0	$\{$all above $+ Sorites\}$

the foraminifera simplices even stands out, let alone are partitioned from the complex. Even at $\theta \geqslant 4$, when the relation λ^{-1} is defined only by "abundant, flood, and rock-forming" amounts of foraminifera, there is little evidence of any assemblage, with the possible exception of {*Textularia, Miliolidae*} between q-levels 22 and 12 (table 6). At $q = 16$, the component {*Cibicides*} occurs, defined by 17 out of the possible 137 grab samples, but it joins {*Textularia, Miliolidae*} at $q = 11$ *after Rotalia* forms part of the complex at $q = 12$. The rest join in approximately the same order as when the complexes are defined by the smaller slicing parameters.

Where are the assemblages? Even traditional 2 x 2 contingency tables cannot find them. On the basis of simple presence and absence in 137 samples ($\theta \geqslant 1$), the ϕ-coefficient for the *Rotalia–Elphidiella* 'assemblage' is zero (0·00011), and it remains zero even when presence and absence are defined only by the very strong relationships with a slicing parameter of $\theta \geqslant 3$. *Elphidiella* does occur sometimes with *Rotalia*, but *Rotalia* is also present in many samples (79/137 when $\theta \geqslant 1$, 46/137 when $\theta \geqslant 3$) when *Elphidiella* is absent. Moreover, when *Elphidiella* is present, it always occurs with *Miliolidae*, *Textularia*, and *Cibicides*. In brief, *Elphidiella* is part of a well-connected structure. As for the *Rotalia–Cibicides* assemblage, its presence–absence contingency table also yields a zero value of ϕ, and even if the 15 samples from the shallow Doha lagoon are removed, on the grounds that they are of a different species (a nice example of not defining rigorously the sets of elements), the value is still zero. Similarly, the *Rotalia–Elphidium* assemblage has a ϕ-coefficient of zero; the *Cibicides* assemblage always occurs with at least four or five other foraminifera; and the *Heterostegina* assemblage "by definition is characterized by the abundant occurrence of *Heterostegina qatarensis*".

Table 6. q-Analysis of $K_F(S; \lambda^{-1})$ sliced at $\theta \geqslant 4$.

q	$K_F(S; \lambda^{-1})$
63–32	{*Miliolidae*}
31–23	{*Miliolidae*}, {*Textularia*}
22–17	{*Miliolidae, Textularia*}
16	{*Miliolidae, Textularia*}, {*Cibicides*}
15–13	{*Miliolidae, Textularia*}, {*Cibicides*}, {*Rotalia*}
12	{*Miliolidae, Textularia, Rotalia*}, {*Cibicides*}
11–9	{all above}
8, 7	{all above}, {*Heterostegina qatarensis*}
6	{all above}, {*Operculina*}
5	{all above}, {*Operculina*}, {*Elphidium*}
4	{all above}, {*Elphidium*}
3	{all above}
2	{all above + *Nodophtalmidium* + *Amphistegina*}
1	{all above + *Heterostegina*}, {*Elphidiella*}
0	{all above + *Globigerina*}

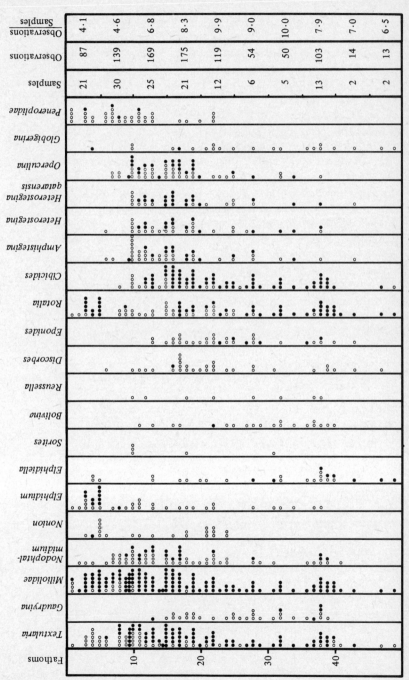

Figure 8. Relation between the presence of foraminifera and depth of water (compiled from Houbolt, 1957).

The fact that this species also always appears with at least five or six other foraminifera seems irrelevant, but such an argument can easily lead us to define an assemblage as any member of the power set $P(S)$ containing 2^n members, according to whether an assemblage contains, or does not contain, some combination of foraminifera. In this way we would have $1\,048\,576$ assemblages, enough to satisfy the most fastidious partitional thinker, but surely in this direction lies madness, not science?

But what of the implied quasifunctional relationship to depth? If the assemblages are not a figment of the imagination, for which no evidence can apparently be marshalled, perhaps there is still some relation between individual foraminifera and the depth of water. Unfortunately, when the data contained in the 137 samples are plotted (figure 8), it is difficult to see an implied 'functional' relationship between anything. The depth ranges are large; rough upper and lower limits can be drawn for about half the foraminifera, but even these are wide (10–40 fathoms); samples are not equal at each depth; and densities do not appear to vary significantly, with the exception of those in very shallow waters. One could, of course, examine directly the relation between the set of foraminifera and a set of depth zones under various density slicing parameters, or we could consider the depths as a zero-order pattern on $K_S(F; \lambda)$, but these approaches hardly seem worthwhile.

The end of reductionism?

Partitional thinking is almost inevitable in the sort of intellectual milieu of reductionism that has prevailed in many areas of science for the past two hundred years. Its causes are many, but not the least is the constrained and simplistic form of mathematics that has been borrowed from 17th–18th century physics, and applied almost unwittingly in areas where it appears less and less appropriate (Gould, 1980a). In addition, Linnaean thinking has saturated many areas of human thought, so that some disciplines appear to consist almost entirely of giving names to things, and then putting the things in boxes. Students in such fields are then examined to see if they can recall all the names of things from memory, and whether they can put them in the 'right' box. The fact that deterministic partitional algorithms have been written for computers should not blind us to the more fundamental fact that we are using such machines to do mechanically the simplistic tasks we have always done in the past.

Take the typical 'problem' for which numerical taxonomic methods are used: information is recorded about a set of objects over another set of variables; an orthogonal, metric space is created in which the objects are given coordinates; and then some sort of rule is applied to cluster objects that are close together. If the objects really do form small and distinct groups in the space, we may have a genuine partition. If this is truly the *structure* of our data sets, a q-analysis will clearly indicate it. In the same way that a function is a restricted form of a relation, so a partition can be considered to be a special case of a cover. But in most empirical examples, we know that we seldom have such a partition: our objects tend to form a globular cluster in the center of the space, with perhaps a few outliers. Where we start cutting things apart with any one of a dozen different routines, each giving different results (Barker, 1974), is virtually arbitrary, and the $(N-1)$-dimensional boundaries of our partitions tend to grow like the random cracks on a glazed pot. The process reminds one of a lunatic hacking apart a pumpkin with a broadaxe, and notice how intellectually seductive the results are: points that are close together in the pieces hacked apart were definitely close together in the pumpkin—no matter where the cuts were made. Because one fact is tautologically indisputable: no matter what clustering routine is applied, points close together in the space (pumpkin) will often appear in the same groups (pieces). But the real question is whether we should be hacking around with pumpkins in the first place. For as Bonham-Carter notes (Bonham-Carter, 1967, page 3):

"If objects are strongly partitioned into groups, each group will appear as a well-defined cluster. More commonly, however, relationships are not so clear, entailing difficult and somewhat subjective decisions ...".

Why do we classify in the first place—apart from the fact that we cannot think of anything more imaginative to do? Presumably because we are driven by a phylogenetic concern to uncover the linkages and connections of evolution (Heywood and McNeill, 1964). But then why not investigate the connections, the *relations*, instead of severing them, destroying the structure, changing the geometry with *t*-forces, and creating obstructions so that our thinking can no longer 'traffic' along the coherent *whole* of the complex formed by the raw material of our observations?

The algebraic topological language of structure allows us, with the prosthetic help of computing machines, to investigate the structure of a relation between data sets. Such an analysis requires that we think hard not only about the relationships we impose, but about the cover sets themselves. Finding the latter is a nontrivial task, and many advances have been made by the imaginative choice of new covers. Mendele'ev, for example, found a new cover for the chemistry of his day, and produced an ordering construct that led to the prediction and search for then unknown elements. Of course, for this imaginative act he was ridiculed by his distinguished peers, and nearly took his own life as a result. Only twenty years ago, I can distinctly remember the same sort of violent opposition to ideas that went against the revealed truth (de Blij, 1959), this time by geologists examining a dissertation in geography (geomorphology) based on the absolutely absurd idea of continental drift.

The approach to our observations through relations, rather than the more restrictive functions, also keeps us much closer to the data themselves. We have all had experience of results in which the arcsin of X_1 is linearly related to the natural logarithm of X_2, the square-root transform of X_3, and the log normal transform of X_4. Everything is statistically significant at the 1% level—except that no one knows what it means. In contrast, the language of structure does not place a linear, or worse, unspecified, mapping between the investigator and the data as a filter destroying the very structure of the set, but allows one to examine, carefully, thoughtfully, and with a full possibility of serendipity effects, the actual connective tissue contained in the relation. So why do we insist on filtering as we do? After all, even Plato in his cave never recommended looking at the shadows through a linear filter darkly.

Finally, a topological approach provides us with one of those rare intellectual perspectives that allow us to see how familiar, and apparently disparate, things are only superficial (literally lying on the surface) manifestations of deeper structures. I do not mean simply that conventional areas of mathematics that are usually taught separately are linked together, although this is certainly true (Gould, 1980b). Rather such approaches force us to ask questions about the nature of spaces themselves. A relation defines a particular topology, and so structures a space. From such a perspective we can no longer think in terms of the container spaces of Newton, but must accept the more radical yet obvious notion of Leibnitz that spaces are defined by relations between objects. Such an intellectual acceptance moves us towards an Einsteinian view of dynamics as a manifestation of what a particular geometry will bear. And in the same way that *space* is defined by relations between objects, so *time* itself may soon be seen as multidimensional and defined by relations between events, rather than the linear continuum that is drilled into us from childhood. Is this not an idea towards which continuum mechanics is presently groping. Let me predict that, with such views of time and space, new geologies will be waiting out there to be discovered. And if we teach, should we not at least encourage the next generation to have a crack at them?

Refences

Ashby W, 1956 *An Introduction to Cybernetics* (Chapman and Hall, London)

Atkin R H, 1965 "Abstract physics" *Il Nuevo Cimento* **38** (1) 496–517

Atkin R H, 1971 "Cohomology of observations" in *Quantum Theory and Beyond* Ed. T Baskin (Cambridge University Press, Cambridge) pp 191–211

Atkin R H, 1972a "From cohomology in physics to Q-connectivity in social science" *International Journal of Man–Machine Studies* **4** 139–167

Atkin R H, 1972b "Urban structure" Urban Structure Research Project Report 1, Department of Mathematics, University of Essex, Colchester, Essex, England

Atkin R H, 1973a "A survey of mathematical theory" Urban Structure Research Project Report 2, Department of Mathematics, University of Essex, Colchester, Essex, England

Atkin R H, 1973b "A study area in Southend-on-Sea" Urban Structure Research Project Report 3, Department of Mathematics, University of Essex, Colchester, Essex, England

Atkin R H, 1974a *Mathematical Structure in Human Affairs* (Heinemann Educational Books, London)

Atkin R H, 1974b "An approach to structure in architectural and urban design.
1. Introduction and mathematical theory" and "2. Algebraic representation and
local structure" *Environment and Planning B* **1** 51-68, 173-192
Atkin R H, 1974c "A community study of the University of Essex" Urban Structure
Research Project Report 4, Department of Mathematics, University of Essex,
Colchester, Essex, England
Atkin R H, 1975a "An approach to structure in architectural and urban design.
3. Illustrative examples *Environment and Planning B* **2** 21-57
Atkin R H, 1975b "Methodology of *Q*-analysis—a study of East Anglia. 1" Regional
Research Project Report 5, Department of Mathematics, University of Essex,
Colchester, Essex, England
Atkin R H, 1977a *Combinatorial Connectivities in Social Systems* (Birkhäuser Verlag,
Basel)
Atkin R H, 1977b "*Q*-analysis theory and practice" Regional Research Project Report 10,
Department of Mathematics, University of Essex, Colchester, Essex, England
Atkin R H, 1981 *Multidimensional Man* (Penguin Books, Harmondsworth, Middx)
Atkin R H, Hartston W, Witten I, 1976 "Fred CHAMP, positional chess analyst"
International Journal of Man-Machine Studies **8** 517-529
Atkin R H, Witten I, 1975 "A multi-dimensional approach to positional chess"
International Journal of Man-Machine Studies **7** 727-750
Barker D, 1974 "Groping through grouping—a comparison of hierarchic and non-
hierarchic strategies" OP-26, Department of Geography, University College London,
Gower Street, London
Beer S, 1974 *Designing Freedom* (CBC Publications, Toronto)
Beer S, 1975 *Platform for Change* (John Wiley, New York)
Bonham-Carter G, 1967 "FORTRAN IV program for *Q*-mode cluster analysis of non-
quantitative data using IBM 7090/7094 computers" Computer Contribution 17,
Kansas State Geological Survey, Lawrence, Kansas
Boyce A, 1964 "The value of some methods of numerical taxonomy with reference to
hominoid classification" in *Phenetic and Phylogenetic Classification* Eds V Heywood,
J McNeill (The Systematics Association, London)
Chadwick G, 1977 "The limits of the plannable—stability and complexity in planning
and planned decisions" *Environment and Planning A* **9** 1189-1192
Chamberlain M, 1976 "A study of Behcet's disease by *Q*-analysis" *International Journal
of Man-Machine Studies* **8**(5) 549-565
de Blij H, 1959 *The Physiographic Provinces and Cyclic Erosion Surfaces of Swaziland*
PhD dissertation, Department of Geography, Northwestern University, Evanston, Ill.
Diffie W, Hellman M, 1976 "New directions in cryptography" *IEEE Transactions on
Information Theory*
Forrester J, 1971 *World Dynamics* (Wright-Allen Press, Cambridge West, Mass)
Gedye J, 1974 "A zygological analysis of a sample of regional cerebral blood flow data"
Department of Clinical Pharmacology and Geriatrics, University College Hospital
Medical School, Gower Street, London
Gould P, 1977 "Concerning a geographic education" in *Invitation to Geography*
Eds D Lanegren, R Palm (McGraw-Hill, New York) pp 202-225
Gould P, 1978a "Signals in the noise" in *Philosophy and Geography* Eds G Olsson,
S Gale (Reidel, Dordrecht) pp 121-154
Gould P, 1978b "How should we classify television programs?" DP-5, International
Television Flows Project, Pennsylvania State University, University Park, Pa
Gould P, 1980a "*Q*-analysis: an introduction for social scientists, geographers and
planners" *International Journal of Man-Machine Studies* **12** 169-199

Gould P, 1980b "Is it necessary to choose? Some technical, hermeneutic and emancipatory thoughts on enquiry" *Conference on Philosophy and Geography: A Search for Common Ground* Bellagio, Italy, August 1-5 (Pion, London) forthcoming

Gould P, Johnson J, 1978 "The structure of television programming—some experiments in the application of *Q*-analysis" and "An experiment in the classification of television programs" DP-1 and DP-3, International Television Flows Project, Pennsylvania State University, University Park, Pa

Gould P, Johnson J, 1980 "National television policy: monitoring structural complexity" *Futures* **12**(3) 178-190

Greenspan D, 1977 "Arithmetic applied mathematics" *Computer and Mathematics with Applications* **3** 253-270

Griffiths J C, 1966 "Future trends in geomathematics" *Mineral Industries* **35**(5) 1-3

Hammer P, 1969 *Advances in Mathematical System Theory* (The Pennsylvania State University Press, University Park)

Heywood and McNeill, 1964 *Phenetic and Phylogenetic Classification* (The Systematics Association, London)

Houbolt J H C, 1957 *Surface Sediments of the Persian Gulf near the Qatar Peninsula* (Mouton, The Hague)

Johnson J 1975 *A Multidimensional Analysis of Urban Road Traffic* PhD thesis, Department of Mathematics, University of Essex, Colchester, Essex, England

Johnson J, 1976 "The *Q*-analysis of road intersections" *International Journal of Man-Machine Studies* **8**(5) 531-548

Johnson J, 1977 "Methodology of *Q*-analysis—a study of road transport" Regional Research Project Report 11, Department of Mathematics, University of Essex, Colchester, Essex, England

Kocklemans J, 1972 "Stegmüller on the relationship between theory and experience" *Philosophy of Science* **39**(3) 397-410

Kuhn T, 1962 *The Structure of Scientific Revolutions* (University of Chicago Press, Chicago)

Laborit H, 1977 *Decoding the Human Message* (Allison and Busby, London); translation of *La Nouvelle Grille* (Robert Laffont, Paris, 1974)

Laing R, 1969 *The Politics of the Family* (CBC Publications, Toronto)

Mulhall D, no date "The representation of personal relationships" Middlesex Hospital Medical School, London

Nielsen C, 1925 *Levende Musik* (Martina Forlag, Kobenhavn)

Open University, 1971 *Relations—Mathematics Foundation Course Unit 19* (The Open University Press, Walton Hall, Milton Keynes)

Schreider J, 1975 *Equality, Resemblance, and Order* (Mir Publishers, Moscow)

Simon H, 1969 *The Science of the Artificial* (MIT Press, Cambridge, Mass)

Spencer-Brown G, 1962 *Laws of Form* (Allen and Unwin, London)

Stegmüller W, 1976 *The Structure and Dynamics of Theories* (Springer, Berlin)

Stokes A, 1973 *A Game that Must be Lost* (Carcanet Press, Manchester)

Wheeler J, Patton C, 1977 "Is physics legislated by cosmogony?" in *The Encyclopedia of Ignorance* Eds R Duncan, M Weston-Smith (Pergamon Press, London)

Zeeman E, 1965 *Topology of the Brain* Mathematics and Computer Science in Biology and Medicine, Medical Research Council, London

Index